T0329875

WIDEBAND RF TECHNOLOGIES AND ANTENNAS IN MICROWAVE FREQUENCIES

WIDEBAND RF TECHNOLOGIES AND ANTENNAS IN MICROWAVE FREQUENCIES

DR. ALBERT SABBAN

WILEY

Published by John Wiley & Sons, Inc., Hoboken, New Jersey

Published simultaneously in Canada

For general information on our other products and services or for technical support, please contact our Customer Care Department within the United States at (800) 762-2974, outside the United States at (317) 572-3993 or fax (317) 572-4002.

Wiley also publishes its books in a variety of electronic formats. Some content that appears in print may not be available in electronic formats. For more information about Wiley products, visit our web site at www.wiley.com.

Library of Congress Cataloging-in-Publication Data:

Names: Sabban, Albert.
Title: Wideband RF technologies and antennas in microwave frequencies / by Albert Sabban.
Description: Hoboken, New Jersey : John Wiley & Sons, Inc., 2016. | Includes
    bibliographical references and index.
Identifiers: LCCN 2016003737 (print) | LCCN 2016015072 (ebook) |
    ISBN 9781119048695 (hardback) | ISBN 9781119048664 (pdf) |
    ISBN 9781119048657 (epub)
Subjects: LCSH: Microwave communication systems. | Microwave receivers. |
    Microwave antennas. | Millimeter wave communication systems. | Broadband
    communication systems. | BISAC: TECHNOLOGY & ENGINEERING / Microwaves.
Classification: LCC TK5103.4833 .S23 2106 (print) | LCC TK5103.4833 (ebook) |
    DDC 621.381/3–dc23
LC record available at http://lccn.loc.gov/2016003737

Set in 10/12pt Times by SPi Global, Pondicherry, India

Printed in the United States of America

10 9 8 7 6 5 4 3 2 1

The book is dedicated to the memory of my father, mother, and sister:
David Sabban, Dolly Sabban, and Aliza Sabban.

CONTENTS

ACKNOWLEDGMENTS

My wife—Mazal Sabban
My daughters—Dolly and Lilach
My son—David Sabban
Grandchildren—Nooa, Avigail, Ido, Shira, and Efrat

Acknowledgements to my engineering colleagues who have helped me through 39 years of my engineering and research career.

AUTHOR BIOGRAPHY

Dr. Albert Sabban
Work Address: *ORT Braude College*, Department of Electrical and Electronic Engineering, P. O. Box 78, Snunit St. 51, Karmiel, 2161002, Israel
Office: EM 426
E-mail: sabban@braude.ac.il

EDUCATION

Ph.D., 1991, Electrical Engineering, Faculty of Electrical and Computer Engineer-
ing, University of Colorado, Boulder, CO, USA
Dissertation: "Multiport Network Model for Evaluating Radiation Loss and
Coupling among Discontinuities in Microstrip Circuits"
M.B.A., 2005, Business and Management, Faculty of Management, University of
Haifa, Israel
M.Sc., 1986, Electrical Engineering, Faculty of Electrical Engineering, Tel Aviv
University, Tel Aviv, Israel, Magna Cum Laude
Thesis: "Spectral Domain Iterative Analysis of Multilayered Microstrip Antennas"
B.Sc., 1976, Electrical Engineering, Faculty of Electrical Engineering, Tel Aviv
University, Tel Aviv, Israel

RESEARCH INTERESTS

- Antennas and microwave, biomedical engineering
- Communication
- System engineer

ACADEMIC APPOINTMENTS

2006 Present senior lecturer, Department of Electrical and Electronic Engineering,
Ort Braude College, Karmiel, Israel

2002–2004 Adjunct lecturer, Faculty of Electrical Engineering, Bar-Ilan University, Ramat Gan, Israel

1988–1989 Research and teaching assistant, Electrical Engineering, Faculty of
Electrical and Computer Engineering, University of Colorado, Boulder,
CO, USA

1979–1984 Teaching assistant, Faculty of Electrical Engineering, Technion –
Israel Institute of Technology (IIT), Haifa, Israel

1976–2007 Research engineer and project leader at RAFAEL, rank A (א' מחקר)

TEACHING EXPERIENCE

A. ORT Braude College of Engineering

Undergraduate Courses

Introduction to Antenna Systems
Introduction to Microwave Engineering
Engineering Design and Ethics
Electromagnetic Waves and Transmission Lines

Solid State Microwave Devices

B. Other Universities or Colleges

University of Colorado at Boulder

Graduate Courses

Product Design and Development

University of Colorado at Boulder:

Teaching assistant (under Prof. C. Johnk) on "Antenna Systems"

Research assistant at MMICAD Center, University of Colorado, Boulder

Technion – IIT

Development of Laboratories

Project manager and teaching assistant in RF and Communications Lab, Electrical Engineering Department, Technion, Haifa, Israel

PROFESSIONAL EXPERIENCE

- 2006 Present senior lecturer, Department of Electrical and Electronic Engineering, Ort Braude College of Engineering, Karmiel, Israel
- 2007–2010 RF and antenna specialist at Given Imaging, Yokneam, Israel
- 2002–2007 Research engineer and millimeter wave project leader at RAFAEL, Haifa, Israel
- At rank A – research
- 2000–2002 Sabbatical at Mini Circuits and RF specialist and consultant at Ravon, Kiryat Bialik, Israel
- 1998–2000 RF modules group manager and project leader at RAFAEL, Haifa, Israel
- 1995–1998 Project leader of Satellite Communications RF Modules
- 1983–1987 Antenna group manager and project leader at RAFAEL, Haifa, Israel
- 1976–1983 Employed by the Armament Development Authority, [RAFAEL], Haifa, Israel. On leave of absence from August 1987 to January 1991. During 1977–1986, employed as a senior research engineer at the Antenna Group as antenna group leader and millimeter wave antenna research leader

Dr. Albert Sabban was RF and antenna specialist at Biomedical Hi-tech Companies. He designed wearable compact antennas to medical systems and was involved in the design of the medical system.

From 1976 to 2007, he was a senior R&D scientist and project leader in RAFAEL, where he passed successfully a system engineering course. During his work in RAFAEL and other institutes and companies, he gained experience in project management, sales, marketing, and training and managed and led groups and projects with more than 20 employees.

Dr. Albert Sabban developed RFIC components on GaAs and silicon substrates from July 2000 to 2001; microwave components by employing LTCC technology from June 2001 to 2002; passive and active microwave components such as power amplifiers, low-noise amplifiers, multipliers, VCOs, power dividers and filters, and RF heads in RAFAEL; 20 W and 100 W power amplifiers at UHF frequencies; a 10 W class C transmitter at L band frequencies with 50% efficiency; a 2 W transmitter at K_a band; and wideband microstrip antenna arrays, dipole antenna arrays, telemetry microstrip antennas, backfire antennas, reflector antennas, couplers, power dividers, and wideband monopulse comparators. From 1979 to 1984, he was a teaching assistant in the Department of Electrical Engineering in Technion, Haifa, Israel. From 1984 to 1987, he was the leader of an antenna R&D group in RAFAEL and, from 1980 to 1987, of several research programs in RAFAEL. In these research programs, he developed wideband millimeter wave microstrip antennas, compact reflector antennas at millimeter wave frequencies, dielectric rod antennas, and passive microwave components. He also developed an iterative spectral domain analysis of single- and double-layered microstrip antennas using conjugate gradient algorithm. From August 1987 to February 1991, he was on leave from RAFAEL and joined the University of Colorado at Boulder where he studied for his Ph.D. degree and worked as a research assistant in the Center for Microwave and Millimeter Wave Computer-Aided Design. His research topic was "Multiport Network Modeling for Evaluating Radiation Loss and Spurious Coupling Among Microstrip Discontinuities in Microstrip Circuits." He also developed a planar lumped model for evaluating spurious coupling and radiation loss among coupled microstrip discontinuities, a spectral domain algorithm to analyze microstrip lines and coupled microstrip lines, and an iterative spectral domain algorithm to analyze wideband microstrip antennas, which he hold a US patent on. Since March 1991, he has been working in RAFAEL as a leading R&D microwave scientist. From 1991 to August 1993, he worked as a leading R&D engineer in developing RF environmental simulation systems and, from August 1993 to August 1994, a compact low-power consumption integrated RF head for Inmarsat-M ground terminal. From August 1995 to August 1998, he worked as a senior R&D scientist and project leader in developing a compact low-power consumption integrated RF head at K_a band for VSAT applications and, from 1988 to July 2000, MMIC RF components and RF heads at Ka band. Since 2002, Dr. A. Sabban has been working as a senior R&D scientist and project leader in developing MMIC RF modules and RF systems at K_a band. In the beginning of 2005, he joined the MEMS group in RAFAEL, where he led and developed projects.

ACADEMIC AND PROFESSIONAL AWARDS AND GRANTS

Ort Braude College best lecturer in education award October 2015.

Member of the winning team of Rafael Best Project Prize 2007.

Member of the winning team of Rafael Best Project Prize 1995.

RAFAEL award for outstanding engineering achievements, December 1984

Second place in student paper competition during IEEE Symposium in Tel-Aviv, 1985

Member of the winning team of Israel Defense Prize, 1984 and 1986

PROFESSIONAL AND RESEARCH ACTIVITIES

A. Invited talks:
- 3-day course. A. Sabban, "Wide Band RF Technologies at Microwave and mm Frequencies" Singapore, 2.2008
- Full-day course. A. Sabban, "Wide Band RF Technologies and Antennas at Microwave and mm Frequencies" Singapore, 12.2009
- Half-day course. A. Sabban, "Wearable Antennas" Orlando, USA, 7.2013
- Half-day course. A. Sabban, "Wide Band RF Technologies and Antennas at Microwave and mm Frequencies" Germany, 10.2013
- Half-day course. A. Sabban, "Wearable Antennas" Memphis, USA, 7.2014
- Half-day course. A. Sabban, "Wearable Antennas" Vancouver, Canada, 7.2015

B. Conference organization
- Organizing Workshops
  - IEEE APMC Conference, Singapore, 2009
  - IEEE APS Conference, Orlando, USA, 7.2013
  - IEEE EuMc Conference, Germany, 10.2013
  - IEEE APS Conference, Memphis, USA, 7.2014
  - IEEE APS Conference, Vancouver, Canada, 7.2015
- Reviewer Board
  - IEEE COMCAS Conference, Israel, 10.2013
  - IET Journal 2012–2015
  - ACES Journal 2013–2015
  - APWL Journal 2013–2015

C. Membership in Professional Societies
- IEEE Antennas and Propagation Society, APS, 1987–2013.
- IEEE Microwave Technology Society, MTT, 1987–2013.
- IEEE Communication Society, 2010–2013.

LIST OF PUBLICATIONS

A. Refereed Journal Papers

[1] A. Sabban "Ultra-Wideband RF Modules for Communication Systems," PARIPEX, Indian Journal of Research, Vol. 5, No. 1, pp. 91–95, January 2016.

[2] A. Sabban, "New Compact Wearable Meta-Material Antennas," Global Journal for Research and Analysis, Vol. IV, pp. 268–271, August 2015.

[3] A. Sabban, "Small Wearable Meta Materials Antennas for Medical Systems," The Applied Computational Electromagnetics Society Journal, Accepted for publication, February 2016.

[4] A. Sabban, "New Wideband Meta Materials Printed Antennas for Medical applications," International Journal of Advance in Medical Science (AMS), Vol. 3, pp. 1–10, April 2015.

[5] A. Sabban, "Wideband RF Modules and Antennas at Microwave and MM Wave Frequencies for Communication Applications," International Journal of Modern Communication Technologies & Research, Vol. 3, pp. 89–97, March 2015.

[6] A. Sabban, "Compact Wearable Tunable Printed Antennas for Medical Applications," Research in Health and Nutrition (RHN) Journal, Vol. 3, pp. 14–19, January 2015.

[7] A. Sabban, "W Band MEMS Detection Arrays," International Journal of Modern Communication Technologies & Research, Vol. 2, pp. 9–13, December 2014.

[8] A. Sabban, "New Wideband Printed Antennas for Medical Applications," IEEE Transactions on Antennas and Propagation, Vol. 61, No. 1, pp. 84–91, January 2013.

[9] A. Sabban, "Comprehensive Study of Printed Antennas on Human Body for Medical Applications," International Journal of Advance in Medical Science (AMS), Vol. 1, pp. 1–10, February 2013.

[10] A. Sabban and K.C Gupta, "A Planar-Lumped Model for Coupled Microstrip Lines and Discontinuities," IEEE Transactions on Microwave Theory and Techniques, Vol. 40, pp. 245–252, February 1992.

[11] A. Sabban and K.C. Gupta, "Characterization of Radiation Loss from Microstrip Discontinuities Using a Multiport Network Modeling Approach," IEEE Transactions on Microwave Theory and Techniques Vol. 39, No. 4, pp. 705–712, April 1991.

[12] A. Sabban and K.C. Gupta, "Effects of Packages on Parasitic Coupling Among Microstrip Discontinuities Using a Multiport Network Modeling Approach," International Journal on Microwave and MM-Wave Computer-Aided Engineering, Vol. 1, No. 4, pp. 403–411, 1991.

[13] R. Kastner, E. Heyman, A. Sabban, "Spectral Domain Iterative Analysis of Single and Double-Layered Microstrip Antennas Using the Conjugate Gradient Algorithm," IEEE Transactions on Antennas and Propagation, Vol. 36, No. 9, pp. 1204–1212, September 1988.

Books

[1] A. Sabban (2015). *"Low Visibility Antennas for Communication Systems."* TAYLOR & FRANCIS GROUP, USA.

[2] A. Sabban (2014). *"Microwave Theory and Transmission Lines,"* Saar Publication, Israel.

Chapters in Books

[1] A. Sabban (2012). Wearable Antennas in *"Advancements in Microstrip and Printed Antennas,"* A. Kishk (Ed.), ISBN 980-953-307-543-8, InTech, Croatia, Available from: http://www.intechopen.com/books/show/title/advancement in microstrip-antennas-with-recent-applications (accessed on March 10, 2016).

[2] A. Sabban (2011). *"Microstrip Antenna Arrays, Microstrip Antennas,"* N. Nasimuddin (Ed.), ISBN: 978-953-307-247-0, InTech, Croatia, Available from: http://www.intechopen.com/articles/show/title/microstrip-antenna-arrays (accessed on March 10, 2016).

Refereed Conference Proceedings, 2001–2014

[1] A. Sabban, "Dually Polarized Tunable Printed Antennas for Medical Applications," 9th IEEE European Conference on Antennas and Propagation Conference, EUCAP 2015, Lisbon, Portugal, May 13–17, 2015.

[2] A. Sabban, "New Microstrip Meta Materials Antennas," IEEE Antennas and Propagation Society International Symposium (APSURSI), Memphis, TN, July 6–11, 2014.

[3] A. Sabban, "Wearable Antennas for Medical Applications," IEEE BodyNet 2013, Boston, MA, October 2013, pp. 1–7.

[4] A. Sabban, "Meta Materials Antennas," New Tech Magazine, Tel Aviv, Israel, June 2013, pp. 16–19.

[5] A. Sabban, "Wideband Tunable Printed Antennas for Medical Applications," IEEE Antennas and Propagation Society International Symposium (APSURSI), Chicago, IL, July 8–14, 2012, pp. 1–2.

[6] A. Sabban, "MM Wave Microstrip Antenna Arrays," New Tech Magazine, Tel Aviv, Israel, June 2012, pp. 16–21.

[7] A. Sabban, "New Compact Wideband Printed Antennas for Medical Applications," IEEE International Symposium on Antennas and Propagation (APSURSI), Spokane, WA, July 3–8, 2011, pp. 251–254.

[8] A. Sabban, "Interaction between New Printed Antennas and Human Body in Medical applications," Asia Pacific Microwave Conference Proceedings (APMC), Yokohama, Japan, December 7–10, 2010, pp. 187–190.

[9] A. Sabban, "Wideband Printed Antennas for Medical Applications," Asia Pacific Microwave Conference, APMC 2009, Singapore, December 7–10, 2009, pp. 393–396.

[10] A. Sabban, "Millimeter Wave Detection Arrays," Asia Pacific Microwave Conference, APMC 2008, Macau, Hong Kong, December 16–20, 2008, pp. 1–4.

[11] A. Sabban, J. Cabiri, E. Carmeli, "18 to 40 GHz Integrated Compact Switched Filter Bank Module," International Symposium on Signals, Systems and Electronics, 2007. ISSSE '07, Montreal, Canada, July 30 to August 2, 2007, pp. 347–350.

[12] A. Sabban, "Applications of MM Wave Microstrip Antenna Arrays," International Symposium on Signals, Systems and Electronics, 2007. ISSSE '07, Montreal, Canada, July 30 to August 2, 2007, pp. 119–122.

[13] A. Sabban, "MM Wave Microstrip Antenna Arrays," New-Tech Microwave Magazine, Tel-Aviv, Israel, June 2012, pp. 28–31.

[14] A. Sabban, "Meta-materials Antennas," New-Tech Microwave Magazine, Tel-Aviv, Israel, December 2012, pp. 28–31.

[15] A. Sabban "RFID Antennas for Medical Applications," New-Tech Microwave Magazine, Tel-Aviv, Israel, December 2011, pp. 22–27.

[16] A. Sabban, "Interaction between Printed Antennas and Human Body in Medical, Applications, New-Tech Magazine, Tel-Aviv, Israel, March 2011, pp. 28–31.

Refereed Conference Proceedings, 1978–2000

[1] A. Sabban, "Ka-Band Compact Integrated High Power Amplifiers for VSAT Satellite Communication Ground Terminal," Asia-Pacific Microwave Conference 2000, Sydney, Australia, December 3–6, 2000, pp. 209–211.

[2] A. Sabban, "Ka Band Microstrip Antenna Arrays with High Efficiency," IEEE Antennas and Propagation Society International Symposium, Orlando, FL, July 11–16, 1999, pp. 2740–2743.

[3] A. Sabban, "Evaluation of Parasitic Coupling in Metallic Package in MIC and MMICS," 29th European Microwave Conference 1999, Tel-Aviv Israel, March 25, 1999.

[4] A. Sabban, "Evaluation and Measurements of Parasitic Coupling in Metallic Package in MIC and MMICS," 29th European Microwave Conference 1999, Munich, Germany, October 1999, pp. 390–393.

[5] A. Sabban, A. Madjar, I. Shapir, "A Ka-band Compact Integrated Transmitter for VSAT Satellite Communication Ground Terminal," 27th European Microwave Conference 1997, Jerusalem, Israel, September 8–12, 1997, pp. 671–675.

[6] A. Sabban, "A Comprehensive Study of Losses in mm-wave Microstrip Antenna Arrays," 27th European Microwave Conference 1997, Jerusalem, Israel, September 8–12, 1997, pp. 163–167.

[7] A. Sabban, A. Britebard, Y. Shemesh, "Development and Fabrication of a Compact Integrated RF-Head for Inmarsat-M Ground Terminal," 27th European Microwave Conference 1997, Jerusalem, Israel, September 8–12, 1997, pp. 1186–1191.

[8] A. Madjar, D. Behar, A. Sabban, I. Shapir, "RF Front End Prototype for a Millimeter Wave (Ka/K) Band VSAT Satellite Communication Earth Terminal," 26th European Microwave Conference 1996, Prague, Czech Republic, September 6–13, 1996, pp. 953–955.

[9] A. Sabban, Y. Shemesh, "A Compact Low Power Consumption Integrated RF-Head for Inmarsat-M Ground Terminal," 25th European Microwave Conference, 1995, Bologna, Italy, September 4, 1995, pp. 81–82.

[10] A. Sabban, "Evaluation of Radiation Loss From Microstrip Line Discontinuities Using A Multiport Network Modeling Approach In MM-Waves Microstrip Antenna Arrays," IEEE URSI Symposium, Seattle, WA, June 1994.

[11] A. Sabban, "Multiport Network Model For Evaluating Spurious Coupling in Microstrip Circuits and MMICS Including Package Effects," RF Technology Symposium, Herzlia, May 1994.

[12] A. Sabban, K.C Gupta, "Evaluation of Parasitic Coupling Among Microstrip Line Discontinuities Using a Multiport Network Modeling Approach," 23rd European Microwave Conference, 1993, Madrid, Spain, September 6–10, 1993, pp. 656–658.

[13] A. Sabban, K.C. Gupta, "Evaluation of Radiation Loss From Coupled Microstrip Line Discontinuities Using a Planar Lumped Model and A Multiport Network Modeling Approach," 22nd European Microwave Conference, 1992, Helsinki, Finland, August 1992, pp. 1375–1380.

[14] A. Sabban, K.C. Gupta, "A Planar-Lumped Model for Coupled Microstrip Discontinuities," IEEE MTT-S International Microwave Symposium Digest, 1990, Dallas, TX, May 8–10, 1990.

[15] A. Sabban, K.C. Gupta, "Multiport Network Model for Evaluating Radiation Loss and Spurious Coupling between Discontinuities in Microstrip Circuits," IEEE MTT-S International Microwave Symposium Digest, 1989, Long-Beach, CA, 1989, pp. 707–710.

[16] R. Kastner, E. Heyman, A. Sabban, "Spectral Domain Iterative Analysis of Single and Double-Layered Microstrip Antennas Using the Conjugate Gradient Algorithm," IEEE Transactions on Antennas and Propagation, Vol. 36, No. 9, September 1988, pp. 1204–1212.

[17] A. Sabban, R. Kastner, E. Heyman, "Spectral Domain Iterative Analysis of Single and Double Layered Microstrip Antennas Using The Conjugate Gradient Algorithm," IEEE Symposium, Tel-Aviv, Israel, April 1987.

[18] A. Sabban, R. Kastner, I. Kotlarenko, "High Efficiency and Gain Printed Antenna Array," IEEE Symposium, Tel-Aviv, Israel, April 1987.

[19] A. Sabban, "Spectral Domain Iterative Analysis of Multilayered Microstrip Antennas," M.Sc. Thesis, Tel-Aviv University, Tel-Aviv, Israel, 1985.

[20] A. Sabban, "A New Broadband Stripline Antenna," IEEE Symposium, Tel-Aviv, Israel, June 1985.

[21] A. Sabban, "A New Wideband Stacked Microstrip Antenna," IEEE Antenna and Propagation Symposium, Houston, TX, June 1983, pp. 63–66.

[22] A. Sabban, E. Navon, "A MM-Waves Microstrip Antenna Array," IEEE Symposium, Tel-Aviv, Israel, March 1983.

[23] A. Sabban, "Wideband Microstrip Antenna Arrays," IEEE Antenna and Propagation Symposium MELCOM, Tel-Aviv, Israel, May 1981.

[24] A. Sabban, "Microstrip Antennas," IEEE Symposium, Tel-Aviv, Israel, October 1979.

International Seminars and Courses with Course Book

[1] A. Sabban, "Wearable Antennas," Course in IEEE Antennas and Propagation Conference, Vancouver, Canada, July 2015.

[2] A. Sabban, "Wearable Antennas," Course in IEEE Antennas and Propagation Conference, Memphis, TN, July 2014.

[3] A. Sabban, "Wideband MM Waves Microwave Technologies," Course in IEEE European Microwave Conference, Nurnberg, Germany, October 2013.

[4] A. Sabban, "Wearable Antennas," Course in IEEE Antennas and Propagation Conference, Orlando, FL, July 2013.

[5] A. Sabban, "Wideband MM Waves Microwave Technologies," Course in IEEE APMC 2009 Conference, Singapore, December 2009.

[6] A. Sabban, "Wideband MM Waves Microwave Technologies," 3 day Course in EEM College, Singapore, February 2008.

Patents

Inventors: A. Sabban, *Microstrip antenna arrays* (US Patent US1986/4,623,893), 1986.

Published and Unpublished Professional Reports (Self-Performed)

[1] Around 35 classified and unclassified research reports at RAFAEL (1976–2007).

[2] Commercial Reports, Presentations and White Papers at Given Imaging (2007–2009).

[3] Research Reports on "Multiport Network Model for Evaluating Radiation Loss and Coupling Among Discontinuities in Microstrip Circuits." Faculty of Electrical and Computer Engineering, University of Colorado Boulder, Boulder, CO.

[4] A. Sabban, "Multiport Network Model for Evaluating Radiation Loss and Coupling Among Discontinuities in Microstrip Circuits," Ph.D. Thesis, University of Colorado Boulder, Boulder, CO, January 1991.

PREFACE

This book's main objective is to present microwave and millimeter wave technologies. However, Chapters 1 and 2 present definitions on communication and RF systems and are written to assist electrical engineers and students in studying basic theory and fundamentals of electromagnetics and antennas. There are many electromagnetic theory books and antenna books for electromagnetic scientist. However, there are few books that help electrical engineers and undergraduate students to study and to understand basic communication and RF systems definitions and electromagnetic and antenna theories and fundamentals with minimum integral and differential equations.

There are several 3D full-wave electromagnetic softwares such as HFSS, ADS, and CST that are used to design and analyze antennas. Modules developed and analyzed in this book were designed by using HFSS and ADS. For almost all RF modules and antennas presented in this book, there was a good agreement between computed and measured results. Only one design and fabrication iteration was needed in the development process of the RF communication modules and antennas. Each chapter covers sufficient details to enable students, scientists from all areas, and electrical and biomedical engineers to follow and understand the topics presented in the book. The book begins with elementary topics on electromagnetics and antennas needed by students and engineers with no background in electromagnetic and antenna theory to study and understand the basic design principle and features of RF and communication systems with communication and medical applications.

Several topics and designs are presented in this book for the first time. These include new designs of RF and communication modules and compact antennas for communication and medical applications. This book may serve as a reference book for students and design engineers. The text contains sufficient mathematical detail and explanations to enable electrical engineering and physics students to understand all topics presented in this book. Design considerations and computed and measured results of the RF modules are also presented.

Chapter 1 presents basic definitions on RF and communication systems. Electromagnetic theory and transmission lines are presented in Chapter 2. An introduction to antennas is given in Chapter 3. MIC and MMIC technologies are presented in Chapter 4. Chapter 5 presents printed antennas for wireless communication systems. MIC and MMIC MM wave receiving channel modules are presented in Chapter 6. Integrated outdoor unit (ODU) for MM wave satellite communication applications is presented in Chapter 7. MIC and MMIC integrated RF heads are presented in Chapter 8. MIC and MMIC components and module design are presented in Chapter 9. Microelectromechanical systems (MEMS) technology is presented in Chapter 10, whereas low temperature co-fired ceramic (LTCC) technology is presented in Chapter 11. Advance antenna technologies for communication system are presented in Chapter 12. Wearable antennas for communication and medical applications are presented in Chapter 13. RF and antenna measurements and measurement setups are presented in Chapter 14.

1

ELECTROMAGNETIC WAVE PROPAGATION AND APPLICATIONS

The purpose of this chapter is to provide a short survey of electromagnetic wave propagation and applications. The electromagnetic spectrum and basic definitions of electromagnetic wave propagation are presented in this chapter. Transmitting and receiving information in microwave frequencies are based on electromagnetic wave propagation.

1.1 ELECTROMAGNETIC SPECTRUM

The electromagnetic spectrum corresponds to electromagnetic waves from the meter range to the mm-wave range. The characteristic feature of this phenomenon is the short wavelength involved. The wavelength is of the same order of magnitude as the circuit devices used. The propagation time from one point of the circuit to another point of the circuit is comparable with period of the oscillating voltages and currents in the circuit. Conventional low circuit analysis based on Kirchhoff's and Ohm's laws could not analyze and describe the variation of fields, voltages, and currents along the length of the components. Components that their dimensions are lower than a 10th of wavelength are called lumped elements. Components that their dimensions are higher than a 10th of wavelength are called distributed elements. Kirchhoff's and Ohm's laws may be applied to lumped elements. However, Kirchhoff's and Ohm's laws cannot be applied to distributed elements.

Wideband RF Technologies and Antennas in Microwave Frequencies, First Edition. Dr. Albert Sabban.
© 2016 John Wiley & Sons, Inc. Published 2016 by John Wiley & Sons, Inc.

TABLE 1.1 Electromagnetic Spectrum and Applications

Band Name	Abbreviation	ITU	Frequency/λ_0	Applications
Tremendously low frequency	TLF		<3 Hz >100,000 km	Natural and artificial EM noise
Extremely low frequency	ELF		3–30 Hz 100,000–10,000 km	Communication with submarines
Super low frequency	SLF		30–300 Hz 10,000–1000 km	Communication with submarines
Ultra low frequency	ULF		300–3000 Hz 1000–100 km	Submarine communication, communication within mines
Very low frequency	VLF	4	3–30 kHz 100–10 km	Navigation, time signals, submarine communication, wireless heart rate monitors, geophysics
Low frequency	LF	5	30–300 kHz 10–1 km	Navigation, clock time signals, AM longwave broadcasting (Europe and parts of Asia), RFID, amateur radio
Medium frequency	MF	6	300–3000 kHz 1–100 m	AM (medium-wave) broadcasts, amateur radio, avalanche beacons
High frequency	HF	7	3–30 MHz 100–10 m	Shortwave broadcasts, radio, amateur radio and aviation, communications, RFID, radar, near-vertical incidence skywave (NVIS) radio communications, marine and mobile radio telephony
Very high frequency	VHF	8	30–300 MHz 10–1 m	FM, television broadcasts and line-of-sight ground-to-aircraft and aircraft-to-aircraft communications, land mobile and maritime mobile communications, amateur radio, weather radio
Ultra high frequency	UHF	9	300–3000 MHz 1–100 mm	Television broadcasts, microwave oven, radio astronomy, mobile phones, wireless LAN, Bluetooth, ZigBee, GPS and two-way radios such as land mobile, FRS, and GMRS radios
Super high frequency	SHF	10	3–30 GHz 100–10 mm	Radio astronomy, wireless LAN, modern radars, communications satellites, satellite television broadcasting, DBS
Extremely high frequency	EHF	11	30–300 GHz 10–1 mm	Radio astronomy, microwave radio relay, microwave remote sensing, directed-energy weapon, scanners
Terahertz or tremendously high frequency	THz or THF	12	300–3000 GHz 1–100 μm	Terahertz imaging, ultrafast molecular dynamics, condensed-matter physics, terahertz time-domain spectroscopy, terahertz computing/communications

To prevent interference and to provide efficient use of the radio spectrum, similar services are allocated in bands (see Refs. [1–3]). Bands are divided at wavelengths of 10^n meters, or frequencies of 3×10^n Hz. Each of these bands has a basic band plan that dictates how it is to be used and shared to avoid interference and to set protocol for the compatibility of transmitters and receivers. In Table 1.1 the electromagnetic spectrum and applications are listed. In Table 1.2 the IEEE Standard for radar frequency bands is listed. In Table 1.3 the International Telecommunication Union (ITU) bands are given. In Table 1.4 radar frequency bands as defined by NATO for ECM systems are listed (see Ref. [4]).

TABLE 1.2   IEEE Standard Radar Frequency Bands

Microwave Frequency Bands—IEEE Standard

Designation	Frequency Range (GHz)
L band	1–2
S band	2–4
C band	4–8
X band	8–12
K_u band	12–18
K band	18–26.5
K_a band	26.5–40
Q band	30–50
U band	40–60
V band	50–75
E band	60–90
W band	75–110
F band	90–140
D band	110–170

TABLE 1.3   The International Telecommunication Union Bands

Band Number	Symbols	Frequency Range	Wavelength Range
4	VLF	3–30 kHz	10–100 km
5	LF	30–300 kHz	1–10 km
6	MF	300–3000 kHz	100–1000 m
7	HF	3–30 MHz	10–100 m
8	VHF	30–300 MHz	1–10 m
9	UHF	300–3000 MHz	10–100 cm
10	SHF	3–30 GHz	1–10 cm
11	EHF	30–300 GHz	1–10 mm
12	THF	300–3000 GHz	0.1–1 mm

TABLE 1.4   Radar Frequency Bands as Defined by NATO for ECM Systems[a]

Band	Frequency Range (GHz)
A band	0–0.25
B band	0.25–0.5
C band	0.5–1.0
D band	1–2
E band	2–3
F band	3–4
G band	4–6
H band	6–8
I band	8–10
J band	10–20
K band	20–40
L band	40–60
M band	60–100

[a] Ref. [4].

1.2   FREE-SPACE PROPAGATION

Consider an isotropic source radiating P_t watts uniformly into free space.

At distance R, the area of the spherical shell with center at the source is $4\pi R^2$. Flux density at distance R is given by Equation 1.1:

$$F = \frac{P_t}{4\pi R^2} \ \text{W/m}^2 \tag{1.1}$$

$$G(\theta) = \frac{P(\theta)}{P_0/4\pi} \tag{1.2}$$

where

$P(\theta)$ is the variation of power with angle

$G(\theta)$ is the gain at the direction θ

P_0 is the total power transmitted

Sphere $= 4\pi$ solid radians

Gain is usually expressed in **decibels** (dB): $G\,[\text{dB}] = 10\log_{10} G$. Gain is realized by focusing power. An isotropic radiator is an antenna that radiates in all directions equally. Effective isotropic radiated power (EIRP) is the amount of power the transmitter would have to produce if it was radiating to all directions equally. The EIRP may vary as a function of direction because of changes in the antenna gain versus angle. We now want to find the power density at the receiver. We know that power is conserved in a lossless medium. The power radiated from a transmitter must pass through a spherical shell on the surface of which is the receiver.

The area of this spherical shell is $4\pi R^2$.

Therefore spherical spreading loss is $1/4\pi R^2$.

We can rewrite the power flux density, as given in Equation 1.3, now considering the transmit antenna gain:

$$F = \frac{\text{EIRP}}{4\pi R^2} = \frac{P_t G_t}{4\pi R^2} \ \text{W/m}^2 \tag{1.3}$$

The power available to a receive antenna of area A_r is given in Equation 1.4:

$$P_r = F \times A_r = \frac{P_t G_t A_r}{4\pi R^2} \tag{1.4}$$

Real antennas have effective flux collecting areas, which are less than the physical aperture area. A_e is defined as the antenna effective aperture area.

Where $A_e = A_{\text{phy}} \times \eta$, η = aperture efficiency.

Antennas have maximum gain G related to the effective aperture area as f as given in Equation 1.5, where A_e is the effective aperture area:

$$G = \text{Gain} = \frac{4\pi A_e}{\lambda^2} \tag{1.5}$$

Aperture antennas (horns and reflectors) have a physical collecting area that can be easily calculated from their dimensions:

$$A_{\text{phy}} = \pi r^2 = \pi \frac{D^2}{4} \tag{1.6}$$

Therefore, using Equations 1.5 and 1.6 we can obtain the formula for aperture antenna gain as given in Equations 1.7 and 1.8:

$$\text{Gain} = \frac{4\pi A_e}{\lambda^2} = \frac{4\pi A_{\text{phy}}}{\lambda^2} \times \eta \tag{1.7}$$

$$\text{Gain} = \left(\frac{\pi D}{\lambda}\right)^2 \times \eta \tag{1.8}$$

$$\text{Gain} \cong \eta \left(\frac{75\pi}{\theta_{3\text{dB}}}\right)^2 = \eta \frac{(75\pi)^2}{\theta_{3\text{dB}H}\theta_{3\text{dB}E}} \tag{1.9}$$

$$\text{where } \theta_{3\text{dB}} \cong \frac{75\lambda}{D}$$

$\theta_{3\text{dB}}$ is the antenna half power beamwidth. Assuming, for instance, a typical aperture efficiency of 0.55 gives

$$\text{Gain} \cong \frac{30,000}{(\theta_{3\text{dB}})^2} = \frac{30,000}{\theta_{3\text{dB}H}\theta_{3\text{dB}E}} \tag{1.10}$$

1.3 FRIIS TRANSMISSION FORMULA

The Friis transmission formula is presented in Equation 1.11:

$$P_r = P_t G_t G_r \left(\frac{\lambda}{4\pi R} \right)^2 \qquad (1.11)$$

Free-space loss (L_p) represents propagation loss in free space (e.g., losses due to attenuation in atmosphere)

where $L_p = (4\pi R/\lambda)^2$. The received power may be given as $P_r = (P_t G_t G_r / L_p)$.

L_a should also be accounted for in the transmission equation. Losses due to polarization mismatch, L_{pol}, should also be accounted. Losses associated with receiving antenna, L_{ra}, and with the receiver, L_r, cannot be neglected in computation of transmission budget. Losses associated with the transmitting antenna can be written as L_{ta}:

$$P_r = \frac{P_t G_t G_r}{L_p L_a L_{ta} L_{ra} L_{pol} L_o L_r} \qquad (1.12)$$

$$P_t = \frac{P_{out}}{L_t}$$

$$\text{EIRP} = P_t G_t$$

where

P_t is the transmitting antenna power
L_t is the loss between power source and antenna
EIRP is the effective isotropic radiated power

$$P_r = \frac{P_t G_t G_r}{L_p L_a L_{ta} L_{ra} L_{pol} L_{other} L_r}$$

$$= \frac{\text{EIRP} \times G_r}{L_p L_a L_{ta} L_{ra} L_{pol} L_{other} L_r} \qquad (1.13)$$

$$= \frac{P_{out} G_t G_r}{L_t L_p L_a L_{ta} L_{ra} L_{pol} L_{other} L_r}$$

where

$$G = 10\log\left(\frac{P_{out}}{P_{in}}\right) \text{dB} \quad \text{gain in dB}$$

$$L = 10\log\left(\frac{P_{in}}{P_{out}}\right) \text{dB} \quad \text{loss in dB}$$

Gain may be derived as given in Equation 1.14:

$$P_{in} = \frac{V_{in}^2}{R_{in}} \quad P_{out} = \frac{V_{out}^2}{R_{out}}$$

$$G = 10\log\left(\frac{P_{out}}{P_{in}}\right) = 10\log\left(\frac{V_{out}^2/R_{out}}{V_{in}^2/R_{in}}\right) \tag{1.14}$$

$$G = 10\log\left(\frac{V_{out}^2}{V_{in}^2}\right) + 10\log\left(\frac{R_{in}}{R_{out}}\right) = 20\log\left(\frac{V_{out}}{V_{in}}\right) + 10\log\left(\frac{R_{in}}{R_{out}}\right)$$

1.3.1 Logarithmic Relations

Important logarithmic operations are listed in Equations 1.15–1.18:

$$\begin{aligned} &10\log_{10}(A \times B) \\ &= 10\log_{10}(A) + 10\log_{10}(B) \\ &= A \text{ dB} + B \text{ dB} \\ &= (A+B) \text{ dB} \end{aligned} \tag{1.15}$$

$$\begin{aligned} &10\log_{10}(A/B) \\ &= 10\log_{10}(A) - 10\log_{10}(B) \\ &= A \text{ dB} - B \text{ dB} \\ &= (A-B) \text{ dB} \end{aligned} \tag{1.16}$$

$$\begin{aligned} &10\log_{10}(A^2) \\ &= 2 \times 10\log_{10}(A) \\ &= 20\log_{10}(A) \\ &= 2 \times (A \text{ in dB}) \end{aligned} \tag{1.17}$$

$$\begin{aligned} &10\log_{10}\left(\sqrt{A}\right) \\ &= \frac{10}{2}\log_{10}(A) \\ &= \frac{1}{2} \times (A \text{ in dB}) \end{aligned} \tag{1.18}$$

In Table 1.5 linear ratio versus logarithmic ratio is listed.

The received power P_r in dBm is given in Equation 1.19. The received power P_r is commonly referred to as "carrier power," C:

$$P_r = \text{EIRP} - L_{ta} - L_p - L_a - L_{pol} - L_{ra} - L_{other} + G_r - L_r \tag{1.19}$$

TABLE 1.5 Linear Ratio versus Logarithmic Ratio

Linear Ratio	dB	Linear Ratio	dB
0.001	−30.0	2.000	3.0
0.010	−20.0	3.000	4.8
0.100	−10.0	4.000	6.0
0.200	−7.0	5.000	7.0
0.300	−5.2	6.000	7.8
0.400	−4.0	7.000	8.5
0.500	−3.0	8.000	9.0
0.600	−2.2	9.000	9.5
0.700	−1.5	10.000	10.0
0.800	−1.0	100.000	20.0
0.900	−0.5	1000.000	30.0
1.000	0.0	18.000	12.6

The surface area of a sphere of radius d is $4\pi d^2$, so that the power flow per unit area W (power flux in W/m^2) at distance d from a transmitter antenna with input power P_T and antenna gain G_T is given in Equation 1.20:

$$W = \frac{P_r G_r}{4\pi d^2} \tag{1.20}$$

The received signal strength depends on the "size" or aperture of the receiving antenna. If the antenna has an effective area A, then the received signal strength is given in Equation 1.21:

$$P_R = P_T G_T \left(\frac{A}{4\pi d^2}\right) \tag{1.21}$$

Define the receiver antenna gain $G_R = 4\pi A/\lambda^2$
where $\lambda = c/f$.

1.4 LINK BUDGET EXAMPLES

$F = 2.4\,\text{GHz} => \lambda = 12.5\,\text{cm}$
At $933\,\text{MHz} => \lambda = 32\,\text{cm}$.
Receiver signal strength: $P_R = P_T\,G_T\,G_R\,(\lambda/4\pi d)^2$
$P_R\,(\text{dBm}) = P_T\,(\text{dBm}) + G_T\,(\text{dBi}) + G_R\,(\text{dBi}) + 10\log_{10}((\lambda/4\pi)^2) - 10\log_{10}(d^2)$
For $F = 2.4\,\text{GHz} => 10\log_{10}((\lambda/4\pi)^2) = -40\text{dB}$
For $F = 933\,\text{MHz} => 10\log_{10}((\lambda/4\pi)^2) = -32\text{dB}$

Mobile phone downlink

$\lambda = 12.5$ cm

$f = 2.4$ GHz

P_R (dBm) $= (P_T \, G_T \, G_R L)$ (dBm) $- 40$dB $+ 10\log_{10} (1/d^2)$

$P_R - (P_T + G_T + G_R + L) - 40$dB $= 10\log_{10} (1/d^2)$

Or $155 - 40 = 10\log_{10} (1/d^2)$

Or $(155 - 40)/20 = \log_{10} (1/d)$

$d = 10\wedge ((155 - 40)/20) = 562$ km

Mobile phone uplink

$d = 10\wedge ((153 - 40)/20) = 446$ km

For standard 802.11

- $P_R - P_T = -113.2$dBm
- 6 Mbps
  - $d = 10^{(113.2-40)/20} = 4500$ m
  - $d = 10^{(113.2-40-3)/20} = 3235$ m with 3dB gain margin
  - $d = 10^{(113.2-40-3-9)/20} = 1148$ m with 3dB gain margin and neglecting antenna gains
- 54 Mbps needing -85dBm
  - $d = 10^{(99.2-40)/20} = 912$ m
  - $d = 10^{(99.2-40-3)/20} = 646$ m with 3dB gain margin
  - $d = 10^{(99.2-40-3-9)/20} = 230$ m with 3dB gain margin and neglecting antenna gains

Signal strength

Measure signal strength in

dBW $= 10\log$ (power in watts)

dBm $= 10\log$ (power in mW)

802.11 can legally transmit at 30dBm.

Most 802.11 PCMCIA cards transmit at 10–20dBm.

Mobile phone base station: 20 W, but 60 users, so 0.3 W/user, but antenna has gain $= 18$dBi.

Mobile phone handset: 21dBm

1.5 NOISE

Noise limits systems ability to process weak signals.

System dynamic range is defined as system capability to detect weak signals in presence of large-amplitude signals.

Noise sources

1. Random noise in resistors and transistors
2. Mixer noise
3. Undesired cross-coupling noise from other transmitters and equipment
4. Power supply noise
5. Thermal noise present in all electronics and transmission media due to thermal agitation of electrons

$$\text{Thermal noise} = kTB \text{ (W)}$$

where k is Boltzmann's constant $= 1.38 \times 10^{-23}$
T is temperature in Kelvin $(C + 273)$
B is bandwidth

Examples

For temperature $= 293°C = > -203\text{dB}, -173\text{dBm/Hz}$
For temperature $= 293°C$ and $22 \text{ MHz} = > -130\text{dB}, -100\text{dBm}$

Random noise

- External noise
- Atmospheric noise
- Interstellar noise

Receiver internal

- Thermal noise
- Flicker noise (low frequency)
- Shot noise

SNR is defined as signal-to-noise ratio. SNR varies with frequency.
SNR = signal power/noise power, and SNR is given in Equation 1.22:

$$\text{SNR} = \frac{S(f)}{N(f)} = \frac{\text{average signal power}}{\text{average noise power}} \tag{1.22}$$

- SNR (dB) $= 10\log_{10}$ (signal power/noise power).

Noise factor, F, is a measure of the degradation of SNR due to the noise added as we process the signal. F is given in Equations 1.23 and 1.24:

$$F = \frac{\text{available output noise power}}{\text{available output noise due to source}} \qquad (1.23)$$

$$\text{Noise figure} = \text{NF} = 10\log(F).$$

Multistage noise figure is given by Equation 1.24:

$$F = F_1 + \frac{F_2 - 1}{G_1} + \frac{F_3 - 1}{G_1 G_2} + \cdots + \frac{F_n - 1}{G_1 G_2 \cdots G_{n-1}} \qquad (1.24)$$

Signal strength is the transmitted power multiplied by a gain minus losses.

Loss sources

• Distance between the transmitter and the receiver.
• The signal passes through rain or fog at high frequencies.
• The signal passes through an object.
• Part of the signal is reflected from an object.
• Signal interferes multipath fading.
• An object not directly in the way impairs the transmission.

The received signal must have a strength that is larger than the receiver sensitivity. SNR of 20dB or larger would be good.

Sensitivity is defined as minimum detectable input signal level for a given output SNR, also called noise floor.

1.6   COMMUNICATION SYSTEM LINK BUDGET

Link budget determines if the received signal is larger than the receiver sensitivity.

A link budget analysis determines if there is enough power at the receiver to recover the information. Link budget must account for effective transmission power. Link budget takes into account the following parameters:

Transmitter

• Transmission power
• Antenna gain
• Losses in cable and connectors

Path losses

• Attenuation
• Ground reflection
• Fading (self-interference)

Receiver

- Receiver sensitivity
- Losses in cable and connectors

1.6.1 Transmitter

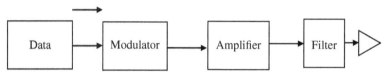

FIGURE 1.1   Transmitter block diagram.

1.6.2 Receiver

Transmitter block diagram is shown in Figure 1.1 (Table 1.6). Receiver block diagram is shown in Figure 1.2 (Table 1.7).

TABLE 1.6   Transmitting Channel Power Budget

Component	Gain (dB)/Loss (dB)	Power (dBm)	Remarks
Input power		−2	
Transmitter gain	40		
Power amplifier output power		38	
Filter loss	1	37	
Cable loss	1	36	
Matching loss	1	35	
Radiated power		35	

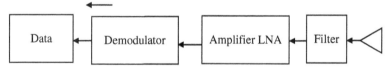

FIGURE 1.2   Receiver block diagram.

TABLE 1.7   Receiving Channel Power Budget

Component	Gain (dB)/Loss (dB)	Power (dBm)	Remarks
Input power		−20	
Receiver gain	23		
Cable loss	1	−21	
Filter loss	1	−22	
Matching loss	1	−23	
LNA amplifier output power		0	

1.7 PATH LOSS

Path loss is a reduction in the signal's power, which is a direct result of the distance between the transmitter and the receiver in the communication path.

There are many models used in the industry today to estimate the path loss, and the most common are the free-space and Hata models. Each model has its own requirements that need to be met in order to be utilized correctly. The free-space path loss is the reference point other models used.

1.7.1 Free-Space Path Loss

$$\text{Free-space path loss (dB)} = 20\log_{10} f + 20\log_{10} d - 147.56$$

where F is frequency in hertz and d is the distance in meters.

Free-space model typically underestimates the path loss experienced for mobile communications. Free-space model predicts point-to-point fixed path loss.

1.7.2 Hata Model

The Hata model is used extensively in cellular communications. The basic model is for urban areas, with extensions for suburbs and rural areas.

The Hata model is valid only for these following ranges:

- Distance 1–20 km
- Base height 30–200 m
- Mobile height 1–10 m
- 150–1500 MHz

The Hata formula for urban areas is

$$L_H = 69.55 + 26.16\log_{10} f_c - 13.82\log_{10} h_b - \text{env}(h_m) + (44.9 - 6.55\log_{10} h_b)\log_{10} R$$

where
 h_b is the base station antenna height in meters
 h_m is the mobile antenna height also measured in meters
 R is the distance from the cell site to the mobile in kilometers
 f_c is the transmit frequency in MHz
 $\text{env}(h_m)$ is an adjustment factor for the type of environment and the height of the mobile ($\text{env}(h_m) = 0$ for urban environments with a mobile height of 1.5 m)

1.8 RECEIVER SENSITIVITY

Sensitivity describes the weakest signal power level that the receiver is able to detect and decode. Sensitivity is determined by the lowest SNR at which the signal can be recovered. Different modulation and coding schemes have different minimum SNRs.

Sensitivity is determined by adding the required SNR to the noise present at the receiver.

Noise sources

- Thermal noise
- Noise introduced by the receiver's amplifier

Thermal noise $= N = kTB$ (W)
where

$k = 1.3803 \times 10^{-23}$ J/K
$T =$ temperature in Kelvin
$B =$ receiver bandwidth

N (dBm) $= 10\log_{10}(kTB) + 30$
Thermal noise is usually very small for reasonable bandwidths.

1.8.1   Basic Receiver Sensitivity Calculation

Sensitivity (W) $= kTB \times NF$ (linear) $\times$ minimum SNR required (linear)
Sensitivity (dBm) $= 10\log_{10}(kTB \times 1000) + NF$ (dB) $+$ minimum SNR required (dB)
Sensitivity (dBm) $= 10\log_{10}(kTB) + 30 + NF$ (dB) $+$ minimum SNR required (dB)

Sensitivity decreases in communication systems when:

- Bandwidth increases.
- Temperature increases.
- Amplifier introduces more noise.
- There are losses in space, rain, and snow.

1.9   RECEIVERS: DEFINITIONS AND FEATURES

Figure 1.3 presents a basic receiver block diagram.

1.9.1   Receiver

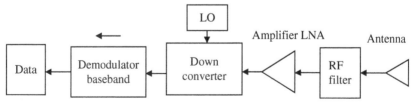

FIGURE 1.3   Basic receiver block diagram.

1.9.2 Receivers: Definitions

RF, IF, and LO frequencies—When a receiver uses a mixer we refer to the input frequency as the RF. The system must provide a signal to mix down the RF, and this is called the local oscillator. The resulting lower frequency is called the intermediate frequency (IF), because it is somewhere between the RF and the baseband frequency.

Baseband frequency—The baseband is the frequency at which the information you want to process is.

Preselector filter—Preselector filter is used to keep undesired radiation from saturating a receiver. For example, we don't want our cell phone to pick up air-traffic control radar.

Amplitude and phase matching versus tracking—In a multichannel receiver (more than one receiver), it is important for the channels to match and track each other over frequency. Amplitude and phase *matching* means that the relative magnitude and phase of signals that pass through the two paths must be almost equal.

Tunable bandwidth versus instantaneous bandwidth—*Instantaneous bandwidth* is what we get with a receiver when we keep the LO at a fixed frequency and sweep the input frequency to measure the response. The resulting bandwidth is a function of the frequency responses of everything in the chain. Their instantaneous bandwidth has a direct effect on the minimum detectable signal. Tunable bandwidth implies that we change the frequency of the LO to track the RF frequency. The bandwidth in this case is only a function of the preselector filter, the LNA, and the mixer. Tunable bandwidth is often many times greater than instantaneous bandwidth.

Gain—The gain of a receiver is the ratio of input signal power to output signal power.

Noise figure—Noise figure of a receiver is a measure of how much the receiver degrades the ratio of signal to noise of the incoming signal. It is related to the minimum detectable signal. If the LO signal has a high AM and/or FM noise, it could degrade the receiver noise figure because, far from the carrier, the AM and FM noise originate from thermal noise. Remember that the effect of LO AM noise is reduced by the balance of the balanced mixer.

1dB compression point—1dB compression point is the power level where the gain of the receiver is reduced by 1dB due to compression.

Linearity—The receiver operates linearly if a 1dB increase in input signal power results in a 1dB increase in IF output signal strength.

Dynamic range—Dynamic range of a receiver is a measurement of the minimum detectable signal to the maximum signal that will start to compress the receiver.

Signal-to-noise ratio (**S/N** or **SNR**)—SNR is a measure of how far a signal is above the noise floor.

Noise factor, noise figure, and noise temperature

- Noise factor is a measure of how the signal to noise ratio is degraded by a device:

    $F =$ noise factor $= (S_{in}/N_{in})/(S_{out}/N_{out})$

    where

    S_{in} is the signal level at the input

    N_{in} is the noise level at the input

    S_{out} is the signal level at the output

    N_{out} is the noise level at the output

- The noise factor of a device is specified with noise from a noise source at room temperature ($N_{in} = KT$), where K is Boltzmann's constant and T is approximately room temperature in Kelvin. KT is somewhere around −174dBm/Hz. Noise figure is the noise factor, expressed in decibels:

    NF (dB) = noise figure = 10log (F).

    $T =$ noise temperature $= 290 \times (F - 1)$.

    1dB NF is about 75 K, and 3dB is 288 K.

    The noise factor contributions of each stage in a four-stage system are given in Equation 1.25:

$$F = F1 + \frac{F2-1}{G1} + \frac{F3-1}{G1G2} + \frac{F4-1}{G1G2G3} \tag{1.25}$$

1.10 TYPES OF RADARS

In **monostatic radars** the transmitting and receiving antennas are colocated. Most radars are monostatic.

In **bistatic radars** the transmitting and receiving antennas are not colocated.

Doppler radar is used to measure the velocity of a target due to its Doppler shift. Police radar is a classic example of Doppler radar.

Frequency-modulated/continuous-wave (FMCW) radar implies that the radar signal is "chirped" or its frequency is varied in time. By varying the frequency in this manner, you can gather both range and velocity information.

Synthetic aperture radar (SAR) uses a moving platform to "scan" the radar in one or two dimensions. Satellite radar images are mostly done using SAR.

1.11 TRANSMITTERS: DEFINITIONS AND FEATURES

Figure 1.4 presents a basic transmitter block diagram.

1.11.1 Transmitter

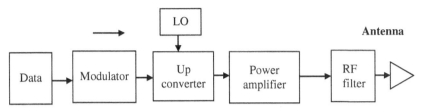

FIGURE 1.4 Transmitter block diagram.

1.11.2 Amplifiers

Class A—The amplifier is biased at close to half of its saturated current. The output conducts during all 360° of phase of the input signal sine wave. Class A does not give maximum efficiency, but provides the best linearity. Drain efficiencies of 50% are possible in class A.

Class B—The power amplifier is biased at a point where it draws nearly 0 DC current; for an FET, this means that it is biased at pinch-off during one half of the input signal sine wave it conducts, but not the other half. Class B amplifier can be very efficient, with theoretical efficiency of 80–85%. However, we are giving up 6dB of gain when we move from class A to class B.

Class C—Class C occurs when the device is biased so that the output conducts for even less than 180° of the input signal. The output power and gain decrease.

Power density—This is a measure of power divided by transistor size. In the case of FETs it is expressed in watts/millimeter. GaN transistors have more than 10 W/mm power density.

Saturated output power (PSAT) is the output power where the P_{in}/P_{out} curve slope goes to zero.

Load pull—This is the process of varying the impedance seen by the *output* of an active device to other than 50 Ω in order to measure performance parameters, in the simplest case, gain. In the case of a power device, a load pull power bench is used to evaluate large signal parameters such as compression characteristics, saturated power, efficiency, and linearity as the output load is varied across the Smith chart.

Harmonic load pull—This is the process of varying the impedance at the output of a device, with separate control of the impedances at F_0, $2 F_0$, $3 F_0$,....

Source pull—This is the process of varying the impedance seen by the input of an active device to other than 50 Ω in order to measure performance parameters. In the case of a low-noise device, source pull is used in a noise parameter extraction setup to evaluate how SNR (noise figure) varies with source impedance.

TABLE 1.8 Power Amplifier Output Power Capabilities

Frequency Band	Solid State	Tube Type
L band through C band	200 W (LDMOS) GaN	
X band	50 W (GaN HEMT device)	3000 W (TWT)
Ka band	6 W (GaAs PHEMT device)	1000 W (klystron)
Q band	4 W (GaAs PHEMT device)	
W band	0.5 W (InP)	1000 W (EIKA) even more! (gyrotron)

Amplifier temperature considerations

In the case of an FET amplifier, the gain drops and the noise figure increases. The gain drop is around −0.006dB/stage/°C.

Noise figure of an LNA increases by +0.006dB/°C. In an LNA, the first stage will dominate the temperature effect.

Power amplifiers

Power amplifiers are used to boost a small signal to a large signal.

Solid-state amplifiers and tube amplifiers are usually employed as power amplifiers.

Power amplifier output power capabilities are listed in Table 1.8.

REFERENCES

[1] ITU. *ITU-R Recommendation V.431: Nomenclature of the Frequency and Wavelength Bands Used in Telecommunications*. Geneva: International Telecommunication Union; 2000.

[2] IEEE Standard 521-2002: Standard Letter Designations for Radar-Frequency Bands. E-ISBN: 0-7381-3356-6; 2003.

[3] AFR 55-44/AR 105-86/OPNAVINST 3430.9A/MCO 3430.1, 27 October 1964 superseded by AFR 55-44/AR 105-86/OPNAVINST 3430.1A/MCO 3430.1A, 6 December 1978: Performing Electronic Countermeasures in the United States and Canada, Attachment 1, ECM Frequency Authorizations.

[4] Belov LA, Smolskiy SM, Kochemasov VN. *Handbook of RF, Microwave, and Millimeter-Wave Components*. Boston: Artech House; 2012. p 27–28.

2

ELECTROMAGNETIC THEORY AND TRANSMISSION LINES FOR RF DESIGNERS

Electromagnetic waves are created by time-varying currents and charges. Propagation of electromagnetic waves in space and in materials obeys Maxwell's equations [1–8]. Electromagnetic fields may be viewed as a combination of electric and magnetic fields. Stationary charges produce electric fields. Moving charges and currents produce magnetic fields.

2.1 DEFINITIONS

Field—A field is a physical quantity that has a value for each point in space and time.

Wavelength—The wavelength is the distance between two sequential equivalent points. Wavelength, λ, is measured in meters.

Wave period—The period is the time, T, for one complete cycle of an oscillation of a wave. The period is measured in seconds.

Frequency—The frequency, f, is the number of periods per unit time (second) and is measured in hertz.

Wideband RF Technologies and Antennas in Microwave Frequencies, First Edition. Dr. Albert Sabban.
© 2016 John Wiley & Sons, Inc. Published 2016 by John Wiley & Sons, Inc.

Phase velocity—The phase velocity, v, of a wave is the rate at which the phase of the wave propagates in space. Phase velocity is measured in meter per second:

$$\lambda = v \times T$$
$$T = \frac{1}{f} \qquad (2.1)$$
$$\lambda = \frac{v}{f}$$

Electromagnetic waves propagate in free space in phase velocity of light:

$$v = 3 \times 10^8 \, \text{m/s} \qquad (2.2)$$

Wavenumber—The wavenumber, k, is the spatial frequency of a wave in radians per unit distance (per meter): $k = 2\pi/\lambda$.

Angular frequency—The angular frequency, ω, represents the frequency in radians per second: $\omega = v \times k = 2\pi f$.

Polarization—A wave is polarized if it oscillates in one direction or plane. The polarization of a transverse wave describes the direction of oscillation in the plane perpendicular to the direction of propagation.

Antenna—Antenna is used to radiate efficiently electromagnetic energy in desired directions. Antennas match radio-frequency systems to space. All antennas may be used to receive or radiate energy. Antennas transmit or receive electromagnetic waves. Antennas convert electromagnetic radiation into electric current, or vice versa. Antennas transmit and receive electromagnetic radiation at radio frequencies. Antennas are a necessary part of all communication links and radio equipment. Antennas are used in systems such as radio and television broadcasting, point-to-point radio communication, wireless LAN, cell phones, radar, medical systems, and spacecraft communication. Antennas are most commonly employed in air or outer space but can also be operated under water, on and inside human body, or even through soil and rock at low frequencies for short distances.

2.2 ELECTROMAGNETIC WAVES

2.2.1 Maxwell's Equations

Maxwell's equations describe how electric and magnetic fields are generated and altered by each other. Maxwell's equations are a classical approximation to the more accurate and fundamental theory of quantum electrodynamics, see Refs. [1–8]. Quantum deviations from Maxwell's equations are usually small. Inaccuracies occur when the particle nature of light is important or when electric fields are strong. Symbols and abbreviations are listed in Tables 2.1 and 2.2.

TABLE 2.1   Symbols and Abbreviations

Dimensions	Parameter	Symbol
Wb/m	Magnetic potential	A
m/s^2	Acceleration	a
Tesla	Magnetic field	B
F	Capacitance	C
V/m	Electric field displacement	D
V/m	Electric field	E
N	Force	F
A/m	Magnetic field strength	H
A	Current	I
A/m^2	Current density	J
H	Self-inductance	L
m	Length	l
H	Mutual inductance	M
C	Charge	q
W/m^2	Poynting vector	$\mathbf{P}$
W	Power	P
Ω	Resistance	R
m^3	Volume	V
m/s^2	Velocity	v
F/m	Dielectric constant	ε
	Relative dielectric constant	ε_r
H/m	Permeability	μ
1/Ω·m	Conductivity	σ
Wb	Magnetic flux	ψ

TABLE 2.2   Symbols and Abbreviations

Parameter	Symbol	Dimensions
Electric field displacement	D	V/m
Electric field	E	V/m
Magnetic field strength	H	A/m
Dielectric constant in space	ε_0	$8.854 \cdot 10^{-12}$ F/m
Permeability in space	μ_0	$\mu_0 = 4\pi \cdot 10^{-7}$ H/m
Volume charge density	ρ_V	C/m^3
Magnetic field	B	Tesla, Wb/m^2
Conductivity	σ	1/Ω·m
Magnetic flux	ψ	Wb
Skin depth	δ_s	m

2.2.2 Gauss's Law for Electric Fields

Gauss's law for electric fields states that the electric flux via any closed surface S is equal to the net charge q divided by the free-space dielectric constant:

$$\oint_S E \times ds = \frac{1}{\varepsilon_0} \int_V \rho dv = \frac{1}{\varepsilon_0} q$$

$$D = \varepsilon E$$

$$\nabla \cdot D = \rho_v$$

$$(2.3)$$

2.2.3 Gauss's Law for Magnetic Fields

Gauss's law for magnetic fields states that the magnetic flux via any closed surface S is equal to zero. There is no magnetic charge in nature:

$$\oint_S B \times ds = 0$$

$$B = \mu H$$

$$\psi_m = \int_S B \times ds$$

$$\nabla \cdot B = 0$$

$$(2.4)$$

2.2.4 Ampère's Law

The original "Ampère's law" stated that magnetic fields can be generated by electrical current. Ampère's law has been corrected by Maxwell who stated that magnetic fields can be generated also by time-variant electric fields. The corrected "Ampère's law" shows that a changing magnetic field induces an electric field and also a time-variant electric field induces a magnetic field:

$$\oint_C \frac{B}{\mu_0} \times dl = \int_S J \times ds + \frac{d}{dt} \int_s \varepsilon_0 E \times ds$$

$$\nabla X H = J + \frac{\partial D}{\partial t}$$

$$J = \sigma E$$

$$i = \int_S J \times ds$$

$$(2.5)$$

2.2.5 Faraday's Law

Faraday's law describes how a propagating time-varying magnetic field through a surface S creates an electric field, as shown in Figure 2.1:

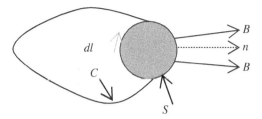

FIGURE 2.1   Faraday's law.

$$\oint_C E \times dl = -\frac{d}{dt}\int_s B \times ds$$

$$\nabla X E = -\frac{\partial B}{\partial t}$$

(2.6)

2.2.6   Wave Equations

The variation of electromagnetic waves as function of time may be written as $e^{j\omega t}$. The derivative as function of time is $j\omega e^{j\omega t}$. Maxwell's equations may be written as

$$\nabla X E = -j\omega\mu H$$
$$\nabla X H = (\sigma + j\omega\varepsilon)E$$

(2.7)

A ∇X (curl) operation on the electric field E results in

$$\nabla X \nabla X E = -j\omega\mu\nabla X H \tag{2.8}$$

By substituting the expression of $\nabla X H$ in equation, we get

$$\nabla X \nabla X E = -j\omega\mu(\sigma + j\omega\varepsilon)E \tag{2.9}$$

$$\nabla X \nabla X E = -\nabla^2 E + \nabla(\nabla \cdot E) \tag{2.10}$$

In free space there are no charges so $\nabla \cdot E = 0$. We get the wave equation for electric field:

$$\nabla^2 E = j\omega\mu(\sigma + j\omega\varepsilon)E = \gamma^2 E$$
$$\gamma = \sqrt{j\omega\mu(\sigma + j\omega\varepsilon)} = \alpha + j\beta$$

(2.11)

where γ is the complex propagation constant, α represents losses in the medium, and β represents the wave phase constant in radian per meter. If we follow after the same procedure on the magnetic field, we will get the wave equation for the magnetic field:

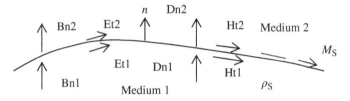

FIGURE 2.2   Fields between two media.

$$\nabla^2 H = j\omega\mu(\sigma + j\omega\varepsilon)H = \gamma^2 H \tag{2.12}$$

The law of conservation of energy imposes boundary conditions on the electric and magnetic fields. When an electromagnetic wave travels from medium 1 to medium 2, the electric and magnetic fields should be continuous as presented in Figure 2.2.

General boundary conditions:

$$n \times (D_2 - D_1) = \rho_S \tag{2.13}$$

$$n \times (B_2 - B_1) = 0 \tag{2.14}$$

$$(E_2 - E_1) \times n = M_S \tag{2.15}$$

$$n \times (H_2 - H_1) = J_S \tag{2.16}$$

Boundary conditions for dielectric medium:

$$n \times (D_2 - D_1) = 0 \tag{2.17}$$

$$n \times (B_2 - B_1) = 0 \tag{2.18}$$

$$(E_2 - E_1) \times n = 0 \tag{2.19}$$

$$n \times (H_2 - H_1) = 0 \tag{2.20}$$

Boundary conditions for conductor:

$$n \times (D_2 - D_1) = \rho_S \tag{2.21}$$

$$n \times (B_2 - B_1) = 0 \tag{2.22}$$

$$(E_2 - E_1) \times n = 0 \tag{2.23}$$

$$n \times (H_2 - H_1) = J_S \tag{2.24}$$

The condition for good conductor is $\sigma \gg \omega\epsilon$:

$$\gamma = \sqrt{j\omega\mu(\sigma + j\omega\varepsilon)} = \alpha + j\beta \approx (1+j)\sqrt{\frac{\omega\mu\sigma}{2}} \tag{2.25}$$

$$\delta_s = \sqrt{\frac{2}{\omega\mu\sigma}} = \frac{1}{\alpha} \qquad (2.26)$$

The conductor skin depth is given as δ_s. The wave attenuation is α.

2.3 TRANSMISSION LINES

Transmission lines are used to transfer electromagnetic energy from one point to another with minimum losses over a wideband of frequencies [1–8]. There are three major types of transmission lines: transmission lines with a cross section that is very small compared to wavelength on which the dominant mode of propagation is the transverse electromagnetic (TEM) mode, closed rectangular and cylindrical conducting tubes on which the dominant modes of propagation are the transverse electric (TE) mode and transverse magnetic (TM) mode, and open boundary structures with cross section greater than 0.1λ that may support surface wave mode of propagation.

For TEM modes $E_z = H_z = 0$. For TE modes $E_z = 0$. For TM modes $H_z = 0$.

Voltages and currents in transmission lines may be derived by using the transmission line equations. Transmission line equations may be derived by employing Maxwell's equations and the boundary conditions on the transmission line section shown in Figure 2.3. Equation 2.27 is the first lossless transmission line equation. l_e is self-inductance per length:

$$\frac{\partial V}{\partial Z} = -\frac{\partial}{\partial t} l_e I = -l_e \frac{\partial I}{\partial t} \qquad (2.27)$$

$$\frac{\partial I}{\partial Z} = -c\frac{\partial V}{\partial t} - gV \qquad (2.28)$$

Equation 2.28 is the second lossless transmission line equation, $c = \dfrac{\Delta C}{\Delta Z}$ F/m:

$$g = \frac{\Delta G}{\Delta Z} \text{ m}/\Omega.$$

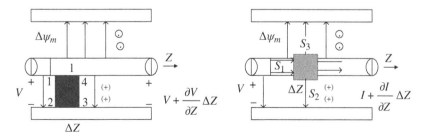

FIGURE 2.3   Transmission line geometry.

Equation 2.29 is the first transmission line equation with losses:

$$-\frac{\partial v}{\partial z} = Ri + L\frac{\partial i}{\partial t} \tag{2.29}$$

Equation 2.30 is the second transmission line equation with losses:

$$-\frac{\partial i}{\partial z} = Gv + C\frac{\partial v}{\partial t} \tag{2.30}$$

where $C = \dfrac{\Delta C(z)}{\Delta Z}$ F/m $\quad L = \dfrac{\Delta L(z)}{\Delta Z}$ H/m $\quad R = \dfrac{1}{G} = \dfrac{\Delta R(z)}{\Delta Z}$ Ω/m

By differentiating Equation 2.29 with respect to z and by differentiating Equation 2.30 with respect to t and adding the result, we get

$$-\frac{\partial^2 v}{\partial z^2} = RGv + (RC + LG)\frac{\partial v}{\partial t} + LC\frac{\partial^2 v}{\partial t^2} \tag{2.31}$$

By differentiating Equation 2.30 with respect to z and by differentiating Equation 2.29 with respect to t and adding the result, we get

$$-\frac{\partial^2 i}{\partial z^2} = RGi + (RC + LG)\frac{\partial i}{\partial t} + LC\frac{\partial^2 i}{\partial t^2} \tag{2.32}$$

Equations 2.31 and 2.32 are analog to the wave equations. The solution of these equations is a superposition of a forward wave, $+z$, and a backward wave, $-z$:

$$V(z,t) = V_+\left(t - \frac{z}{v}\right) + V_-\left(t + \frac{z}{v}\right)$$

$$I(z,t) = Y_0\left\{V_+\left(t - \frac{z}{v}\right) - V_-\left(t + \frac{z}{v}\right)\right\} \tag{2.33}$$

where Y_0 is the characteristic admittance of the transmission line $Y_0 = 1/Z_0$.

The variation of electromagnetic waves as function of time may be written as $e^{j\omega t}$. The derivative as function of time is $j\omega e^{j\omega t}$. By using these relations we may write phasor transmission line equations:

$$\frac{dV}{dZ} = -ZI$$

$$\frac{dI}{dZ} = -YV$$

$$\frac{d^2V}{dz^2} = \gamma^2 V \tag{2.34}$$

$$\frac{d^2I}{dz^2} = \gamma^2 I$$

where

$$Z = R + j\omega L \quad \Omega/\text{m}$$
$$Y = G + j\omega C \quad \text{m}/\Omega \qquad (2.35)$$
$$\gamma = \alpha + j\beta = \sqrt{ZY}$$

The solution of the transmission line equations in harmonic steady state is

$$v(z,t) = \text{Re } V(z)e^{j\omega t}$$
$$i(z,t) = \text{Re } I(z)e^{j\omega t} \qquad (2.36)$$

$$V(z) = V_+ e^{-\gamma z} + V_- e^{\gamma z}$$
$$I(z) = I_+ e^{-\gamma z} + I_- e^{\gamma z} \qquad (2.37)$$

For a lossless transmission line, we may write

$$\frac{dV}{dZ} = -j\omega L I$$
$$\frac{dI}{dZ} = -j\omega C V$$
$$\frac{d^2 V}{dz^2} = -\omega^2 LCV \qquad (2.38)$$
$$\frac{d^2 I}{dz^2} = -\omega^2 LCI$$

The solution of the lossless transmission line equations is

$$V(z) = e^{j\omega t}\left(V^+ e^{-j\beta z} + V^- e^{j\beta z}\right)$$
$$I(z) = Y_0\left(e^{j\omega t}\left(V^+ e^{-j\beta z} - V^- e^{j\beta z}\right)\right) \qquad (2.39)$$
$$v_p = \frac{\omega}{\beta} = \frac{1}{\sqrt{LC}} = \frac{1}{\sqrt{\mu\varepsilon}}$$

where v_p is the phase velocity.

$$Z_0 = \frac{V_+}{I_+} = \frac{V_-}{I_-} = \sqrt{\frac{(R+j\omega L)}{(G+j\omega C)}}$$
$$\text{For } R = 0 \quad G = 0 \qquad (2.40)$$
$$Z_0 = \frac{V_+}{I_+} = \frac{V_-}{I_-} = \sqrt{\frac{L}{C}}$$

where Z_0 is the characteristic impedance of the transmission line.

2.3.1 Waves in Transmission Lines

A load Z_L is connected, at $z = 0$, to a transmission line with impedance Z_0. The voltage on the load is V_L. The current on the load is I_L, see Figure 2.4.

$$V(0) = V_L = I(0) \cdot Z_L$$
$$I(0) = I_L \tag{2.41}$$

For $z = 0$ we can write

$$V(0) = I(0) \cdot Z_L = V_+ + V_-$$
$$I(0) = Y_0 (V_+ - V_-) \tag{2.42}$$

By substituting $I(0)$ in $V(0)$, we get

$$Z_L = Z_0 \frac{V_+ + V_-}{(V_+ - V_-)}$$
$$Z_L = Z_0 \frac{1 + (V_-/V_+)}{1 - (V_-/V_+)} \tag{2.43}$$

The ratio V_-/V_+ is defined as reflection coefficient, $\Gamma_L = V_-/V_+$:

$$Z_L = Z_0 \frac{1 + \Gamma_L}{1 - \Gamma_L} \tag{2.44}$$

$$\Gamma_L = \frac{(Z_L/Z_0) - 1}{(Z_L/Z_0) + 1} \tag{2.45}$$

The reflection coefficient as function of z may be written as

$$\Gamma(Z) = \frac{V_-}{V_+} = \frac{V_- e^{j\beta z}}{V_+ e^{-j\beta z}} = \Gamma_L e^{2j\beta z} \tag{2.46}$$

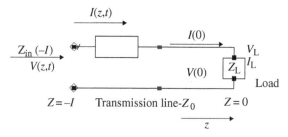

FIGURE 2.4 Transmission line with load.

The input impedance as function of z may be written as

$$Z_{\text{in}}(z) = \frac{V(z)}{I(z)} = \frac{V_+ e^{-j\beta z} + V_- e^{j\beta z}}{\left(V_+ e^{-j\beta z} - V_- e^{j\beta z}\right) Y_0}$$
$$= Z_0 \frac{1 + \Gamma_L e^{2j\beta z}}{1 - \Gamma_L e^{2j\beta z}} \tag{2.47}$$

$$Z_{\text{in}}(-l) = Z_L \frac{\cos\beta l + j(Z_0/Z_L)\sin\beta l}{\cos\beta l + j(Z_L/Z_0)\sin\beta l} \tag{2.48}$$

The voltage and current as function of z may be written as

$$V(z) = V_+ e^{-j\beta z}(1 + \Gamma(z))$$
$$I(z) = Y_0 V_+ e^{-j\beta z}(1 - \Gamma(z)) \tag{2.49}$$

The ratio between the maximum and minimum voltage along a transmission line is called voltage standing wave ratio (VSWR) S:

$$S = \left|\frac{V(z)\text{max}}{V(z)\text{min}}\right| = \frac{1 + |\Gamma(z)|}{1 - |\Gamma(z)|} \tag{2.50}$$

2.4 MATCHING TECHNIQUES

Usually in communication systems the load impedance is not the same as the impedance of commercial transmission lines. We will get maximum power transfer from source to load if impedances are matched, see Refs. [1–8]. A perfect impedance match corresponds to a VSWR 1 : 1. A reflection coefficient magnitude of zero is a perfect match: a value of 1 is perfect reflection. The reflection coefficient (Γ) of a short circuit has a value of -1 (1 at an angle of 180°). The reflection coefficient of an open circuit is one at an angle of 0°. The **return loss** of a load is merely the magnitude of the reflection coefficient expressed in decibels. The correct equation for return loss is return loss $= -20 \times \log[\text{mag}(\Gamma)]$.

For a maximum voltage V_m in a transmission line, the maximum power will be

$$P_{\text{max}} = \frac{V_m^2}{Z_0} \tag{2.51}$$

For an unmatched transmission line, the maximum power will be

$$P_{\text{max}} = \frac{\left(1 - |\Gamma|^2\right) V_+^2}{Z_0}$$
$$V_{\text{max}} = (1 + |\Gamma|) V_+ \tag{2.52}$$
$$V_+^2 = \frac{V_{\text{max}}^2}{(1 + |\Gamma|)^2}$$

$$P_{max} = \frac{1-|\Gamma|}{1+|\Gamma|} \cdot \frac{V_{max}^2}{Z_0} = \frac{V_{max}^2}{VSWR \cdot Z_0} \tag{2.53}$$

A $2:1$ VSWR will result in half of the maximum power transferred to the load. The reflected power may cause damage to the source.

Equation 2.47 indicates that there is a one-to-one correspondence between the reflection coefficient and input impedance. A movement of distance z along a transmission line corresponds to a change of $e^{-2j\beta z}$, which represent a rotation via an angle of $2\beta z$. In the reflection coefficient plane, we may represent any normalized impedance by contours of constant resistance, r, and contours of constant reactance, x. The corresponding impedance moves on a constant radius circle via an angle of $2\beta z$ to a new impedance value. Those relations may be presented by a graphical aid called Smith chart. Those relations may be presented by the following set of equations:

$$\Gamma_L = \frac{(Z_L/Z_0)-1}{(Z_L/Z_0)+1}$$

$$z(l) = \frac{Z_L}{Z_0} = r + jx \tag{2.54}$$

$$\Gamma_L = \frac{r+jx-1}{r+jx+1} = p + jq$$

$$\frac{Z(z)}{Z_0} = \frac{1+\Gamma(z)}{1-\Gamma(z)} = r + jx$$

$$\Gamma(z) = u + jv \tag{2.55}$$

$$\frac{1+u+jv}{1-u-jv} = r + jx$$

$$\left(u - \frac{r}{1+r}\right)^2 + v^2 = \frac{1}{(1+r)^2}$$

$$\left(u-1\right)^2 + \left(v - \frac{1}{x}\right)^2 = \frac{1}{x^2} \tag{2.56}$$

Equation 2.56 presents two families of circles in the reflection coefficient plane. The first family is contours of constant resistance, r, and the second family is contours of constant reactance, x. The center of the Smith chart is $r = 1$. Moving away from the load corresponds to moving around the chart in a clockwise direction. Moving away from the generator toward the load corresponds to moving around the chart in a counterclockwise direction, as presented in Figure 2.5. A complete revolution around the chart in a clockwise direction corresponds to a movement of half wavelength away from the load. The Smith chart may be employed to calculate the reflection coefficient and the input impedance for a given transmission line and load impedance. If we are at a matched impedance condition at the center of the Smith Chart, any length of transmission line with impedance Z_0 does nothing to the input match. But if the reflection coefficient of the network (S_{11}) is at some nonideal impedance, adding transmission

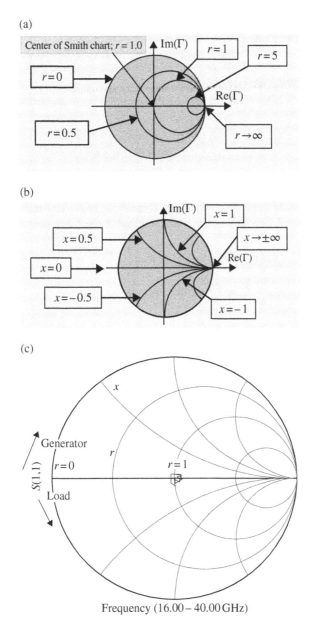

FIGURE 2.5   (a) R circles, (b) X circles, and (c) the Smith chart.

line between the network and the reference plane rotates the observed reflection coefficient clockwise about the center of the Smith chart. Further, the rotation occurs at a fixed radius (and VSWR magnitude) if the transmission line has the same characteristic impedance as the source impedance Z_0. Adding a quarter wavelength means a

180° phase rotation. Adding one quarter wavelength from a short circuit moves us 180° to the right side of the chart to an open circuit.

2.4.1 The Smith Chart Guidelines

The Smith chart contains almost all possible complex impedances within one circle:

- The Smith chart horizontal centerline represents resistance/conductance.
- Zero resistance is located on the left end of the horizontal centerline.
- Infinite resistance is located on the right end of the horizontal centerline.
- Horizontal centerline is the resistive/conductive horizontal scale of the chart.
    Impedances in the Smith chart are normalized to the characteristic impedance of the transmission line and are independent of the characteristic impedance of the transmission.
- The center of the line and also of the chart is 1.0 point, where $R = Z_0$ or $G = Y_0$.

At point $r = 1.0$, $Z = Z_0$ and no reflection will occur.

2.4.2 Quarter-Wave Transformers

A quarter-wave transformer may be used to match a device with impedance Z_L to a system with impedance Z_0, as shown in Figure 2.6. A quarter-wave transformer is a matching network with bandwidth somewhat inversely proportional to the relative mismatch we are trying to match. For a single-stage quarter-wave transformer, the correct transformer impedance is the geometric mean between the impedances of the load and the source. If we substitute to Equation 2.47 $l = \lambda/4$, $\beta = 2\pi/\lambda$ we get

$$\bar{Z}(-l) = \frac{(Z_L/Z_{02})\cos\beta(\lambda/4) + j\sin\beta(\lambda/4)}{\cos\beta(\lambda/4) + j(Z_L/Z_{02})\sin\beta(\lambda/4)} = \frac{Z_{02}}{Z_L}$$

$$\bar{Z}(-l) = \frac{Z_{01}}{Z_{02}} = \frac{Z_{02}}{Z_L} \tag{2.57}$$

We will achieve matching when

$$Z_{02} = \sqrt{Z_L Z_{01}} \tag{2.58}$$

For complex Z_L values Z_{02} also will be complex impedance. However, standard transmission lines have real impedance values. To match a complex

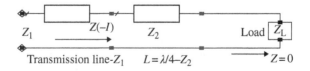

FIGURE 2.6 Quarter-wave transformer.

$Z_L = R + jX$, we transform Z_L to a real impedance $Z_{L1} - jX$ to Z_L. Connecting a capacitor $-jX$ to Z_L is not practical at high frequencies. A capacitor at high frequencies has parasitic inductance and resistance. A practical method to transform Z_L to a real impedance Z_{L1} is to add a transmission line with impedance Z_0 and length l to get a real value Z_{L1}.

2.4.3   Wideband Matching: Multisection Transformers

Multisection quarter-wave transformers are employed for wideband applications. Responses such as Chebyshev (equiripple) and maximally flat are possible for multisection transformers. Each section brings us to intermediate impedance. In four section transformers from 25 to 50 Ω, intermediate impedances are chosen by using an arithmetic series. For an arithmetic series the steps are equal, $\Delta Z = 6.25\,\Omega$, so the impedances are 31.25, 37.5, and 43.75 Ω. Solving for the transformers yields $Z_1 = 27.951$, $Z_2 = 34.233$, $Z_3 = 40.505$, and $Z_4 = 46.771\,\Omega$. A second solution to multisection transformers involves a geometric series from impedance Z_L to impedance Z_S. Here the impedance from one section to the next adjacent section is a constant ratio.

2.4.4   Single-Stub Matching

A device with admittance Y_L can be matched to a system with admittance Y_0 by using a shunt or series single stub. At a distance l from the load, we can get a normalized admittance $\bar{Y}_{in} = 1 + j\bar{B}$. By solving Equation 2.59 we can calculate l:

$$\bar{Y}(l) = \frac{1 + j(Z_L/Z_0)\tan\beta l}{(Z_L/Z_0) + j\tan\beta l} = 1 + j\bar{B} \tag{2.59}$$

At this location we can add a shunt stub with normalized input susceptance $-j\bar{B}$ to yield $\bar{Y}_{in} = 1$ as presented in Equation 2.60. $\bar{Y}_{in} = 1$ represents a matched load. The stub can be short-circuited line or open-circuited line. The susceptance $\bar{B}$ is given in Equation 2.60:

$$\begin{aligned} \bar{Y}_{in} &= 1 + j\bar{B} \\ \bar{Y}_{1n} &= -j\bar{B} \\ \bar{B} &= ctg\beta l_1 \end{aligned} \tag{2.60}$$

The length l_1 of the short-circuited line, as shown in Figure 2.7, may be calculated by solving Equation 2.61:

$$\mathrm{Im}\left\{ \frac{1 + j\frac{Z_L}{Z_0}\tan\beta l}{\frac{Z_L}{Z_0} + j\tan\beta l} \right\} = ctg\beta l_1 \tag{2.61}$$

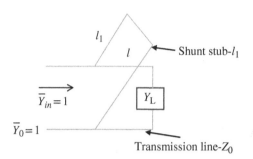

FIGURE 2.7   Single-stub matching.

2.5   COAXIAL TRANSMISSION LINE

A coax transmission line consists of two round conductors in which one completely surrounds the other, with the two separated by a continuous solid dielectric [1–6]. The desired propagation mode is TEM. The major advantage of coax over microstrip line is that the transmission line does not radiate. The disadvantage is that coax lines are more expensive. Coax lines are usually employed up to 18 GHz. Coax lines are highly expensive at frequencies higher than 18 GHz. To obtain good performance at higher frequencies, small diameter cables are required to stay below the cutoff frequency. Maxwell's laws are employed to compute the electric and magnetic fields in the coax transmission line. Cross section of the coaxial transmission line is shown in Figure 2.8:

$$\oint_S E \cdot ds = \frac{1}{\varepsilon_0} \int_V \rho dv = \frac{1}{\varepsilon_0} q \tag{2.62}$$

$$E_r = \frac{\rho_L}{2\pi r \varepsilon}$$

$$V = \int_a^b E \cdot dr = \int_a^b \frac{\rho_L}{2\pi r \varepsilon} \cdot dr = \frac{\rho_L}{2\pi \varepsilon} \ln \frac{b}{a} \tag{2.63}$$

Ampère's law is employed to calculate the magnetic field:

$$\oint H \cdot dl = 2\pi r H_\phi = I \tag{2.64}$$

$$H_\phi = \frac{I}{2\pi r}$$

$$V = \int_a^b E_r \cdot dr = -\int_a^b \eta H_\phi \cdot dr = \frac{I\eta}{2\pi} \ln \frac{b}{a} \tag{2.65}$$

$$Z_0 = \frac{V}{I} = \frac{\eta}{2\pi} \ln \frac{b}{a} \quad \eta = \sqrt{\mu/\varepsilon}$$

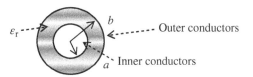

FIGURE 2.8 Coaxial transmission line.

TABLE 2.3 Industry-Standard Coax Cables

Cable Type	Outer Diameter (Inches)	$2b$ (Inches)	$2a$ (Inches)	Z_0 (Ω)	f_c (GHz)
RG-8A	0.405	0.285	0.089	50	14.0
RG-58A	0.195	0.116	0.031	50	35.3
RG-174	0.100	0.060	0.019	50	65.6
RG-196	0.080	0.034	0.012	50	112
RG-214	0.360	0.285	0.087	50	13.9
RG-223	0.216	0.113	0.037	50	34.6
SR-085	0.085	0.066	0.0201	50	60.2
SR-141	0.141	0.1175	0.0359	50	33.8
SR-250	0.250	0.210	0.0641	50	18.9

The power flow in the coaxial transmission line may be calculated by using the Poynting vector:

$$P = (E \times H) \times n = EH$$

$$P = \frac{VI}{2\pi r^2; \ln(b/a)}$$

$$W = \int_s P \times ds = \int_a^b \frac{VI}{2\pi r^2; \ln(b/a)} 2\pi r \, dr = VI$$

(2.66)

Table 2.3 presents several industry-standard coax cables. The cable dimensions, cable impedance, and cable cutoff frequency are given in Table 2.3. RG cables are flexible cables. SR cables are semirigid cables.

2.5.1 Cutoff Frequency and Wavelength of Coax Cables

f_c—cutoff frequency

The criterion for cutoff frequency is that the circumference at the midpoint inside the dielectric must be less than a wavelength. Therefore the cutoff wavelength for the TE_{01} mode is $\lambda_c = \pi(a+b)\sqrt{\mu_r \varepsilon_r}$.

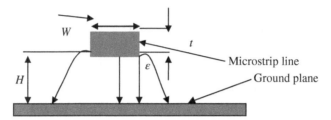

FIGURE 2.9   Microstrip line cross section.

2.6   MICROSTRIP LINE

Microstrip is a planar printed transmission line. Microstrip is the most popular RF transmission line over the last 20 years [3, 7]. Microstrip transmission lines consist of a conductive strip of width "W" and thickness "t" and a wider ground plane, separated by a dielectric layer of thickness "H." In practice, microstrip line is usually made by etching circuitry on a substrate that has a ground plane on the opposite face. Cross section of the microstrip line is shown in Figure 2.9. The major advantage of microstrip over stripline is that all components can be mounted on top of the board. The disadvantage is that when high isolation is required such as in a filter or switch, some external shielding is needed. Microstrip circuits may radiate, causing unintended circuit response. Microstrip is dispersive, meaning that signals of different frequencies travel at slightly different speeds. Other microstrip line configurations are offset stripline and suspended air microstrip line. For microstrip line not all of the fields are constrained to the same dielectric. At the line edges the fields pass via air and dielectric substrate. The effective dielectric constant should be calculated.

2.6.1   Effective Dielectric Constant

Part of the fields in the microstrip line structure exists in air and the other part of the fields exists in the dielectric substrate. The effective dielectric constant is somewhat less than the substrate's dielectric constant.

The effective dielectric constant of the microstrip line is calculated by

$$\text{For} \left(\frac{W}{H} \right) < 1 \tag{2.67a}$$

$$\varepsilon_e = \frac{\varepsilon_r + 1}{2} + \frac{\varepsilon_r - 1}{2} \left[\left(1 + 12 \left(\frac{H}{W} \right) \right)^{-0.5} + 0.04 \left(1 - \left(\frac{W}{H} \right) \right)^2 \right]$$

$$\text{For} \left(\frac{W}{H} \right) \geq 1 \tag{2.67b}$$

$$\varepsilon_e = \frac{\varepsilon_r + 1}{2} + \frac{\varepsilon_r - 1}{2} \left[\left(1 + 12 \left(\frac{H}{W} \right) \right)^{-0.5} \right]$$

This calculation ignores strip thickness and frequency dispersion, but their effects are negligible.

2.6.2   Characteristic Impedance

The characteristic impedance Z_0 is a function of the ratio of the height to the width W/H of the transmission line and also has separate solutions depending on the value of W/H. The characteristic impedance Z_0 of microstrip is calculated by

$$\text{For} \left(\frac{W}{H}\right) < 1 \tag{2.68a}$$

$$Z_0 = \frac{60}{\sqrt{\varepsilon_e}} \ln \left[8\left(\frac{H}{W}\right) + 0.25\left(\frac{H}{W}\right) \right] \Omega$$

$$\text{For} \left(\frac{W}{H}\right) \geq 1 \tag{2.68b}$$

$$Z_0 = \frac{120\pi}{\sqrt{\varepsilon_e} \left[\left(\frac{H}{W}\right) + 1.393 + 0.66 \times \ln\left(\frac{H}{W} + 1.444\right) \right]} \Omega$$

We can calculate Z_0 by using Equation 2.68 for a given (W/H). However, to calculate (W/H) for a given Z_0, we first should calculate ε_e. However, to calculate ε_e we should know (W/H). We first assume that $\varepsilon_e = \varepsilon_r$ and compute (W/H) for this value of (W/H); we compute ε_e. Then we compute a new value of (W/H). Two to three iterations are needed to calculate accurate values of (W/H) and ε_e. We may calculate (W/H) with around 10% accuracy by using Equation 2.69:

$$\frac{W}{H} = 8\frac{\sqrt{\left[e^{\frac{Z_0}{42.4}\sqrt{(\varepsilon_r + 1)}} - 1\right]\left[\frac{7 + (4/\varepsilon_r)}{11}\right] + \left[\frac{1 + (1/\varepsilon_r)}{0.81}\right]}}{\left[e^{\frac{Z_0}{42.4}\sqrt{(\varepsilon_r + 1)}} - 1\right]} \tag{2.69}$$

2.6.3   Higher-Order Transmission Modes in Microstrip Line

In order to prevent higher-order transmission modes, we should limit the thickness of the microstrip substrate to 10% of a wavelength. The cutoff frequency of the higher-order transmission mode is given as $f_c = \frac{c}{4H\sqrt{\varepsilon - 1}}$.

Examples: Higher-order modes will not propagate in microstrip lines printed on alumina substrate 15 mil thick up to 18 GHz. Higher-order modes will not propagate in microstrip lines printed on GaAs substrate 4 mil thick up to 80 GHz. Higher-order modes will not propagate in microstrip lines printed on quartz

TABLE 2.4   Examples of Microstrip Line Parameters

Substrate	W/H	Impedance (Ω)
Alumina ($\varepsilon_r = 9.8$)	0.95	50
GaAs ($\varepsilon_r = 12.9$)	0.75	50
$\varepsilon_r = 2.2$	3	50

substrate 5 mil thick up to 120 GHz. Examples of microstrip line parameters are listed in Table 2.4.

2.6.4   Losses in Microstrip Line

Losses in microstrip line are due to conductor loss, radiation loss, and dielectric loss.

2.6.5   Conductor Loss

Conductor loss may be calculated by using Equation 2.70:

$$\alpha_c = 8.686 \log\left(\frac{R_S}{(2WZ_0)}\right) \text{ dB/length}$$

$$R_S = \sqrt{\pi f \mu \rho} \quad \text{skin resistance}$$

(2.70)

Conductor losses may also be calculated by defining an equivalent loss tangent δ_c, given by $\delta_c = \delta_s / h$, where $\delta_s = \sqrt{2/\omega\mu\sigma}$ and where σ is the strip conductivity, h is the substrate height, and μ is the substrate permeability.

2.6.6   Dielectric Loss

Dielectric loss may be calculated by using Equation 2.71:

$$\alpha_d = 27.3 \frac{\varepsilon_r}{\sqrt{\varepsilon_{eff}}} \frac{\varepsilon_{eff}-1}{\varepsilon_r-1} \frac{tg\delta}{\lambda_0} \text{ dB/cm}$$

$$tg\delta = \text{dielectric loss coefficent}$$

(2.71)

Losses in microstrip lines are presented in Tables 2.5, 2.6, and 2.7 for several microstrip line structures. For example, total loss of a microstrip line presented in Table 2.6 at 40 GHz is 0.5dB/cm, and total loss of a microstrip line presented in Table 2.7 at 40 GHz is 1.42dB/cm. We may conclude that losses in microstrip lines limit the applications of microstrip technology at mm-wave frequencies.

TABLE 2.5   Microstrip Line Losses for Alumina Substrate 10 mil Thick[a]

Frequency (GHz)	Tangent Loss (dB/cm)	Metal Loss (dB/cm)	Total Loss (dB/cm)
10	−0.005	−0.12	−0.124
20	−0.009	−0.175	−0.184
30	−0.014	−0.22	−0.23
40	−0.02	−0.25	−0.27

[a] Line parameters. Alumina, $H = 254\,\mu m$ (10 mils), $W = 247\,\mu m$, $\varepsilon^r = 9.9$, $\tan^0\delta = 0.0002$, 3 μm gold, conductivity 3.5E7 mhos/m.

TABLE 2.6   Microstrip Line Losses for Alumina Substrate 5 mil Thick[a]

Frequency (GHz)	Tangent Loss (dB/cm)	Metal Loss (dB/cm)	Total Loss (dB/cm)
10	−0.004	−0.23	−0.23
20	−0.009	−0.333	−0.34
30	−0.013	−0.415	−0.43
40	−0.018	−0.483	−0.5

[a] Alumina, $H = 127\,\mu m$ (5 mils), $W = 120\,\mu m$, $\varepsilon^r = 9.9$, $\tan^0\delta = 0.0002$, 3 μm gold, conductivity 3.5E7 mhos/m.

TABLE 2.7   Microstrip Line Losses for GaAs Substrate 2mil Thick[a]

Frequency (GHz)	Tangent Loss (dB/cm)	Metal Loss (dB/cm)	Total Loss (dB/cm)
10	−0.010	−0.66	−0.67
20	−0.02	−0.96	−0.98
30	−0.03	−1.19	−1.22
40	−0.04	−1.38	−1.42

[a] GaAs, $H = 50\,\mu m$ (2 mils), $W = 34\,\mu m$, $\varepsilon^r = 12.88$, $\tan^0\delta = 0.0004$, 3 μm gold, conductivity 3.5E7 mhos/m.

2.7   MATERIALS

In Table 2.8 hard materials are presented. Alumina is the most popular hard substrate in microwave integrated circuits (MICs). Gallium arsenide is the most popular hard substrate in monolithic microwave integrated circuit (MIMIC) technology at microwave frequencies.

In Table 2.9 soft materials are presented. Duroid is the most popular soft substrate in MIC circuits and in printed antenna industry. Dielectric losses in duroid are significantly lower than dielectric losses in FR4 substrate. However, the cost of FR4 substrate is significantly lower than the cost of duroid. Commercial MIC devices use usually FR4 substrate. Duroid is the most popular soft substrate used in the development of printed antennas with high efficiency at microwave frequencies.

TABLE 2.8 Hard Materials: Ceramics

Material	Symbol or Formula	Dielectric Constant	Dissipation Factor (tan δ)	Coefficient of Thermal Expansion (ppm/°C)	Thermal Cond. (W/m °C)	Mass Density (g/cm^3)
Alumina 99.5%	Al_2O_3	9.8	0.0001	8.2	35	3.97
Alumina 96%	Al_2O_3	9.0	0.0002	8.2	24	3.8
Aluminum nitride	AlN	8.9	0.0005	7.6	290	3.26
Beryllium oxide	BeO	6.7	0.003	6.05	250	
Gallium arsenide	GaAs	12.88	0.0004	6.86	46	5.32
Indium phosphide	InP	12.4				
Quartz		3.8	0.0001	0.6	5	
Sapphire		9.3, 11.5				
Silicon (high resistivity)	Si (HRS)	4		2.5	138	2.33
Silicon carbide	SiC	10.8	0.002	4.8	350	3.2

TABLE 2.9 Soft Materials

Manufacturer and Material	Symbol or Formula	Relative Dielectric Constant (ϵ_r)	Tolerance on Dielectric Constant	Tan δ	Mass Density (g/cm^3)	Thermal Conductivity (W/m·°C)	Coefficient of Thermal Expansion (ppm/°C x/y/z)
Rogers duroid 5870	PTFE/random glass	2.33	0.02	0.0012	2.2	0.26	22/28/173
Rogers duroid 5880	PTFE/random glass	2.2	0.02	0.0012	2.2	0.26	31/48/237
Rogers duroid 6002	PTFE/random glass	2.94	0.04	0.0012	2.1	0.44	16/16/24
Rogers duroid 6006	PTFE/random glass	6	0.15	0.0027	2.7	0.48	38/42/24
Rogers duroid 6010	PTFE/random glass	10.2–10.8	0.25	0.0023	2.9	0.41	24/24/24
FR4	Glass/epoxy	4.8		0.022		0.16	
Polyethylene		2.25					
Polyflon CuFlon	PTFE	2.1		0.00045			12.9
Polyflon PolyGuide	Polyolefin	2.32		0.0005			108
Polyflon NorCLAD	Thermoplastic	2.55		0.0011			53
Polyflon Clad Ultem	Thermoplastic	3.05		0.003	1.27		56
PTFE	PTFE	2.1		0.0002	2.1	0.2	
Rogers R/flex 3700	Thermally stable thermoplastic	2.0		0.002			8

(*Continued*)

TABLE 2.9 (*Continued*)

Manufacturer and Material	Symbol or Formula	Relative Dielectric Constant (ϵ_r)	Tolerance on Dielectric Constant	Tan δ	Mass Density (g/cm³)	Thermal Conductivity (W/m Â°C)	Coefficient of Thermal Expansion (ppm/Â°C x/y/z)
Rogers RO3003	PTFE ceramic	3	0.04	0.0013	2.1	0.5	17/17/24
Rogers RO3006	PTFE ceramic	6.15	0.15	0.0025	2.6	0.61	17/17/24
Rogers RO3010	PTFE ceramic	10.2	0.3	0.0035	3	0.66	17/17/24
Rogers RO3203	PTFE ceramic	3.02					
Rogers RO3210	PTFE ceramic	10.2					
Rogers RO4003	Thermoset plastic ceramic glass	3.38	0.05	0.0027	1.79	0.64	11/14/46
Rogers RO4350B	Thermoset plastic ceramic glass	3.48	0.05	0.004	1.86	0.62	14/16/50
Rogers TMM 3	Ceramic/thermoset	3.27	0.032	0.002	1.78	0.7	15/15/23
Rogers TMM 10	Ceramic/thermoset	9.2	0.23	0.0022	2.77	0.76	21/21/20

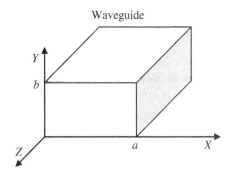

FIGURE 2.10   Rectangular waveguide structure.

2.8   WAVEGUIDES

Waveguides are low-loss transmission lines.

Waveguides may be rectangular or circular. Rectangular waveguide structure is presented in Figure 2.10. Waveguide structure is uniform in the z direction. Fields in waveguides are evaluated by solving the Helmholtz equation. Wave equation is given in Equation 2.72. Wave equation in rectangular coordinate system is given in Equation 2.73:

$$\nabla^2 E = \omega^2 \mu \varepsilon E = -k^2 E$$
$$\nabla^2 H = \omega^2 \mu \varepsilon H = -k^2 H \tag{2.72}$$
$$k = \omega \sqrt{\mu \varepsilon} = \frac{\omega}{v} = \frac{2\pi}{\lambda}$$

$$\nabla^2 E = \frac{\partial^2 E_i}{\partial x^2} + \frac{\partial^2 E_i}{\partial y^2} + \frac{\partial^2 E_i}{\partial z^2} + k^2 E_i = 0 \quad i = x, y, z$$
$$\nabla^2 H = \frac{\partial^2 H_i}{\partial x^2} + \frac{\partial^2 H_i}{\partial y^2} + \frac{\partial^2 H_i}{\partial z^2} + k^2 H_i = 0 \tag{2.73}$$

The wave equation solution may be written as $E = f(z)g(x, y)$. Field variation in the z direction may be written as $e^{-j\beta z}$. The derivative of this expression in the direction may be written as $-j\beta e^{-j\beta z}$. Maxwell's equation may be presented as written in Equations 2.74 and 2.75. A field may be represented as a superposition of waves in the transverse and longitudinal directions:

$$E(x, y, z) = e(x, y)e^{-j\beta z} + e_z(x, y)e^{-j\beta z}$$
$$H(x, y, z) = h(x, y)e^{-j\beta z} + h_z(x, y)e^{-j\beta z}$$
$$\nabla X E = (\nabla_t - j\beta a_z) \times (e + e_z)e^{-j\beta z} = -j\omega\mu(h + h_z)e^{-j\beta z}$$
$$\nabla_t \times e - j\beta a_z \times e + \nabla_t \times e_z - j\beta a_z \times e_z = -j\omega\mu(h + h_z) \tag{2.74}$$
$$a_z \times e_z = 0$$
$$\nabla_t \times e = -j\omega\mu h_z$$
$$-j\beta a_z \times e + \nabla_t \times e_z = -a_z \times \nabla_t e_z - j\beta a_z \times e = -j\omega\mu h$$

$$\nabla_t \times h = -j\omega e e_z$$
$$a_z \times \nabla_t h_z + j\beta a_z \times h = -j\omega e e \tag{2.75}$$

$$\nabla \times \mu H = (\nabla_t - j\beta a_z) \times (h + h_z)\mu e^{-j\beta z} = 0$$
$$\nabla_t \times h - j\beta a_z \times h_z = 0 \tag{2.76}$$
$$\nabla \times \varepsilon E = 0 \quad \nabla_t \times e - j\beta a_z \times e_z = 0$$

Waves may be characterized as TEM, TE, or TM waves. In TEM waves $e_z = h_z = 0$. In TE waves $e_z = 0$. In TM waves $h_z = 0$.

2.8.1   TE Waves

In TE waves $e_z = 0$. h_z is given as the solution of Equation 2.77. The solution to Equation 2.77 may be written as $h_z = f(x)g(y)$:

$$\nabla_t^2 h_z = \frac{\partial^2 h_z}{\partial x^2} + \frac{\partial^2 h_z}{\partial y^2} + k_c^2 h_z = 0 \tag{2.77}$$

By applying $h_z = f(x)g(y)$ to Equation 2.77 and dividing by fg, we get Equation 2.78:

$$\frac{f''}{f} + \frac{g''}{g} + k_c^2 = 0 \tag{2.78}$$

where f is a function that varies in the x direction and g is a function that varies in the y direction. The sum of f and g may be equal to zero only if they are equal to a constant. These facts are written in Equation 2.79:

$$\frac{f''}{f} = -k_x^2; \frac{g''}{g} = -k_y^2$$
$$k_x^2 + k_y^2 = k_c^2 \tag{2.79}$$

The solutions for f and g are given in Equation 2.80. A_1, A_2, B_1, B_2 are derived by applying h_z boundary conditions to Equation 2.78:

$$f = A_1 \cos k_x x + A_2 \sin k_x x$$
$$g = B_1 \cos k_y y + B_2 \sin k_y y \tag{2.80}$$

h_z boundary conditions are written in Equation 2.81:

$$\frac{\partial h_z}{\partial x} = 0 \quad at \quad x = 0, a$$
$$\frac{\partial h_z}{\partial y} = 0 \quad at \quad y = 0, b \tag{2.81}$$

By applying h_z boundary conditions to Equation 2.80, we get the relations written in Equation 2.82:

$$-k_x A_1 \sin k_x x + k_x A_2 \cos k_x x = 0$$
$$-k_y B_1 \sin k_y y + k_y B_2 \cos k_y y = 0$$
$$A_2 = 0 \quad k_x a = 0 \quad k_x = \frac{n\pi}{a} \quad n = 0,1,2 \qquad (2.82)$$
$$B_2 = 0 \quad k_y b = 0 \quad k_y = \frac{m\pi}{b} \quad m = 0,1,2$$

The solution for h_z is given in Equation 2.83:

$$h_z = A_{nm} \cos \frac{n\pi x}{a} \cos \frac{m\pi y}{b}$$
$$n = m \neq 0 \quad n = 0,1,2 \quad m = 0,1,2 \qquad (2.83)$$
$$k_{c,nm} = \left[\left(\frac{n\pi}{a}\right)^2 + \left(\frac{m\pi}{b}\right)^2 \right]^{1/2}$$

Both n and m cannot be zero. The wavenumber at cutoff is $k_{c,nm}$ and depends on the waveguide dimensions. The propagation constant γ_{nm} is given in Equation 2.84:

$$\gamma_{nm} = j\beta_{nm} = j\left(k_0^2 - k_c^2\right)^{1/2}$$
$$= j\left[\left(\frac{2\pi}{\lambda_0}\right)^2 - \left(\frac{n\pi}{a}\right)^2 - \left(\frac{m\pi}{b}\right)^2 \right]^{1/2} \qquad (2.84)$$

For $k_0 > k_{c,nm}$, β is real and wave will propagate. For $k_0 < k_{c,nm}$, β is imaginary and the wave will decay rapidly. Frequencies that define propagating and decaying waves are called cutoff frequencies. We may calculate cutoff frequencies by using Equation 2.85:

$$f_{c,nm} = \frac{c}{2\pi} k_{c,nm} = \frac{c}{2\pi} \left[\left(\frac{n\pi}{a}\right)^2 + \left(\frac{m\pi}{b}\right)^2 \right]^{1/2} \qquad (2.85)$$

For $a = 2$ cutoff wavelength is computed by using Equation 2.86:

$$\lambda_{c,nm} = \frac{2ab}{[n^2 b^2 + m^2 a^2]^{1/2}} = \frac{2a}{[n^2 + 4m^2]^{1/2}}$$
$$\lambda_{c,10} = 2a \quad \lambda_{c,01} = a \quad \lambda_{c,11} = 2a/\sqrt{5} \qquad (2.86)$$
$$\frac{c}{2a} \langle f_{01} \rangle \frac{c}{a}$$

For $\dfrac{c}{2a} \langle f_{01} \rangle \dfrac{c}{a}$ the dominant is TE_{10}.

By using Equations 2.74–2.76 we can derive the electromagnetic fields that propagate in the waveguide as given in Equation 2.87:

$$H_z = A_{nm} \cos\frac{n\pi x}{a} \cos\frac{m\pi y}{b} e^{\pm j\beta_{nm}z}$$

$$H_x = \pm j\frac{n\pi\beta_{nm}}{ak_{c,nm}^2} A_{nm} \sin\frac{n\pi x}{a} \cos\frac{m\pi y}{b} e^{\pm j\beta_{nm}z}$$

$$H_y = \pm j\frac{m\pi\beta_{nm}}{bk_{c,nm}^2} A_{nm} \cos\frac{n\pi x}{a} \sin\frac{m\pi y}{b} e^{\pm j\beta_{nm}z}$$

$$E_X = Z_{h,nm} j\frac{m\pi\beta_{nm}}{bk_{c,nm}^2} A_{nm} \cos\frac{n\pi x}{a} \sin\frac{m\pi y}{b} e^{\pm j\beta_{nm}z} \qquad (2.87)$$

$$E_Y = -jZ_{h,nm} \frac{n\pi\beta_{nm}}{ak_{c,nm}^2} A_{nm} \sin\frac{n\pi x}{a} \cos\frac{m\pi y}{b} e^{\pm j\beta_{nm}z}$$

The impedance of the nm modes is given as $Z_{h,nm} = \dfrac{e_x}{h_y} = \dfrac{k_0}{\beta_{nm}}\sqrt{\dfrac{\mu_0}{\varepsilon_0}}$.

The power of the nm mode is computed by using the Poynting vector calculation as shown in Equation 2.88:

$$P_{nm} = 0.5\mathrm{Re}\int_0^a\int_0^b E \times H^* \cdot a_z dxdy = 0.5\mathrm{Re}\, Z_{h,nm}\int_0^a\int_0^b \left(H_y H_Y^* + H_x H_x^*\right)dxdy$$

$$\int_0^a\int_0^b \cos^2\frac{n\pi x}{a}\sin^2\frac{m\pi y}{b}dxdy = \frac{ab}{4}\quad n\neq 0\ \ m\neq 0 \qquad (2.88)$$

$$\text{or}\ \ \frac{ab}{2}\quad n\ \ \text{or}\ \ m=0$$

$$P_{nm} = \frac{|A_{nm}|^2}{2\varepsilon_{0n}\varepsilon_{0m}}\left(\frac{\beta_{nm}}{k_{c,nm}}\right)^2 Z_{h,nm}\quad \varepsilon_{0n}=1,\, n=0\ \ \varepsilon_{0n}=2,\, n>0$$

The TE mode with the lowest cutoff frequency in rectangular waveguide is TE_{10}. TE_{10} fields in rectangular waveguide are shown in Figure 2.11.

2.8.2 TM Waves

In TM waves $h_z = 0$. e_z is given as the solution of Equation 2.89. The solution to Equation 2.13 may be written as $e_z = f(x)g(y)$. e_z should be zero at the metallic walls. e_z boundary conditions are written in Equation 2.90. The solution for e_z is given in Equation 2.91:

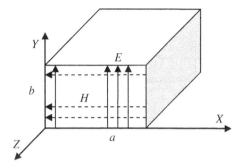

FIGURE 2.11   TE$_{10}$ mode.

$$\nabla_t^2 e_z = \frac{\partial^2 e_z}{\partial x^2} + \frac{\partial^2 e_z}{\partial y^2} + k_c^2 e_z = 0 \tag{2.89}$$

$$\begin{aligned} e_z = 0 \quad at \quad x = 0, a \\ e_z = 0 \quad at \quad y = 0, b \end{aligned} \tag{2.90}$$

$$\begin{aligned} e_z &= A_{nm} \sin\frac{n\pi x}{a} \sin\frac{m\pi y}{b} \\ n &= m \neq 0 \quad n = 0,1,2 \quad m = 0,1,2 \\ k_{c,nm} &= \left[\left(\frac{n\pi}{a}\right)^2 + \left(\frac{m\pi}{b}\right)^2 \right]^{1/2} \end{aligned} \tag{2.91}$$

The first propagating TM mode is TM$_{11, \, n=m=1}$. By using Equations 2.74–2.76, and 2.89, we can derive the electromagnetic fields that propagate in the waveguide as given in Equation 2.92:

$$\begin{aligned} E_z &= \sin\frac{n\pi x}{a} \sin\frac{m\pi y}{b} e^{\pm j\beta_{nm} z} \\ E_x &= -j\frac{n\pi\beta_{nm}}{ak_{c,nm}^2} \cos\frac{n\pi x}{a} \sin\frac{m\pi y}{b} e^{\pm j\beta_{nm} z} \\ E_y &= -j\frac{m\pi\beta_{nm}}{bk_{c,nm}^2} A_{nm} \sin\frac{n\pi x}{a} \cos\frac{m\pi y}{b} e^{\pm j\beta_{nm} z} \\ H_X &= \frac{-E_y}{Z_{e,nm}} \\ H_Y &= \frac{E_x}{Z_{e,nm}} \end{aligned} \tag{2.92}$$

The impedance of the nm modes is $Z_{e,nm} = \frac{\beta_{nm}}{k_0} \sqrt{\frac{\mu_0}{\varepsilon_0}}$.

The TM mode with the lowest cutoff frequency in rectangular waveguide is TM$_{11}$. TM$_{11}$ fields in rectangular waveguide are shown in Figure 2.12.

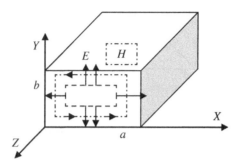

FIGURE 2.12 TM$_{11}$ mode.

2.9   CIRCULAR WAVEGUIDE

Circular waveguide is used to transmit electromagnetic wave in circular polarization. At high frequencies attenuation of several modes in circular waveguide is lower than in rectangular waveguide. Circular waveguide structure is uniform in the z direction. Fields in waveguides are evaluated by solving the Helmholtz equation in cylindrical coordinate system. A circular waveguide in cylindrical coordinate system is presented in Figure 2.13. Wave equation is given in Equation 2.93. Wave equation in cylindrical coordinate is given in Equation 2.94:

$$\nabla^2 E = \omega^2 \mu \varepsilon E = -k^2 E$$
$$\nabla^2 H = \omega^2 \mu \varepsilon H = -k^2 H \tag{2.93}$$
$$k = \omega\sqrt{\mu\varepsilon} = \frac{\omega}{v} = \frac{2\pi}{\lambda}$$

$$\nabla^2 E = \frac{1}{r}\frac{\partial}{\partial r}\left(r\frac{\partial E_i}{\partial r}\right) + \frac{1}{r^2}\frac{\partial^2 E_i}{\partial \phi^2} + \frac{\partial^2 E_i}{\partial z^2} - \gamma^2 E_i = 0 \quad i = r, \phi, z$$
$$\nabla^2 H = \frac{1}{r}\frac{\partial}{\partial r}\left(r\frac{\partial H_i}{\partial r}\right) + \frac{1}{r^2}\frac{\partial^2 H_i}{\partial \phi^2} + \frac{\partial^2 H_i}{\partial z^2} - \gamma^2 H_i = 0 \tag{2.94}$$

The solution to Equation 2.94 may be written as $E = f(r)g(\phi)h(z)$. By applying E to Equation 2.94 and dividing by $f(r)g(\phi)h(z)$, we get Equation 2.95:

$$\nabla^2 E = \frac{1}{rf}\frac{\partial}{\partial r}\left(r\frac{\partial f}{\partial r}\right) + \frac{1}{r^2 g}\frac{\partial^2 g}{\partial \phi^2} + \frac{\partial^2 h}{h\partial z^2} - \gamma^2 = 0 \tag{2.95}$$

where f is a function that varies in the r direction, g is a function that varies in the ϕ direction, and h is a function that varies in the z direction. The sum of f, g, and h may be equal to zero only if they are equal to a constant. The solution for h is written in Equation 2.96. The propagation constant is γ_g:

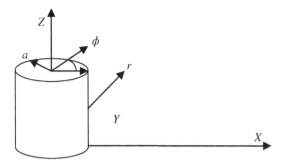

FIGURE 2.13   Circular waveguide structure.

$$\frac{\partial^2 h}{h \partial z^2} - \gamma_g^2 = 0$$

$$h = A e^{-\gamma_g z} + B e^{\gamma_g z}$$
(2.96)

$$\frac{r}{f}\frac{\partial}{\partial r}\left(r\frac{\partial f}{\partial r}\right) + \frac{1}{g}\frac{\partial^2 g}{\partial \phi^2} - \left(\gamma^2 - \gamma_g^2\right) = 0$$
(2.97)

The solution for g is written in Equation 2.98:

$$\frac{\partial^2 g}{g \partial \phi^2} + n^2 = 0$$

$$g = A_n \sin n\phi + B_n \cos n\phi$$
(2.98)

$$r\frac{\partial}{\partial r}\left(r\frac{\partial f}{\partial r}\right) + \left((k_c r)^2 - n^2\right)f = 0$$

$$k_c^2 + \gamma^2 = \gamma_g^2$$
(2.99)

Equation 2.99 is a Bessel equation. The solution of this equation is written in Equation 2.100. $J_n(k_c r)$ is a Bessel equation with order n and represents a standing wave. The wave varies a cosine function in the circular waveguide. $N_n(k_c r)$ is a Bessel equation with order n and represents a standing wave. The wave varies a sine function in the circular waveguide:

$$f = C_n J_n(k_c r) + D_n N_n(k_c r)$$
(2.100)

The general solution for the electric fields in the circular waveguide is given in Equation 2.101. For $r = 0 N_n(k_c r)$ goes to infinity, so $D_n = 0$:

$$E(r,\phi,z) = (C_n J_n(k_c r) + D_n N_n(k_c r))(A_n \sin n\phi + B_n \cos n\phi)e^{\pm \gamma_g z}$$
(2.101)

$$A_n \sin n\phi + B_n \cos n\phi = \sqrt{A_n^2 + B_n^2} \cos\left(n\phi + \tan^{-1}\left(\frac{A_n}{B_n}\right)\right) = F_n \cos n\phi \qquad (2.102)$$

The general solution for the electric fields in the circular waveguide is given in Equation 2.103:

$$E(r,\phi,z) = E_0(J_n(k_c r))(\cos n\phi)e^{\pm \gamma_g z} \quad \text{If} \quad \alpha = 0$$
$$E(r,\phi,z) = E_0(J_n(k_c r))(\cos n\phi)e^{\pm \beta_g z} \qquad (2.103)$$
$$\beta_g = \pm \sqrt{\omega^2 \mu \varepsilon - k_c^2}$$

2.9.1   TE Waves in Circular Waveguide

In TE waves $e_z = 0$. H_z is given as the solution of Equation 2.104. The solution to Equation 2.104 may be written as given in Equation 2.105:

$$\nabla^2 H_z = \gamma^2 H_z \qquad (2.104)$$

$$H_z = H_{0z}(J_n(k_c r))(\cos n\phi)e^{\pm j\beta_g z} \qquad (2.105)$$

The electric and magnetic fields are solutions of Maxwell's equations as written in Equations 2.106 and 2.107:

$$\nabla X E = -j\omega\mu H \qquad (2.106)$$

$$\nabla \times H = j\omega\varepsilon E \qquad (2.107)$$

Field variation in the z direction may be written as $e^{-j\beta z}$. The derivative of this expression in the direction may be written as $-j\beta e^{j\beta z}$. The electric and magnetic field components are solutions of Equations 2.108 and 2.109:

$$E_r = -\frac{j\omega\mu}{k_c^2}\frac{1}{r}\left(\frac{\partial H_z}{\partial \phi}\right)$$
$$E_\phi = \frac{j\omega\mu}{k_c^2}\left(\frac{\partial H_z}{\partial r}\right) \qquad (2.108)$$

$$H_\phi = -\frac{j\beta_g}{k_c^2}\frac{1}{r}\left(\frac{\partial H_z}{\partial \phi}\right)$$
$$H_r = \frac{-j\beta_g}{k_c^2}\left(\frac{\partial H_z}{\partial r}\right) \qquad (2.109)$$

TABLE 2.10 Circular Waveguide TE Modes

p	$(n=0)\ X'_{np}$	$(n=1)\ X'_{np}$	$(n=2)\ X'_{np}$	$(n=3)\ X'_{np}$	$(n=4)\ X'_{np}$	$(n=5)\ X'_{np}$
1	3.832	1.841	3.054	4.201	5.317	6.416
2	7.016	5.331	6.706	8.015	9.282	10.52
3	10.173	8.536	9.969	11.346	12.682	13.987
4	13.324	11.706	13.17			

H_z, H_r, and E_ϕ boundary conditions are written in Equation 2.110:

$$\frac{\partial H_z}{\partial r}=0 \quad at \quad r=a$$
$$H_r=0 \quad at \quad r=a \qquad (2.110)$$
$$E_\phi=0 \quad at \quad r=a$$

By applying the boundary conditions to Equation 2.105, we get the relations written in Equation 2.111. The solutions of Equation 2.111 are listed in Table 2.10:

$$\frac{\partial H_z}{\partial r}|r=a=H_{0z}\left(J'_n(k_ca)\right)(cosn\phi)e^{-\beta_g z}=0$$
$$J'_n(k_ca)=0 \qquad (2.111)$$

The wavenumber at cutoff is $k_{c,np}$. $k_{c,np}$ depends on the waveguide dimensions. The propagation constant $\gamma_{g,np}$ is given in Equation 2.112:

$$\gamma_{g,np}=j\beta_{g,np}=j\left(k_0^2-k_c^2\right)^{1/2}$$
$$=j\left[\left(\frac{2\pi}{\lambda_0}\right)^2-\left(\frac{X'_{np}}{a}\right)^2\right]^{1/2} \qquad (2.112)$$

For $k_0>k_{c,nm}$, β is real and wave will propagate. For $k_0<k_{c,nm}$, β is imaginary and the wave will decay rapidly. Frequencies that define propagating and decaying waves are called cutoff frequencies. We may calculate cutoff frequencies by using Equation 2.113:

$$f_{c,nm}=\frac{cX'_{np}}{2\pi a} \qquad (2.113)$$

We get the field components by solving Equations 2.108 and 2.109. Field components are written in Equation 2.114:

$$H_z = H_{0z}\left(J_n\left(\frac{X'_{np}r}{a}\right)\right)(\cos n\phi)e^{-j\beta_g z}$$

$$H_\phi = \frac{E_{0r}}{Z_g}\left(J_n\left(\frac{X'_{np}r}{a}\right)\right)(\sin n\phi)e^{-j\beta_g z}$$

$$H_r = \frac{E_{0\phi}}{Z_g}\left(J'_n\left(\frac{X'_{np}r}{a}\right)\right)(\cos n\phi)e^{-j\beta_g z} \qquad (2.114)$$

$$E_\phi = E_{0\phi}\left(J'_n\left(\frac{X'_{np}r}{a}\right)\right)(\cos\phi)e^{-j\beta_g z}$$

$$E_r = E_{0r}\left(J_n\left(\frac{X'_{np}r}{a}\right)\right)(\sin\phi)e^{-j\beta_g z}$$

The impedance of the np modes is written in Equation 2.115:

$$Z_{g,np} = \frac{E_r}{H_\phi} = \frac{\omega\mu}{\beta_{g,np}} = \frac{\eta}{\sqrt{1-(f_c/f)^2}} \qquad (2.115)$$

2.9.2   TM Waves in Circular Waveguide

In TM waves $h_z = 0$. e_z is given as the solution of Equation 2.116. The solution to Equation 2.116 is written in Equation 2.117:

$$\nabla^2 E_z = \gamma^2 E_z \qquad (2.116)$$

$$E_z = E_{0z}(J_n(k_c r))(\cos n\phi)e^{\pm j\beta_g z} \qquad (2.117)$$

The electric and magnetic fields are solutions of Maxwell's equations as written in Equations 2.118 and 2.119:

$$\nabla X E = -j\omega\mu H \qquad (2.118)$$

$$\nabla \times H = j\omega\varepsilon E \qquad (2.119)$$

Field variation in the z direction may be written as $e^{-j\beta z}$. The derivative of this expression in the direction may be written as $-j\beta e^{j\beta z}$. The electric and magnetic field components are solutions of Equations 2.120 and 2.121:

$$H_r = \frac{j\omega\varepsilon}{k_c^2}\frac{1}{r}\left(\frac{\partial H_z}{\partial\phi}\right)$$

$$\qquad\qquad (2.120)$$

$$H_\phi = \frac{-j\omega\varepsilon}{k_c^2}\left(\frac{\partial H_z}{\partial r}\right)$$

$$E_\phi = -\frac{-j\beta_g}{k_c^2}\frac{1}{r}\left(\frac{\partial E_z}{\partial \phi}\right)$$

$$E_r = \frac{-j\beta_g}{k_c^2}\left(\frac{\partial E_z}{\partial r}\right)$$

(2.121)

where E_r is the boundary condition written in Equation 2.122:

$$E_z = 0 \quad \text{at} \quad r = a \qquad (2.122)$$

By applying the boundary conditions to Equation 2.117, we get the relations written in Equation 2.123. The solutions of Equation 2.123 are listed in Table 2.11:

$$E_z|(r=a) = H_{0z}(J_n(k_c a))(\cos n\phi)e^{-j\beta_g z} = 0$$

$$J_n(k_c a) = 0$$

(2.123)

The wavenumber at cutoff is $k_{c,np}$. $k_{c,np}$ depends on the waveguide dimensions. The propagation constant $\gamma_{g,np}$ is given in Equation 2.124:

$$\gamma_{g,np} = j\beta_{g,np} = j\left(k_0^2 - k_c^2\right)^{1/2}$$

$$= j\left[\left(\frac{2\pi}{\lambda_0}\right)^2 - \left(\frac{X_{np}}{a}\right)^2\right]^{1/2}$$

(2.124)

For $k_0 > k_{c,nm}$, β is real and wave will propagate. For $k_0 < k_{c,nm}$, β is imaginary and the wave will decay rapidly. Frequencies that define propagating and decaying waves are called cutoff frequencies. We may calculate cutoff frequencies by using Equation 2.125:

$$f_{c,nm} = \frac{cX_{np}}{2\pi a} \qquad (2.125)$$

We get the field components by solving Equations 2.120 and 2.121. Field components are written in Equation 2.126:

TABLE 2.11   Circular Waveguide TM Modes

p	$n=0, X_{np}$	$n=1, X_{np}$	$n=2, X_{np}$	$n=3, X_{np}$	$n=4, X_{np}$	$n=5, X_{np}$
1	2.405	3.832	5.136	6.38	7.588	8.771
2	5.52	7.106	8.417	9.761	11.065	12.339
3	8.645	10.173	11.62	13.015	14.372	
4	11.792	13.324	14.796			

TE$_{11}$ fields

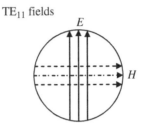

FIGURE 2.14   TE$_{11}$ fields in circular waveguide.

$$E_z = E_{0z}\left(J_n\left(\frac{X_{np}r}{a}\right)\right)(\cos n\phi)e^{-j\beta_g z}$$

$$H_\phi = \frac{E_{0r}}{Z_g}\left(J_n'\left(\frac{X_{np}r}{a}\right)\right)(\cos n\phi)e^{-j\beta_g z}$$

$$H_r = \frac{E_{0\phi}}{Z_g}\left(J_n\left(\frac{X_{np}r}{a}\right)\right)(\sin n\phi)e^{-j\beta_g z} \qquad (2.126)$$

$$E_\phi = E_{0\phi}\left(J_n\left(\frac{X_{np}r}{a}\right)\right)(\sin\phi)e^{-j\beta_g z}$$

$$E_r = E_{0r}\left(J_n'\left(\frac{X_{np}r}{a}\right)\right)(\cos\phi)e^{-j\beta_g z}$$

The impedance of the *np* modes is written in Equation 2.127:

$$Z_{g,np} = \frac{E_r}{H_\varphi} = \frac{\beta_{g,np}}{\omega\varepsilon} = \eta\sqrt{1-\left(\frac{f_c}{f}\right)^2} \qquad (2.127)$$

The mode with the lowest cutoff frequency in circular waveguide is TE$_{11}$.
TE$_{11}$ fields in circular waveguide are shown in Figure 2.14.

REFERENCES

[1] Ramo S, Whinnery JR, Van Duzer T. *Fields and Waves in Communication Electronics.* 3rd ed. New York: John Wiley & Sons, Inc.; 1994.

[2] Collin RE. *Foundation for Microwave Engineering.* Auckland: McGraw-Hill; 1966.

[3] Balanis CA. *Antenna Theory: Analysis and Design.* 2nd ed. New York: John Wiley & Sons, Inc.; 1996.

[4] Godara LC, editor. *Handbook of Antennas in Wireless Communications.* Boca Raton: CRC Press LLC; 2002.

[5] Kraus JD, Marhefka RJ. *Antennas for all Applications.* 3rd ed. New York: McGraw Hill; 2002.

[6] Ulaby FT. *Electromagnetics for Engineers.* Upper Saddle River: Pearson Education; 2005.

[7] James JR, Hall PS, Wood C. *Microstrip Antenna Theory and Design.* New York: Peregrinus on behalf of the Institution of Electrical Engineers; 1981.

[8] Sabban A. *RF Engineering, Microwave and Antennas.* Tel Aviv: Saar Publication; 2014.

3

BASIC ANTENNAS FOR COMMUNICATION SYSTEMS

3.1 INTRODUCTION TO ANTENNAS

Antennas are major component in communication systems, see Refs. [1–11]. Antenna is used to radiate efficiently electromagnetic energy in desired directions. Antennas match radio-frequency systems, sources of electromagnetic energy, to space. All antennas may be used to receive or radiate energy. Antennas transmit or receive electromagnetic waves. Antennas convert electromagnetic radiation into electric current, or vice versa. Antennas transmit and receive electromagnetic radiation at radio frequencies. Antennas are a necessary part of all communication links and radio equipment. Antennas are used in systems such as radio and television broadcasting, point-to-point radio communication, wireless LAN, cell phones, radar, medical systems, and spacecraft communication. Antennas are most commonly employed in air or outer space but can also be operated underwater, on and inside the human body, or even through soil and rock at low frequencies for short distances. Physically, an antenna is an arrangement of one or more conductors. In transmitting mode, an alternating current is created in the elements by applying a voltage at the antenna terminals, causing the elements to radiate an electromagnetic field. In receiving mode, an electromagnetic field from another source induces an alternating current in the elements and a corresponding voltage at the antenna's terminals. Some receiving antennas (such as parabolic and horn) incorporate shaped reflective surfaces to receive the radio waves striking them and direct or focus them onto the actual conductive elements.

Wideband RF Technologies and Antennas in Microwave Frequencies, First Edition. Dr. Albert Sabban.
© 2016 John Wiley & Sons, Inc. Published 2016 by John Wiley & Sons, Inc.

3.2   ANTENNA PARAMETERS

Radiation pattern—A radiation pattern is the antenna-radiated field as function of the direction in space. It is a way of plotting the radiated power from an antenna. This power is measured at various angles at a constant distance from the antenna.

Radiator—This is the basic element of an antenna. An antenna can be made up of multiple radiators.

Boresight—The direction in space to which the antenna radiates maximum electromagnetic energy.

Range—Antenna range is the radial range from an antenna to an object in space.

Azimuth (Az)—This is the angle from left to right from a reference point, from 0° to 360°.

Elevation (El)—El angle is the angle from the horizontal (x, y) plane, from $-90°$ (down) to $+90°$ (up).

Main beam—The main beam is the region around the direction of maximum radiation, usually the region that is within 3dB of the peak of the main lobe.

Beamwidth—Beamwidth is the angular range of the antenna pattern in which at least half of the maximum power is emitted. This angular range, of the major lobe, is defined as the points at which the field strength falls around 3dB regarding to the maximum field strength.

Side lobes—Side lobes are smaller beams that are away from the main beam. Side lobes present radiation in undesired directions. The side-lobe level is a parameter used to characterize the antenna radiation pattern. It is the maximum value of the side lobes away from the main beam and is expressed usually in decibels.

Radiated power—The total radiated power when the antenna is excited by a current or voltage of known intensity.

Isotropic radiator—Theoretical lossless radiator that radiates, or receives, equal electromagnetic energy in free space to all directions.

Directivity—The ratio between the amounts of energy propagating in a certain direction compared to the average energy radiated to all directions over a sphere:

$$D = \frac{P(\theta,\phi)\text{maximal}}{P(\theta,\phi)\text{average}} = 4\pi \frac{P(\theta,\phi)\text{maximal}}{P_{\text{rad}}} \tag{3.1}$$

$$P(\theta,\phi)\text{average} = \frac{1}{4\pi} \int\int P(\theta,\phi)\sin\theta\, d\theta d\phi = \frac{P_{\text{rad}}}{4\pi} \tag{3.2}$$

$$D \sim \frac{4\pi}{\theta E \times \theta H} \tag{3.3}$$

θE – Beamwidth in radian in EL plane

θH – Beamwidth in radian in AZ plane

Antenna effective area (A_{eff})—The antenna area that contributes to the antenna directivity:

$$D = \frac{4\pi A_{\text{eff}}}{\lambda^2} \qquad (3.4)$$

Antenna gain (G)—The ratio between the amounts of energy propagating in a certain direction compared to the energy that would be propagating in the same direction if the antenna were not directional, isotropic radiator, is known as its gain.

Radiation efficiency (α)—The radiation efficiency is the ratio of power radiated to the total input power. The efficiency of an antenna takes into account losses and is equal to the total radiated power divided by the radiated power of an ideal lossless antenna:

$$G = \alpha D \qquad (3.5)$$

A comparison of directivity and gain values for several antennas is given in Table 3.1:

$$\text{For small antennas} \left(l < \frac{\lambda}{2} \right) G \cong \frac{41,000}{\theta E^\circ \times \theta H^\circ} \qquad (3.6a)$$

$$\text{For medium size antennas} (\lambda < l < 8\lambda) G \cong \frac{31,000}{\theta E^\circ \times \theta H^\circ} \qquad (3.6b)$$

$$\text{For big antennas} (8\lambda < l) G \cong \frac{27,000}{\theta E^\circ \times \theta H^\circ} \qquad (3.6c)$$

Antenna impedance—Antenna impedance is the ratio of voltage at any given point along the antenna to the current at that point. Antenna impedance depends upon the height of the antenna above the ground and the influence of surrounding objects. The impedance of a quarter-wave monopole near a perfect ground is

TABLE 3.1 Antenna Directivity versus Antenna Gain

Antenna Type	Directivity (dBi)	Gain (dBi)
Isotropic radiator	0	0
Dipole $\lambda/2$	2	2
Dipole above ground plane	6–4	6–4
Microstrip antenna	7–8	6–7
Yagi antenna	6–18	5–16
Helix antenna	7–20	6–18
Horn antenna	10–30	9–29
Reflector antenna	15–60	14–58

approximately 36 Ω. The impedance of a half-wave dipole is approximately 75 Ω.

Phased arrays—Phased array antennas are electrically steerable. The physical antenna can be stationary. Phased arrays, smart antennas, incorporate active components for beam steering.

Steerable antennas—These vary from arrays with switchable elements and partially mechanically and arrays with fully electronically steerable arrays (hybrid antenna systems). Such systems can be equipped with phase and amplitude shifters for each element, or the design can be based on digital beamforming (DBF). This technique, in which the steering is performed directly on a digital level, allows the most flexible and powerful control of the antenna beam.

3.3 DIPOLE ANTENNA

Dipole antenna is a small wire antenna. Dipole antenna consists of two straight conductors excited by a voltage fed via a transmission line as shown in Figure 3.1. Each side of the transmission line is connected to one of the conductors. The most common dipole is the half-wave dipole, in which each of the two conductors is approximately quarter wavelength long, so the length of the antenna is half wavelength.

We can calculate the fields radiated from the dipole by using the potential function. The electric potential function is ϕ_1. The electric potential function is A. The potential function is given in Equation 3.7:

$$\phi_1 = \frac{1}{4\pi\varepsilon_0} \int_c \frac{\rho_1 e^{j(\omega t - \beta R)}}{R} dl$$

$$A_1 = \frac{\mu_0}{4\pi} \int_c \frac{i e^{j(\omega t - \beta R)}}{R} dl \tag{3.7}$$

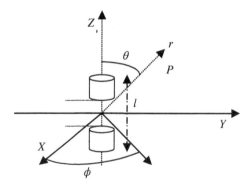

FIGURE 3.1 Dipole antenna.

3.3.1 Radiation from a Small Dipole

The length of a small dipole is small compared to wavelength and is called elementary dipole. We may assume that the current along the elementary dipole is uniform. We can solve the wave equation in spherical coordinates by using the potential function given in Equation 3.7. The electromagnetic fields in a point $P(r, \theta, \phi)$ are given in Equation 3.8. The electromagnetic fields in Equation 3.2 vary as $\frac{1}{r}, \frac{1}{r^2}, \frac{1}{r^3}$. For $r \ll 1$, the dominant component of the field varies as $1/(r)^3$ and is given in Equation 3.9. These fields are the dipole near fields. In this case the waves are standing waves, and the energy oscillates in the antenna near zone and is not radiated to the open space. The real part of the Poynting vector is equal to zero. For $r \gg 1$, the dominant component of the field varies as $1/r$ as given in Equation 3.10. These fields are the dipole far fields:

$$E_r = \eta_0 \frac{l I_0 \cos\theta}{2\pi r^2}\left(1 - \frac{j}{\beta r}\right) e^{j(\omega t - \beta r)}$$

$$E_\theta = j\eta_0 \frac{\beta l I_0 \sin\theta}{4\pi r}\left(1 - \frac{j}{\beta r} - \frac{1}{(\beta r)^2}\right) e^{j(\omega t - \beta r)}$$

$$H_\phi = j\frac{\beta l I_0 \sin\theta}{4\pi r}\left(1 - \frac{j}{\beta r}\right) e^{j(\omega t - \beta r)} \qquad (3.8)$$

$$H_r = 0 \quad H_\theta = 0 \quad E_\phi = 0$$

$$I = I_0 \cos\omega t$$

$$E_r = -j\eta_0 \frac{l I_0 \cos\theta}{2\pi\beta r^3} e^{j(\omega t - \beta r)}$$

$$E_\theta = -j\eta_0 \frac{l I_0 \sin\theta}{4\pi\beta r^3} e^{j(\omega t - \beta r)} \qquad (3.9)$$

$$H_\phi = \frac{l I_0 \sin\theta}{4\pi r^2} e^{j(\omega t - \beta r)}$$

$$E_r = 0$$

$$E_\theta = j\eta_0 \frac{l \beta I_0 \sin\theta}{4\pi r} e^{j(\omega t - \beta r)} \qquad (3.10)$$

$$H_\phi = j\frac{l \beta I_0 \sin\theta}{4\pi r} e^{j(\omega t - \beta r)}$$

$$\frac{E_\theta}{H_\phi} = \eta_0 = \sqrt{\frac{\mu_0}{\varepsilon_0}} \qquad (3.11)$$

In the far fields the electromagnetic fields vary as $1/r$ and $\sin\theta$. Wave impedance in free space is given in Equation 3.11.

3.3.2 Dipole Radiation Pattern

The antenna radiation pattern represents the radiated fields in space at a point $P(r, \theta, \phi)$ as function of θ, ϕ. The antenna radiation pattern is three-dimensional. When ϕ is constant and θ vary we get the E-plane radiation pattern. When ϕ vary and θ is constant, usually $\theta = \pi/2$, we get the H-plane radiation pattern.

3.3.3 Dipole E-Plane Radiation Pattern

The dipole E-plane radiation pattern is given in Equation 3.12 and presented in Figure 3.2:

$$|E_\theta| = \eta_0 \frac{I\beta l_0 |\sin\theta|}{4\pi r} \tag{3.12}$$

At a given point $P(r, \theta, \phi)$, the dipole E-plane radiation pattern is given in Equation 3.13:

$$|E_\theta| = \eta_0 \frac{I\beta l_0 |\sin\theta|}{4\pi r} = A|\sin\theta|$$
$$\text{Choose } A = 1 \tag{3.13}$$
$$|E_\theta| = |\sin\theta|$$

Dipole E-plane radiation pattern in a spherical coordinate system is shown in Figure 3.3.

3.3.4 Dipole H-Plane Radiation Pattern

For $\theta = \pi/2$ the dipole H-plane radiation pattern is given in Equation 3.8 and presented in Figure 3.4:

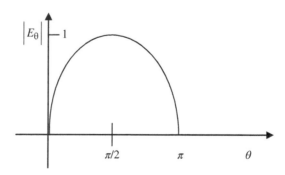

FIGURE 3.2 Dipole E-plane radiation pattern.

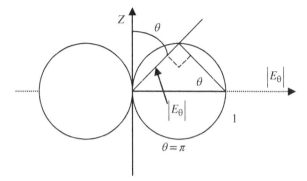

FIGURE 3.3 Dipole E-plane radiation pattern in a spherical coordinate system.

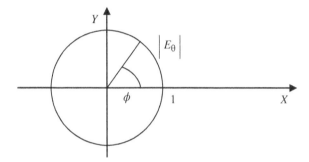

FIGURE 3.4 Dipole H-plane radiation pattern for $\theta = \pi/2$.

$$|E_\theta| = \eta_0 \frac{I\beta l_0}{4\pi r} \tag{3.14}$$

At a given point $P(r, \theta, \phi)$, the dipole H-plane radiation pattern is given in Equation 3.15:

$$|E_\theta| = \eta_0 \frac{I\beta l_0 |\sin\theta|}{4\pi r} = A$$

Choose $A = 1$ $\tag{3.15}$

$$|E_\theta| = 1$$

The dipole H-plane radiation pattern in the x–y plane is a circle with $r = 1$.

The radiation pattern of a vertical dipole is omnidirectional. It radiates equal power in all azimuthal directions perpendicular to the axis of the antenna. The dipole H-plane radiation pattern in spherical coordinate system is shown in Figure 3.4.

3.3.5   Antenna Radiation Pattern

A typical antenna radiation pattern is shown in Figure 3.5. The antenna main beam is measured between the points that the maximum relative field intensity E decays to 0.707E. Half of the radiated power is concentrated in the antenna main beam. The antenna main beam is called 3dB beamwidth. Radiation to the undesired direction is concentrated in the antenna side lobes.

For a dipole the power intensity varies as $(\sin^2\theta)$. At $\theta=45°$ and $\theta=135°$ the radiated power equals to half the power radiated toward $\theta=90°$. The dipole beamwidth is $\theta=(135-45)=90°$.

3.3.6   Dipole Directivity

Directivity is defined as the ratio between the amounts of energy propagating in a certain direction compared to the average energy radiated to all directions over a sphere as written in Equations 3.16 and 3.17:

$$D=\frac{P(\theta,\phi)\text{maximal}}{P(\theta,\phi)\text{average}}=4\pi\frac{P(\theta,\phi)\text{maximal}}{P_{\text{rad}}} \tag{3.16}$$

$$P(\theta,\phi)\text{average}=\frac{1}{4\pi}\int\int P(\theta,\phi)\sin\theta d\theta d\phi=\frac{P_{\text{rad}}}{4\pi} \tag{3.17}$$

The radiated power from a dipole is calculated by computing the Poynting vector P as given in Equation 3.18:

$$P=0.5(E\times H^*)=\frac{15\pi I_0^2 l^2\sin^2\theta}{r^2\lambda^2}$$

$$W_T=\int_s P\times ds=\frac{15\pi I_0^2 l^2}{\lambda^2}\int_0^\pi\sin^3\theta d\theta\int_0^{2\pi}d\phi=\frac{40\pi^2 I_0^2 l^2}{\lambda^2} \tag{3.18}$$

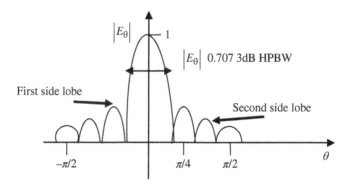

FIGURE 3.5   Antenna typical radiation pattern.

The overall radiated energy is W_T. W_T is computed by integration of the power flow over an imaginary sphere surrounding the dipole. The power flow of an isotropic radiator is equal to W_T divided by the surrounding area of the sphere, $4\pi r^2$, as given in Equation 3.19. The dipole directivity at $\theta = 90°$ is 1.5 or 1.76 dB as shown in Equation 3.20:

$$\oint_s ds = r^2 \int_0^{\pi} \sin\theta d\theta \int_0^{2\pi} d\phi = 4\pi r^2$$

$$P_{iso} = \frac{W_T}{4\pi r^2} = \frac{10\pi I_0^2 l^2}{r^2 \lambda^2} \tag{3.19}$$

$$D = \frac{P}{P_{iso}} = 1.5\sin^2\theta$$

$$G_{dB} = 10\log_{10} G = 10\log_{10} 1.5 = 1.76 \text{dB} \tag{3.20}$$

For small antennas or for antennas without losses, $D = G$, losses are negligible. For a given θ and ϕ for small antennas, the approximate directivity is given by Equation 3.21:

$$D = \frac{41,253}{\theta_{3dB}\phi_{3dB}} \tag{3.21}$$

$$G = \xi D \quad \xi = \text{Efficency}$$

Antenna losses degrade the antenna efficiency. Antenna losses consist of conductor loss, dielectric loss, radiation loss, and mismatch losses. For resonant small antennas $\xi = 1$. For reflector and horn antennas, the efficiency varies from $\xi = 0.5$ to $\xi = 0.7$.

3.3.7 Antenna Impedance

Antenna impedance determines the efficiency of transmitting and receiving energy in antennas. The dipole impedance is given in Equation 3.22:

$$R_{rad} = \frac{2W_T}{I_0^2}$$

$$\text{For a dipole}: R_{rad} = \frac{80\pi^2 l^2}{\lambda^2} \tag{3.22}$$

3.3.8 Impedance of a Folded Dipole

A folded dipole is a half-wave dipole with an additional wire connecting its two ends. If the additional wire has the same diameter and cross section as the dipole, two nearly identical radiating currents are generated. The resulting far-field emission pattern is nearly identical to the one for the single-wire dipole described in Equation 3.10, but at resonance its feed point impedance R_{rad-f} is four times the radiation resistance of a dipole. This is because for a fixed amount of power, the total radiating current I_0 is equal to twice the current in each wire and thus equal to twice the current at the feed point. Equating the average radiated power to the average power delivered at the feed point, we obtain that $R_{rad-f} = 4\,R_{rad} = 300\ \Omega$. The folded dipole has a wider bandwidth than a single dipole.

3.4   BASIC APERTURE ANTENNAS

Reflector and horn antennas are defined as aperture antennas. The longest dimension of an aperture antenna is higher than several wavelengths.

3.4.1   The Parabolic Reflector Antenna

The **parabolic reflector** antenna [3] consists of a radiating feed that is used to illuminate a reflector that is curved in the form of an accurate parabolic with diameter D as presented in Figure 3.6. This shape enables a very beam to be obtained. To provide the optimum illumination of the reflecting surface, the level of the parabola illumination should be greater by 10dB in the center than that at the parabola edges. The parabolic reflector antenna gain may be calculated by using Equation 3.23. α is the parabolic reflector antenna efficiency.

3.4.1.1   *Parabolic Reflector Antenna Gain*

$$G \cong 10\log_{10}\left(\alpha\frac{(\pi D)^2}{\lambda^2}\right) \tag{3.23}$$

Reflector geometry is presented in Figure 3.7. The following relations given in Equations 3.24–3.27 can be derived from the reflector geometry:

$$PQ = r'\cos\theta' \tag{3.24}$$

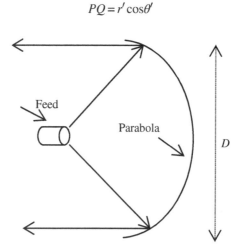

FIGURE 3.6   Parabolic antenna.

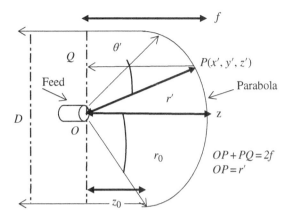

FIGURE 3.7 Reflector geometry.

$$2f = r'(1 + \cos\theta') \tag{3.25}$$

$$2f = r' + r'\cos\theta' = \sqrt{(x')^2 + (y')^2 + (z')^2} + z' \tag{3.26}$$

The relation between the reflector diameter D and θ is given in Equations 3.27–3.30:

$$\theta_0 = \tan^{-1}\frac{D/2}{z_0} \tag{3.27}$$

$$z_0 = f - \frac{(D/2)^2}{4f} \tag{3.28}$$

$$\theta_0 = \tan^{-1}\left|\frac{D/2}{z_0}\right| = \tan^{-1}\left|\frac{D/2}{f - \left((D/2)^2/4f\right)}\right| = \tan^{-1}\left|\frac{f/2D}{(f/D)^2 - (1/16)}\right| \tag{3.29}$$

$$f = \frac{D}{4}\cot\left(\frac{\theta_0}{2}\right) \tag{3.30}$$

3.4.2 Reflector Directivity

Reflector directivity is a function of the reflector geometry and feed radiation characteristics as given in Equations 3.31 and 3.32:

$$D_0 = \frac{4\pi U_{\max}}{P_{\mathrm{rad}}} = \frac{16\pi^2}{\lambda^2}f^2\left|\int_0^{\theta_0}\sqrt{G_{\mathrm{F}}(\theta')}\tan\left(\frac{\theta'}{2}\right)d\theta'\right|^2 \tag{3.31}$$

$$D_0 = \frac{(\pi D)^2}{\lambda^2} \left[\cot^2\left(\frac{\theta_0}{2}\right) \left| \int_0^{\theta_0} \sqrt{G_F(\theta')} \tan\left(\frac{\theta'}{2}\right) d\theta' \right|^2 \right] \qquad (3.32)$$

The reflector aperture efficiency is given in Equation 3.33. The feed radiation pattern may be presented as in Equation 3.34:

$$\epsilon_{ap} = \left[\cot^2\left(\frac{\theta_0}{2}\right) \left| \int_0^{\theta_0} \sqrt{G_F(\theta')} \tan\left(\frac{\theta'}{2}\right) d\theta' \right|^2 \right] \qquad (3.33)$$

$$G_F(\theta') = G_0^n \cos^n(\theta') \quad \text{for } 0 \le \theta' \le \frac{\pi}{2} \qquad (3.34)$$

$$G_F(\theta') = 0 \ \text{ for } \frac{\pi}{2} < \theta' \le \pi$$

$$\int_0^{\pi/2} G_0^n \cos^n(\theta') \sin(\theta') d\theta' = 2$$

where $G_0^n = 2(n+1)$.

Uniform illumination of the reflector aperture may be achieved if $G_F(\theta')$ is given by Equation 3.35:

$$G_F(\theta') = \sec^4\left(\frac{\theta'}{2}\right) \quad \text{for } 0 \le \theta' \le \frac{\pi}{2} \qquad (3.35)$$

$$G_F(\theta') = 0 \ \text{ for } \frac{\pi}{2} < \theta' \le \pi$$

The reflector aperture efficiency is computed by multiplying all the antenna efficiencies due to spillover, blockage, taper, phase error, cross-polarization losses, and random error over the reflector surface, $\epsilon_{ap} = \epsilon_s \epsilon_t \epsilon_b \epsilon_x \epsilon_p \epsilon_r$, where

ϵ_s – spillover efficiency, written in Equation 3.36

ϵ_t – taper efficiency, written in Equation 3.37

ϵ_b – blockage efficiency; ϵ_p – phase efficiency;

ϵ_x – cross polarization efficiency;

ϵ_r – random error over the reflector surface efficiency

$$\epsilon_s = \frac{\displaystyle\int_0^{\theta_0} G_F(\theta') \sin\theta' \, d\theta'}{\displaystyle\int_\pi^{\theta_0} G_F(\theta') \sin\theta' \, d\theta'} \qquad (3.36)$$

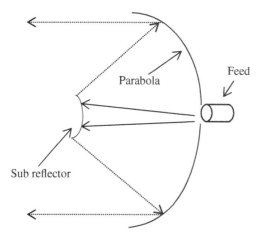

FIGURE 3.8 Cassegrain feed system.

$$\epsilon_t = 2\cot^2\left(\frac{\theta_0}{2}\right)\frac{\left|\int_0^{\theta_0}\sqrt{G_F(\theta')}\tan\left(\frac{\theta'}{2}\right)d\theta'\right|^2}{\int_0^{\theta_0}G_F(\theta')\sin\theta'd\theta'} \qquad (3.37)$$

In the literature [1] we can find graphs that present the reflector antenna efficiencies as function of the reflector antenna geometry and feed radiation pattern. However Equations 3.32–3.38 give us a good approximation of the reflector directivity:

$$D_0 = \frac{(\pi D)^2\epsilon_{ap}}{\lambda^2} \qquad (3.38)$$

3.4.3 Cassegrain Reflector

The parabolic reflector or dish antenna consists of a radiating element that may be a simple dipole or a waveguide horn antenna. This is placed at the center of the metallic parabolic reflecting surface as shown in Figure 3.8. The energy from the radiating element is arranged so that it illuminates the subreflecting surface. The energy from the subreflector is arranged so that it illuminates the main reflecting surface. Once the energy is reflected, it leaves the antenna system in a narrow beam.

3.5 HORN ANTENNAS

Horn antennas are used as a feed element for radio astronomy, satellite tracking and communication reflector antennas, and phased arrays radiating elements used in antenna calibration and measurements. Figure 3.9a presents *E*-plane sectoral horn.

(a)   (b)

(c)   (d)

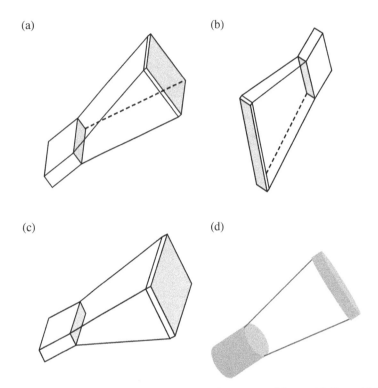

FIGURE 3.9 (a) *E*-plane sectoral horn. (b) *H*-plane sectoral horn. (c) Pyramidal horn. (d) Conical horn.

Figure 3.9b presents *H*-plane sectoral horn. Figure 3.9c presents a pyramidal horn. Figure 3.9d presents a conical horn.

3.5.1 *E*-Plane Sectoral Horn

Figure 3.10 presents an *E*-plane sectoral horn. Horn antennas are fed by a waveguide. The excited mode is TE_{10}. Field expressions over the horn aperture are similar to the fields of a TE_{10} mode in rectangular waveguide with the aperture dimensions of a, b_1. The fields in the antenna aperture are given in Equations 3.39–3.41:

$$E_y'(x',y') \approx E_1 \cos\left(\frac{\pi}{a}x'\right)e^{-j\left[ky'^2/(2\rho_1)\right]} \tag{3.39}$$

$$H_x'(x',y') \approx \frac{E_1}{\eta}\cos\left(\frac{\pi}{a}x'\right)e^{-j\left[ky'^2/(2\rho_1)\right]} \tag{3.40}$$

$$H_z'(x',y') \approx jE_1\left(\frac{\pi}{ka\eta}\right)\sin\left(\frac{\pi}{a}x'\right)e^{-j\left[ky'^2/(2\rho_1)\right]} \tag{3.41}$$

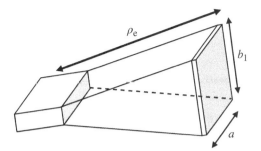

FIGURE 3.10   *E-plane sectoral horn.*

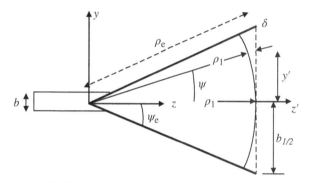

FIGURE 3.11   *E-plane sectoral horn geometry.*

The horn length is ρ_1, as shown in Figure 3.11. The extra distance along the aperture sides compared with the distance to the center is δ and is given by Equation 3.42:

$$\delta = \rho_e - \sqrt{\rho_e - \left(b_1/2\right)^2} = \rho_e\left(1 - \sqrt{1 - \left(b_1/2\rho_e\right)^2}\right) = b_1^2 \bigg/ 8\rho_e \qquad (3.42)$$

$$\frac{\delta}{\lambda} = S = \frac{b_1^2}{8\lambda\rho_e} \qquad (3.43)$$

where S represents the quadratic phase distribution as given in Equation 3.43:

$$\left(\delta(y') + \rho_1\right)^2 = \rho_1{}^2 + (y')^2 \qquad (3.44)$$

$$\delta(y') = -\rho_1 + \sqrt{\rho_1{}^2 + (y')^2} = -\rho_1 + \rho_1\sqrt{1 + \left(\frac{y'}{\rho_1}\right)^2} \qquad (3.45)$$

The maximum phase deviation at the aperture $\varnothing_{max}$ is given by Equation 3.46:

$$\varnothing_{max} = k\delta(y'): \left(y' = \frac{b_1}{2}\right) = \frac{kb_1^2}{8\rho_1} \tag{3.46}$$

The total flare angle of the horn, $2\psi_e$, is given in Equation 3.47:

$$2\psi_e = 2\tan^{-1}\left(\frac{b_1}{2\rho_1}\right) \tag{3.47}$$

3.5.1.1 *Directivity of the E-Plane Horn*   The maximum radiation is given by Equation 3.48:

$$U_{max} = \frac{r^2}{2\eta}|E|_{max}^2$$

$$U_{max} = \frac{r^2}{2\eta}|E|_{max}^2 \tag{3.48}$$

$$U_{max} = \frac{2ka^2\rho_1}{\pi^3\eta}|E|^2|F(t)|^2 \tag{3.49}$$

$$|F(t)|^2 = \left[C^2\left(\frac{b_1}{\sqrt{2\lambda\rho_1}}\right) + S^2\left(\frac{b_1}{\sqrt{2\lambda\rho_1}}\right)\right] \tag{3.50}$$

C and S are Fresnel integers and are given in Table 3.2.
The total radiated power by the horn is given in Equation 3.51:

$$P_{rad} = \frac{ab_1}{4\eta}|E|^2 \tag{3.51}$$

The directivity of E-plane horn D_E is given in Equation 3.52:

$$D_E = \frac{4\pi U_{max}}{P_{rad}} = \frac{64a\rho_1}{\pi\lambda b_1}\left[C^2\left(\frac{b_1}{\sqrt{2\lambda\rho_1}}\right) + S^2\left(\frac{b_1}{\sqrt{2\lambda\rho_1}}\right)\right] \tag{3.52}$$

TABLE 3.2 Fresnel Integers

x	$C_1(x)$	$S_1(x)$	$C(x)$	$S(x)$
0.0	0.62666	0.62666	0.0	0.0
0.1	0.52666	0.62632	0.10000	0.00052
0.2	0.42669	0.62399	0.19992	0.00419
0.3	0.32690	0.61766	0.29940	0.01412
0.4	0.22768	0.60536	0.39748	0.03336
0.5	0.12977	0.58518	0.49234	0.06473
0.6	0.03439	0.55532	0.58110	0.11054
0.7	−0.05672	0.51427	0.65965	0.17214
0.8	−0.14119	0.46092	0.72284	0.24934
0.9	−0.21606	0.39481	0.76482	0.33978
1.0	−0.27787	0.31639	0.77989	0.43826
1.1	−0.32285	0.22728	0.76381	0.53650
1.2	−0.34729	0.13054	0.71544	0.62340
1.3	−0.34803	0.03081	0.63855	0.68633
1.4	−0.32312	−0.06573	0.54310	0.71353
1.5	−0.27253	−0.15158	0.44526	0.69751
1.6	−0.19886	−0.21861	0.36546	0.63889
1.7	−0.10790	−0.25905	0.32383	0.54920
1.8	−0.00871	−0.26682	0.33363	0.45094
1.9	0.08680	−0.23918	0.39447	0.37335
2.0	0.16520	−0.17812	0.48825	0.34342
2.1	0.21359	−0.09141	0.58156	0.37427
2.2	0.22242	0.00743	0.63629	0.45570
2.3	0.18833	0.10054	0.62656	0.55315
2.4	0.11650	0.16879	0.55496	0.61969
2.5	0.02135	0.19614	0.45742	0.61918
2.6	−0.07518	0.17454	0.38894	0.54999
2.7	−0.14816	0.10789	0.39249	0.45292
2.8	−0.17646	0.01329	0.46749	0.39153
2.9	−0.15021	−0.08181	0.56237	0.41014
3.0	−0.07621	−0.14690	0.60572	0.49631
3.1	0.02152	−0.15883	0.56160	0.58181
3.2	0.10791	−0.11181	0.46632	0.59335
3.3	0.14907	−0.02260	0.40570	0.51929
3.4	0.12691	0.07301	0.43849	0.42965
3.5	0.04965	0.13335	0.53257	0.41525
3.6	−0.04819	0.12973	0.58795	0.49231
3.7	−0.11929	0.06258	0.54195	0.57498
3.8	−0.12649	−0.03483	0.44810	0.56562
3.9	−0.06469	−0.11030	0.42233	0.47521
4.0	0.03219	−0.12048	0.49842	0.42052
4.1	0.10690	−0.05815	0.57369	0.47580
4.2	0.11228	0.03885	0.54172	0.56320
4.3	0.04374	0.10751	0.44944	0.55400
4.4	−0.05287	0.10038	0.43833	0.46227
4.5	−0.10884	0.02149	0.52602	0.43427

(Continued)

TABLE 3.2 (*Continued*)

x	$C_1(x)$	$S_1(x)$	$C(x)$	$S(x)$
4.6	−0.08188	−0.07126	0.56724	0.51619
4.7	0.00810	−0.10594	0.49143	0.56715
4.8	0.08905	−0.05381	0.43380	0.49675
4.9	0.09277	0.04224	0.50016	0.43507
5.0	0.01519	0.09874	0.56363	0.49919
5.1	−0.07411	0.06405	0.49979	0.56239
5.2	−0.09125	−0.03004	0.43889	0.49688
5.3	−0.01892	−0.09235	0.50778	0.44047
5.4	0.07063	−0.05976	0.55723	0.51403
5.5	0.08408	0.03440	0.47843	0.55369
5.6	0.00641	0.08900	0.45171	0.47004
5.7	−0.07642	0.04296	0.53846	0.45953
5.8	−0.06919	−0.05135	0.52984	0.54604
5.9	0.01998	−0.08231	0.44859	0.51633
6.0	0.08245	−0.01181	0.49953	0.44696
6.1	0.03946	0.07180	0.54950	0.51647
6.2	−0.05363	0.06018	0.46761	0.53982
6.3	−0.07284	−0.03144	0.47600	0.45555
6.4	0.00835	−0.07765	0.54960	0.49649
6.5	0.07574	−0.01326	0.48161	0.54538
6.6	0.03183	0.06872	0.46899	0.46307
6.7	−0.05828	0.04658	0.54674	0.49150
6.8	−0.05734	−0.04600	0.48307	0.54364
6.9	0.03317	−0.06440	0.47322	0.46244
7.0	0.06832	0.02077	0.54547	0.49970
7.1	−0.00944	0.06977	0.47332	0.53602
7.2	−0.06943	0.00041	0.48874	0.45725
7.3	−0.00864	−0.06793	0.53927	0.51894
7.4	0.06582	−0.01521	0.46010	0.51607
7.5	0.02018	0.06353	0.51601	0.46070
7.6	−0.06137	0.02367	0.51564	0.53885
7.7	−0.02580	−0.05958	0.46278	0.48202
7.8	0.05828	−0.02668	0.53947	0.48964
7.9	0.02638	0.05752	0.47598	0.53235
8.0	−0.05730	0.02494	0.49980	0.46021
8.1	−0.02238	−0.05752	0.52275	0.53204
8.2	0.05803	−0.01870	0.46384	0.48589
8.3	0.01387	0.05861	0.53775	0.49323
8.4	−0.05899	0.00789	0.47092	0.52429
8.5	−0.00080	−0.05881	0.51417	0.46534
8.6	0.05767	0.00729	0.50249	0.53693
8.7	−0.01616	0.05515	0.48274	0.46774
8.8	−0.05079	−0.02545	0.52797	0.52294
8.9	0.03461	−0.04425	0.46612	0.48856
9.0	0.03526	0.04293	0.53537	0.49985
9.1	−0.04951	0.02381	0.46661	0.51042
9.2	−0.01021	−0.05338	0.52914	0.48135

TABLE 3.2 (*Continued*)

x	$C_1(x)$	$S_1(x)$	$C(x)$	$S(x)$
9.3	0.05354	0.00485	0.47628	0.52467
9.4	−0.02020	0.04920	0.51803	0.47134
9.5	−0.03995	−0.03426	0.48729	0.53100
9.6	0.04513	−0.02599	0.50813	0.46786
9.7	0.00837	0.05086	0.49549	0.53250
9.8	−0.04983	−0.01094	0.50192	0.46758
9.9	0.02916	−0.04124	0.49961	0.53215
10.0	0.02554	0.04298	0.49989	0.46817
10.1	−0.04927	0.00478	0.49961	0.53151
10.2	0.01738	−0.04583	0.50186	0.46885
10.3	0.03233	0.03621	0.49575	0.53061
10.4	−0.04681	0.01094	0.50751	0.47033
10.5	0.01360	−0.04563	0.48849	0.52804

Figure 3.12 presents the *H*-plane horn radiation pattern as function of *S*, where

$$S = \frac{b_1^2}{8\lambda\rho_e}.$$

3.5.2 *H*-Plane Sectoral Horn

H-plane sectoral horn is shown in Figure 3.13. *H*-plane sectoral horn geometry is shown in Figure 3.14.

Field expressions over the horn aperture are similar to the fields of a TE_{10} mode in rectangular waveguide with the aperture dimensions of a, b_1. The fields in the antenna aperture are written in Equations 3.53 and 3.54:

$$E_y'(x',y') \approx E_2 \cos\left(\frac{\pi}{a}x'\right)e^{-j\left[kx'^2/(2\rho_2)\right]} \tag{3.53}$$

$$H_x'(x',y') \approx \frac{E_2}{\eta}\cos\left(\frac{\pi}{a}x'\right)e^{-j\left[kx'^2/(2\rho_2)\right]} \tag{3.54}$$

The horn length is ρ_2. The extra distance along the aperture sides compared with the distance to the center is δ and is given by Equations 3.55 to 3.58.

$$\delta = \rho_h - \sqrt{\rho_h - \left(\frac{a_1}{2}\right)^2} = \rho_h\left(1 - \sqrt{1 - \left(\frac{a_1}{2\rho_h}\right)^2}\right) = \frac{a_1^2}{8\rho_h} \tag{3.55}$$

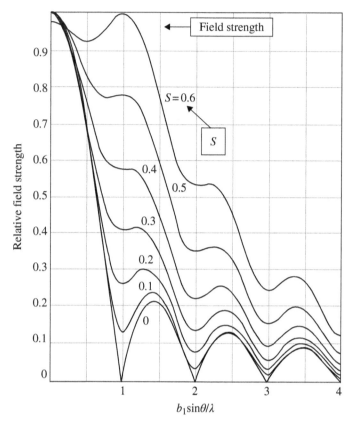

FIGURE 3.12   *H*-plane horn radiation pattern as function of S, $S = b_1^2/8\lambda\rho_e$.

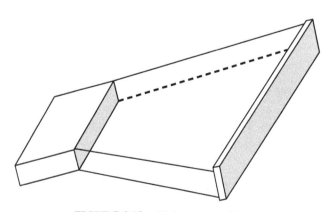

FIGURE 3.13   *H*-plane sectoral horn.

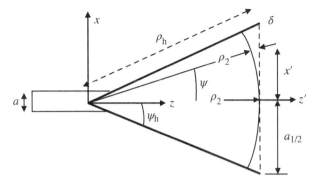

FIGURE 3.14 *E*-plane sectoral horn geometry.

$$\frac{\delta}{\lambda} = S = \frac{a_1^2}{8\lambda\rho_h} \tag{3.56}$$

where S represents the quadratic phase distribution as written in Equation 3.56:

$$\left(\delta(x') + \rho_2\right)^2 = \rho_2^2 + \left(x'\right)^2 \tag{3.57}$$

$$\delta(x') = -\rho_2 + \sqrt{\rho_2^2 + (x')^2} = -\rho_2 + \rho_1\sqrt{1 + \left(\frac{x'}{\rho_2}\right)^2} \tag{3.58}$$

The maximum phase deviation at the aperture $\varnothing_{max}$ is given by Equation 3.59:

$$\varnothing_{max} = k\delta(x') : \left(x' = \frac{a_1}{2}\right) = \frac{ka_1^2}{8\rho_2} \tag{3.59}$$

The total flare angle of the horn, $2\psi_e$, is given in Equation 3.60:

$$2\psi_h = 2\tan^{-1}\left(\frac{a_1}{2\rho_2}\right) \tag{3.60}$$

3.5.2.1 Directivity of the E-Plane Horn The maximum radiation is given by Equations 3.61 and 3.62:

$$U_{max} = \frac{r^2}{2\eta}|E|_{max}^2 \tag{3.61}$$

$$U_{max} = \frac{b^2\rho_2}{4\lambda\eta}|E|^2|F(t)|^2 \tag{3.62}$$

$$|F(t)|^2 = \left[(C(u) - C(v))^2 + (S(u) - S(v))^2\right] \tag{3.63}$$

where $u = \dfrac{1}{\sqrt{2}}\left(\dfrac{\sqrt{\rho_2\lambda}}{a_1} - \dfrac{a_1}{\sqrt{\rho_2\lambda}}\right)$ and $v = \dfrac{1}{\sqrt{2}}\left(\dfrac{\sqrt{\rho_2\lambda}}{a_1} + \dfrac{a_1}{\sqrt{\rho_2\lambda}}\right)$.

C and S are Fresnel integers.

The total radiated power by the horn is given in Equation 3.64:

$$P_{\text{rad}} = \frac{ab_1}{4\eta}|E|^2 \tag{3.64}$$

The directivity of H-plane horn D_H is given in Equation 3.65:

$$D_H = \frac{4\pi U_{\max}}{P_{\text{rad}}} = \frac{4b\pi\rho_2}{\lambda a_1}\left[(C(u)-C(v))^2 + (S(u)-S(v))^2\right] \tag{3.65}$$

H-plane horn radiation pattern as function of S, $S = a_1^2/8\lambda\rho_h$, is shown in Figure 3.15.

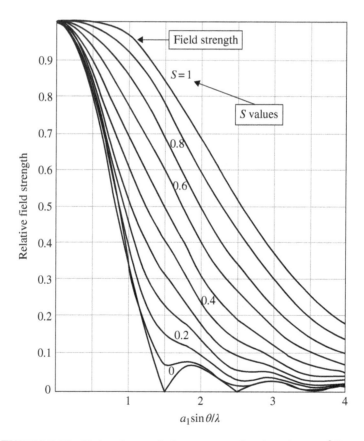

FIGURE 3.15 H-plane horn radiation pattern as function of S, $S = a_1^2/8\lambda\rho_h$.

3.5.3 Pyramidal Horn Antenna

The pyramidal horn antenna is a combination of the E and H horns as shown in Figure 3.16. The pyramidal horn antenna is realizable only if $\rho_h = \rho_e$.

The directivity of pyramidal horn antenna D_P is given in Equation 3.66:

$$D_P = \frac{4\pi U_{max}}{P_{rad}} = \frac{\pi\lambda^2}{32ab}D_E D_H \tag{3.66}$$

Kraus [3] gives the following approximation for pyramidal horn beamwidth:

$$\theta^e_{3dB} = \frac{56}{A_{e\lambda}} \tag{3.67}$$

$$\theta^h_{3dB} = \frac{67}{A_{h\lambda}} \tag{3.68}$$

$$\theta^e_{10dB} = \frac{100.8}{A_{e\lambda}} \tag{3.69}$$

$$\theta^h_{10dB} = \frac{120.6}{A_{h\lambda}} \tag{3.70}$$

$A_{e\lambda}$ is the aperture dimensions in wavelength in the E plane. $A_{h\lambda}$ is the aperture dimensions in wavelength in the H plane. The pyramidal horn gain is given by Equation 3.71:

$$G = 10\log_{10}4.5A_{h\lambda}A_{e\lambda} \text{ dBi} \tag{3.71}$$

The relative power at any angle, $P^h_{dB}(\theta)$, is given approximately by Equation 3.72:

$$P^h_{dB}(\theta) = 10\left(\frac{\theta}{\theta^h_{10dB}}\right)^2 \tag{3.72}$$

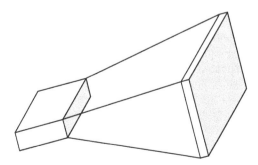

FIGURE 3.16 Pyramidal horn antenna.

3.6 ANTENNA ARRAYS FOR COMMUNICATION SYSTEMS

3.6.1 Introduction

An array of antenna elements is a set of antennas used for transmitting or receiving electromagnetic waves. An array of antennas is a collection of N similar radiators with the same three-dimensional radiation pattern. All the radiating elements are fed with the same frequency and with a specified amplitude and phase relationship for the drive voltage of each element. The array functions as a single antenna, generally with higher antenna gain than would be obtained from the individual elements. In antenna arrays electromagnetic wave interference is used to enhance electromagnetic signal in one desired direction at the expense of other directions. It may also be used to null the radiation pattern in one particular direction. Antenna array theory is presented in Ref. [1]. Several printed arrays and gain limitation of printed arrays are presented in Refs. [4–10]. Analysis and computations of losses in microstrip lines are given in Refs. [4–10]. Millimeter-wave arrays are presented in Refs. [6–10].

3.6.2 Array Radiation Pattern

The polar radiation pattern of a single element is called the element pattern (EP). The array pattern is the polar radiation pattern that would result if the elements were replaced by isotropic radiators, with the same amplitude and phase of excitation as the actual elements, and spaced at points on a grid corresponding to the far-field phase centers of the radiators. If we assume that all the polar radiation patterns of the elements taken individually are identical (within a certain tolerance) and that the patterns are all aligned in the same direction in azimuth and elevation, then the total array antenna pattern is got by multiplying the array factor (AF) by the EP. The total array antenna pattern, ET, is $ET = AF \times EP$.

The radiated field strength at a certain point in space in the far field is calculated by adding the contributions of each element to the total radiated fields. The summation of the contribution of each element to the total radiated fields is called the AF. The field strengths fall off as $1/r$ where r is the distance from the antenna to the field point as shown in Figure 3.17. We must take into account the amplitude and phase angle of the radiator excitation and also the phase delay, which is due to the time it takes the signal to get from the source to the field point. This phase delay is expressed as $2\pi r/\lambda$ where λ is the free-space wavelength of the electromagnetic wave.

Contours of equal field strength may be interpreted as an amplitude polar radiation pattern. Contours of the squared modulus of the field strength may be interpreted as a power polar radiation pattern. Figure 3.17 presents four-element array. The array factor of N element array is given in Equation 3.73. β presents the phase difference between the elements in the array.

Figure 3.18 presents a two-element array:

$$AF = 1 + e^{j\varphi} + e^{j2\varphi} + \cdots + e^{j\varphi(N-1)} = \sum_{n=1}^{N} e^{j\varphi(n-1)} \qquad (3.73)$$

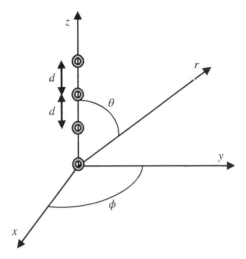

FIGURE 3.17   Coordinate system for external field calculations.

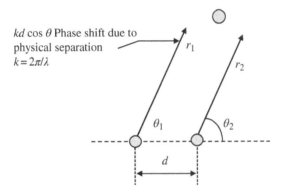

FIGURE 3.18   Two-element array.

where $\varphi = kd\cos\theta + \beta$ and $k = 2\pi/\lambda$.

The series summation is given in Equation 3.74:

$$AF = \frac{\sin(N\varphi/2)}{\sin(\varphi/2)} \tag{3.74}$$

The array factor will be zero when $\sin(N\varphi/2) = 0$. The array nulls will occur as given in Equation 3.75:

$$\theta_n = \cos^{-1}\left[\frac{\lambda}{2\pi d}\left(-\beta \pm \frac{2n}{N}\pi\right)\right] \quad n \neq N = 1,2,3\ldots \tag{3.75}$$

For $\varphi = \pm 2m\pi$ the array maximum level is given in Equation 3.76:

$$\theta_m = \cos^{-1}\left[\frac{\lambda}{2\pi d}(-\beta \pm 2m\pi)\right] \quad m = 0,1,2,3\ldots \tag{3.76}$$

The array 3dB beamwidth is given in Equation 3.77:

$$\theta_{3dB} = \cos^{-1}\left[\frac{\lambda}{2\pi d}\left(-\beta \pm \frac{2.782}{N}\right)\right] \tag{3.77}$$

The peak value of the side lobe is given in Equation 3.78:

$$\theta_{SL} = \cos^{-1}\left[\frac{\lambda}{2\pi d}\left(-\beta \pm \frac{3\pi}{N}\right)\right] \tag{3.78}$$

The side-lobe level for $\beta = 0$ is -13.46dB, as calculated in Equation 3.79:

$$AF_{SL} = 20\log_{10}\left(\frac{2}{3\pi}\right) = -13.46\text{dB} \tag{3.79}$$

3.6.3 Broadside Array

In a broadside array the main beam is perpendicular to the array. The array will radiate maximum energy perpendicular to the array if all elements in the array are fed with the same amplitude and phase level, $\beta = 0$.

The array factor will be zero when $\sin(N\varphi/2)$. The array nulls will occur as given in Equation 3.80:

$$\theta_n = \cos^{-1}\left(\pm\frac{n\lambda}{Nd}\right) n \neq N = 1,2,3\ldots \tag{3.80}$$

For $\varphi = \pm 2m\pi$ the array maximum level is given in Equation 3.81:

$$\theta_m = \cos^{-1}\left(\frac{m\lambda}{d}\right) m = 0,1,2,3\ldots \tag{3.81}$$

The array 3dB beamwidth is given in Equation 3.82 when $\pi d/\lambda \ll 1$:

$$\theta_{3dB} = \cos^{-1}\left(\frac{1.391\lambda}{\pi Nd}\right) \pi d/\lambda \ll 1 \tag{3.82}$$

The peak value of the side lobe is given in Equation 3.83:

$$\theta_{SL} = \cos^{-1}\left[\frac{\lambda}{2d}\left(\pm\frac{(2s+1)}{N}\right)\right] s = 0,1,2,3\ldots \tag{3.83}$$

3.6.4 End-Fire Array

In end-fire array the main beam is in the direction of the array axis. The array will radiate maximum energy in the direction of the array axis. The array will radiate maximum energy in the direction $\theta = 0$ if $\beta = -kd$. The array will radiate maximum energy in the direction $\theta = 180$ if $\beta = kd$.

The array nulls will occur as given in Equation 3.84:

$$\theta_n = \cos^{-1}\left(1 - \frac{n\lambda}{Nd}\right) n \neq N = 1, 2, 3 \ldots \tag{3.84}$$

For $\varphi = \pm 2m\pi$ the array maximum level is given in Equation 3.85:

$$\theta_m = \cos^{-1}\left(1 - \frac{m\lambda}{d}\right) m = 0, 1, 2, 3 \ldots \tag{3.85}$$

The array 3dB beamwidth is given in Equation 3.86 when $\pi d / \lambda \ll 1$:

$$\theta_{3\mathrm{dB}} = \cos^{-1}\left(1 - \frac{1.391\lambda}{\pi Nd}\right) \pi d / \lambda \ll 1 \tag{3.86}$$

The peak value of the side lobe is given in Equation 3.87:

$$\theta_{\mathrm{SL}} = \cos^{-1}\left[\frac{\lambda}{2d}\left(1 - \frac{(2s+1)}{N}\right)\right] s = 0, 1, 2, 3 \ldots \tag{3.87}$$

3.6.5 Printed Arrays

In Figure 3.19a parallel feed network of microstrip antenna array is shown. In Figure 3.19b parallel–series feed network of microstrip antenna array is presented.

Array directivity, D, may be written as $D = D_0 = U_{\max} / U_0 \sim \mathrm{AF}_{\max}^2 = N$.

Half power beamwidth may be written as $(\mathrm{HPBW}) = 2 \times (90 - \arccos (1.39\lambda / \pi Nd))$.

In Table 3.3 array directivity and beamwidth as function of number of element are listed. The radiation pattern of a 16 broadside element array is shown in Figure 3.20. The array directivity is 12dB.

3.6.6 Stacked Microstrip Antenna Arrays

A stacked Ku-band 16-element microstrip antenna array was designed at 14.5 GHz as shown in Figure 3.21. The resonator and the feed network were printed on a substrate with relative dielectric constant of 2.5 with thickness of 0.5 mm. The resonator is a circular microstrip resonator with diameter $a = 4.2$ mm. The radiating element was

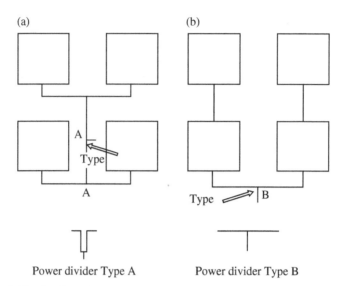

Power divider Type A   Power divider Type B

FIGURE 3.19 Configuration of microstrip antenna array. (a) Parallel feed network. (b) Parallel–series feed network.

TABLE 3.3   Array Directivity as Function of Number of Element

N Elements	D_0 (dB)	HPBW—θ_{3dB}
4	6	26°
8	9	10°
12	10.8	9°
16	12	7°
32	15	3.5°
64	18.1	1.75°

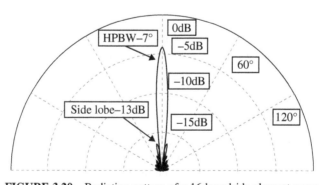

FIGURE 3.20   Radiation pattern of a 16 broadside element array.

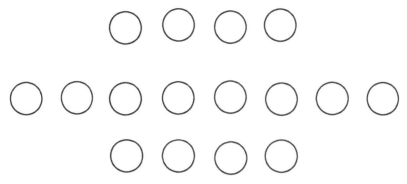

FIGURE 3.21   Stacked Ku-band 16-element microstrip antenna arrays.

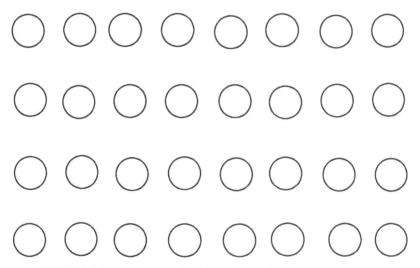

FIGURE 3.22   Stacked Ku-band 32-element microstrip antenna array.

printed on a substrate with relative dielectric constant of 2.2 with thickness of 0.5 mm. The distance between the radiating elements is around 0.75λ. The array dimensions are $125 \times 40 \times 1$ mm. The antenna bandwidth is 10% for VSWR better than $2:1$. The antenna beamwidth is around $10° \times 25°$. The measured antenna gain is around 18.5dBi. Losses in feed network are around 1dB.

A stacked Ku-band 32-element microstrip antenna array was designed at 14.5 GHz as shown in Figure 3.22. The resonator and the feed network were printed on a substrate with relative dielectric constant of 2.5 with thickness of 0.5 mm. The resonator is a circular microstrip resonator with diameter $a = 4.2$ mm. The radiating element was printed on a substrate with relative dielectric constant of 2.2 with thickness of 0.5 mm. The distance between the radiating elements is around 0.75λ. The array dimensions

TABLE 3.4   Measured Results of Stacked Microstrip Antenna Array

Array	F (GHz)	Bandwidth (%)	Beamwidth (°)	Gain (dBi)	Dimensions (mm)
16	14.5	10	10×25	18.5	$125 \times 40 \times 1$
32	14.5	10	10×20	20.5	$125 \times 125 \times 1$
16	10	13.5	23×23	17.0	$60 \times 60 \times 1.6$
64	34	10	10×10	23.5	$55 \times 55 \times 0.5$

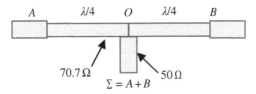

FIGURE 3.23   Power combiner/splitter.

are $125 \times 125 \times 1$ mm. The antenna bandwidth is 10% for VSWR better than 2 : 1. The antenna beamwidth is around $10° \times 20°$. The measured antenna gain is around 20.5dBi. Losses in feed network are around 1.5dB. Measured results of several stacked microstrip antenna arrays are listed in Table 3.4. The measured gain of the 64-element array at 34 GHz is around 23.5dBi.

A basic configuration of a power combiner/splitter is shown in Figure 3.23. This configuration of a power combiner/splitter is used in the feed network of several printed arrays. The power combiner/splitter consists of two sections of quarter-wavelength transformer. The distance from A to O is $\lambda/4$. The impedance at point O is 100 Ω. For an equal-split splitter, the impedance of the quarter-wavelength transformer is $1.41 \times Z_0$, or 70.7 Ω for $Z_0 = 50$ Ω. For an input signal V, the outputs at ports A and B are equal in magnitude and in phase.

3.6.7   Ka-band Microstrip Antenna Arrays

Microstrip antenna arrays with integral feed networks may be broadly divided into arrays fed by parallel feeds and series fed arrays. Usually series fed arrays are more efficient than parallel fed arrays. However, parallel fed arrays have a well-controlled aperture distribution. Two Ka-band microstrip antenna arrays that consist of 64 radiating elements have been designed on a 10 mil duroid substrate with $\varepsilon_r = 2.2$. The first array uses a parallel feed network and the second uses a parallel–series feed network as shown in Figure 3.24a and b. Comparison of the performance of the arrays is given in Table 3.5. Results given in Table 3.5 verify that the parallel–series fed array is more efficient than the parallel fed array due to minimization of the number of discontinuities in the parallel–series feed network.

The parallel–series fed array has been modified by using a 5 cm coaxial line to replace the same length of microstrip line. Results given in Table 3.5 indicate that

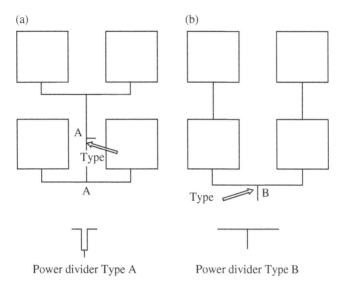

Power divider Type A   Power divider Type B

FIGURE 3.24 Configure of 64-element microstrip antenna arrays. (a) Parallel feed network. (b) Parallel–series feed network.

TABLE 3.5 Performance of 64-Element Microstrip Antenna Arrays

Parameter	Corporate Feed	Parallel Feed	Corporate Feed Mix
Number of elements	64	64	64
Beamwidth°	8.5	8.5	8.5
Computed gain (dBi)	26.3	26.3	26.3
Microstrip line loss (dB)	1.1	1.2	0.5
Radiation losses (dB)	0.7	1.3	0.7
Mismatch loss (dB)	0.5	0.5	0.5
Expected gain (dBi)	24.0	23.3	24.6
Efficiency (%)	58.9	50.7	67.6

the efficiency of the parallel–series fed array that incorporates coaxial line in the feed network is around 67.6% due to minimization of the microstrip line length. Two microstrip antenna arrays that consist of 256 radiating elements have been designed. In the first array, type A as shown in Figure 3.25a, using power divider, Type A minimizes the number of microstrip discontinuities. The second array, type B as shown in Figure 3.25b, incorporates more bend discontinuities in the feeding network. Comparison of the performance of the arrays is given in Table 3.6. The type A array with 256 radiating elements has been modified by using a 10 cm coaxial line to replace the same length of microstrip line. The performance comparison of the arrays given in Table 3.6 shows that the gain of the modified array has increased by 1.6dB. Results given in Table 3.6 verify that the type A array is more efficient than the type B array due to minimization of the number of bend discontinuities in the type A array feed network.

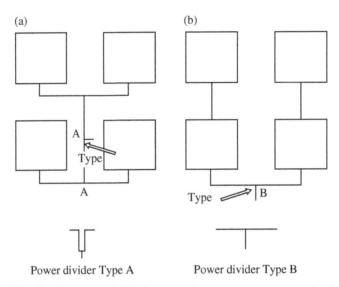

FIGURE 3.25 Configuration of 256-element microstrip antenna arrays. (a) Parallel feed network. (b) Parallel–series feed network.

TABLE 3.6 Performance of 256-Element Microstrip Antenna Arrays

Parameter	Type A	Type B	Type C
Number of elements	256	256	256
Beamwidth°	4.2	4.2	4.2
Computed gain (dBi)	32	32	32
Microstrip line loss (dB)	3.1	3.1	1.5
Radiation losses (dB)	1	1.9	1
Mismatch loss (dB)	0.5	0.5	0.5
Expected gain (dBi)	27.43	26.5	29.03
Efficiency (%)	34.9	28.2	50.47

The measured gain results are very close to the computed gain results and verify the loss computation.

REFERENCES

[1] Balanis CA. *Antenna Theory: Analysis and Design*. 2nd ed. New York: John Wiley & Sons, Inc.; 1996.

[2] Godara LC, editor. *Handbook of Antennas in Wireless Communications*. Boca Raton: CRC Press LLC; 2002.

[3] Kraus JD, Marhefka RJ. *Antennas for all Applications*. 3rd ed. New York: McGraw Hill; 2002.

[4] James JR, Hall PS, Wood C. *Microstrip Antenna Theory and Design*. London: Peregrinus on behalf of the Institution of Electrical Engineers; 1981.

[5] Sabban A, Gupta KC. Characterization of radiation loss from microstrip discontinuities using a multiport network modeling approach. IEEE Trans Microwave Theory Tech 1991;39 (4):705–712.

[6] Sabban A. Multiport network model for evaluating radiation loss and coupling among discontinuities in microstrip circuits [Ph.D. thesis]. University of Colorado at Boulder; January 1991.

[7] Kathei PB, Alexopoulos NG. Frequency-dependent characteristic of microstrip, discontinuities in millimeter-wave integrated circuits. IEEE Trans Microwave Theory Tech 1985;33:1029–1035.

[8] Sabban A. A new wideband stacked microstrip antenna. IEEE Antenna and Propagation Symposium; June 1983; Houston, TX.

[9] Sabban A. *Low Visibility Antennas for Communication Systems*. Boca Raton: Taylor & Francis Group; 2015.

[10] Sabban A. Wideband Microstrip Antenna Arrays. IEEE Antenna and Propagation Symposium; June 1981; MELCOM, Tel-Aviv.

[11] Sabban A. *RF Engineering, Microwave and Antennas*. Tel Aviv: Saar Publications; 2014.

4

MIC AND MMIC MICROWAVE AND MILLIMETER WAVE TECHNOLOGIES

4.1 INTRODUCTION

Communication and radar industry in microwave and mm-wave frequencies is currently in continuous growth. Radio frequency modules such as front end, filters, power amplifiers, antennas, passive components, and limiters are important modules in radar and communication links, see Refs. [1–10]. The electrical performance of the modules determines if the system will meet the required specifications. Moreover, in several cases the modules performance limits the system performance. Minimization of the size and weight of the RF modules is achieved by employing monolithic microwave integrated circuits (MMIC) and microwave integrated circuits (MIC) technology. However, integration of MIC and MMIC components and modules raise several technical challenges. Design parameters that may be neglected at low frequencies cannot be ignored in the design of wideband integrated RF modules. Powerful RF design software, such as ADS and HFSS, are required to achieve accurate design of RF modules in mm-wave frequencies. Accurate design of mm-wave RF modules is crucial. It is an impossible mission to tune mm-wave RF modules in the fabrication process.

Microwave and mm-wave technologies

- MIC, microwave integrated circuits
- MMIC, monolithic microwave integrated circuits

Wideband RF Technologies and Antennas in Microwave Frequencies, First Edition. Dr. Albert Sabban.
© 2016 John Wiley & Sons, Inc. Published 2016 by John Wiley & Sons, Inc.

- MEMS
- LTCC

4.2   MICROWAVE INTEGRATED CIRCUITS MODULES

Traditional microwave systems consists of connectorized components (such us amplifier, filters, and mixers) connected by coaxial cables. These modules have big dimensions and suffer from high losses. Dimension and losses may be minimized by using MIC technology. There are three types of MIC circuits: HMIC, standard MIC, and miniature HMIC. HMIC is a hybrid microwave integrated circuit. Solid state and passive elements are bonded to the dielectric substrate. The passive elements are fabricated by using thick or thin film technology. Standard MIC uses a single-level metallization for conductors and transmission lines. Miniature HMIC uses multilevel process in which passive elements such as capacitors and resistors are batch deposited on the substrate. Semiconductor devices such as amplifiers and diodes are bonded on the substrate. Figure 4.1 presents a MIC module. Figure 4.2 presents the layout of the MIC receiving link. The receiving channel consists of a low-noise amplifier, filters, dielectric resonant oscillators, and a diode mixer.

4.3   DEVELOPMENT AND FABRICATION OF A COMPACT
INTEGRATED RF HEAD FOR INMARSAT-M GROUND TERMINAL

This section describes the design, performance, and fabrication of a compact and low-cost MIC RF head for Inmarsat-M applications. Surface-mount MIC technology is employed to fabricate the RF head.

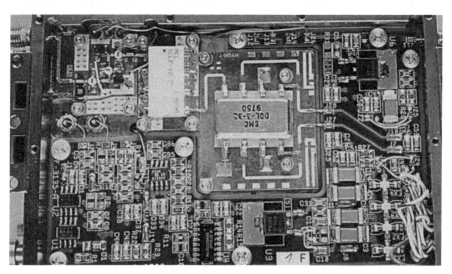

FIGURE 4.1   A MIC module.

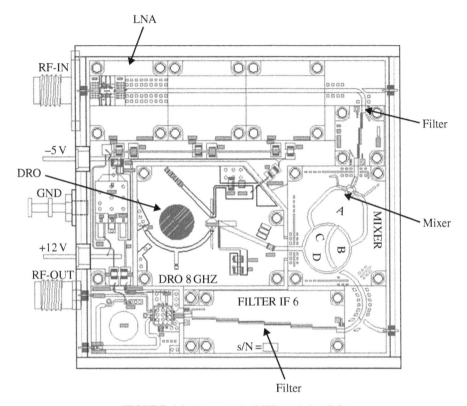

FIGURE 4.2 Layout of a MIC receiving link.

4.3.1   Introduction

Mobile telecommunication industry is currently in continuous growth. Moreover the great public demand for cellular and cordless telephones has stimulated a wide interest in new mobile services such as portable satellite communication terminals.

The Inmarsat-M system provides digital communications between the public switched terrestrial networks and mobile users.

Communication links to and from mobile installations are established via Inmarsat geostationary satellite and the associated ground station.

The RF head includes receiving and transmitting channels, RF controller, synthesizers, modem, and a DC supply unit. The RF head size is $30 \times 20 \times 2.5$ cm and weighs 1 kg. The transmitting channel may be operated in high-power mode to transmit 10 W or in low-power mode to transmit 4 W. The transmitted power level is controlled by automatic leveling control unit to ensure low power consumption over all the frequency and temperature range. The RF head vent is set to on and off automatically by the RF controller. The gain of the receiving channel is 76dB and is temperature compensated by using a temperature sensor and a voltage-controlled attenuator. Surface-mount technology is employed to fabricate the RF head.

4.3.2 Description of the Receiving Channel

Block diagram of the receiving channel is shown in Figure 4.3.

The receiving channel consists of low-noise amplifiers, filters, active mixer, saw filter, temperature sensor, voltage-controlled attenuator, and IF amplifiers.

The low-noise amplifier has 10dB gain and 0.9dB noise figure for frequencies ranging from 1.525 to 1.559 GHz. The total gain of the receiving channel is 76dB.

The low-noise amplifier employs a $1.8 gallium arsenide (GaAS) FET. The receiving channel noise figure and power budget calculation are given in Table 4.1. The channel gain is temperature compensated by connecting the output port of a temperature sensor to the reference voltage port of a voltage-controlled attenuator. The variation of the sensor output voltage as function of temperature varies the attenuation level of the attenuator. Gain stability as function of temperature is less than 1dB. The receiving channel noise figure is less than 2.2dB, including 1dB diplexer losses. The receiving channel rejects out of band signals with power level lower than −20dBm.

Receiving Channel Specifications

Frequency range: 1525–1559 MHz

Local oscillator frequency: 1355–1389 MHz

I.F. frequency: 170 MHz

Channel 1dB bandwidth: 50 kHz

Noise figure: 3dB

Input signals: −105 to −135dBm

Output signals: −30 to −60dBm

Receiving sensitivity: $C/N = 41$dB/Hz for −130dBm input signal.

4.3.3 Receiving Channel Design and Fabrication

A major parameter in the receiving channel design was to achieve a very low target price in manufacturing hundreds of units. The components were selected to meet the electrical requirements for a given low target price assigned to each component. Low production cost of the RF head is achieved by using SMT technology to manufacture the RF head. Trimming is not required in the fabrication procedure of the receiving channel. The gain and noise figure values of the receiving channel are measured in each RF head. Around 300 receiving channels have been manufactured up to date.

4.3.4 Description of the Transmitting Channel

Block diagram of the transmitting channel is shown in Figure 4.4. The transmitting channel consists of low-power amplifiers, passband filters, voltage-controlled

System1

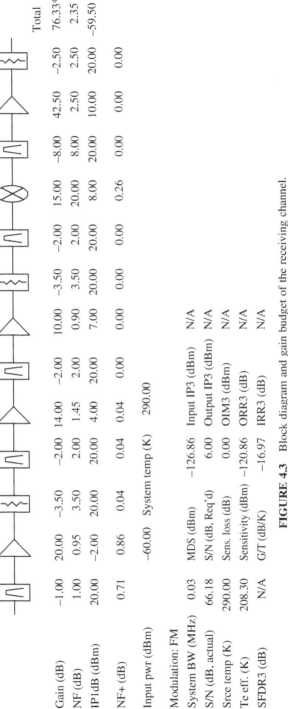

	Diplexer	LNA	VCA	Filter	Matched LNA	Filter	LNA	VCA	Filter	Mixer magnum	Saw filter	IF amp	Losses	Total
Gain (dB)	-1.00	20.00	-3.50	-2.00	14.00	-2.00	10.00	-3.50	-2.00	15.00	-8.00	42.50	-2.50	76.33*
NF (dB)	1.00	0.95	3.50	2.00	1.45	2.00	0.90	3.50	2.00	20.00	8.00	2.50	2.50	2.35
IP1dB (dBm)	20.00	-2.00	20.00	20.00	4.00	20.00	7.00	20.00	20.00	8.00	20.00	10.00	20.00	-59.50
NF+ (dB)	0.71	0.86	0.04	0.00	0.04	0.00	0.00	0.00	0.00	0.26	0.00	0.00	0.00	

Input pwr (dBm)   -60.00   System temp (K)   290.00

Modulation: FM

System BW (MHz)	0.03	MDS (dBm)	-126.86	Input IP3 (dBm)	N/A
S/N (dB, actual)	66.18	S/N (dB, Req'd)	6.00	Output IP3 (dBm)	N/A
Srce temp (K)	290.00	Sens. loss (dB)	0.00	OIM3 (dBm)	N/A
Te eff. (K)	208.30	Sensitivity (dBm)	-120.86	ORR3 (dB)	N/A
G/T (dB/K)	N/A	IRR3 (dB)	-16.97	IRR3 (dB)	N/A
SFDR3 (dB)	N/A				

FIGURE 4.3   Block diagram and gain budget of the receiving channel.

TABLE 4.1 Noise Figure and Gain Calculation

Component	Noise Figure (dB)	Gain (dB)	P_{out} (dBm)
Diplexer	1	−1	−131
LNA	0.95	20	−111
VCA	3.5	−3.5	−114.5
Filter	2	−2	−116.5
Matched LNA unit	1.45	14	−102.5
Filter	2	−2	−104.5
LNA	0.9	10	−94.5
VCA	3.5	−3.5	−98
Filter	2	−2	−100
Mixer	20	15	−85
Saw filter	8	−8	−93
IF amplifiers	2.5	42.5	−50.5
Transmission line losses	2.5	−2.5	−53
Total	2.35	77	−53

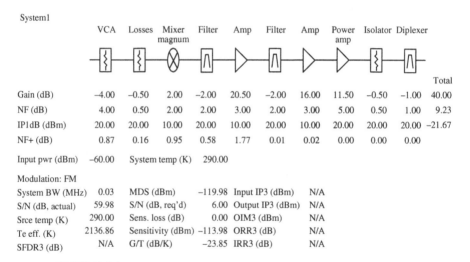

FIGURE 4.4 Block diagram and gain budget of the transmitting channel.

attenuator, active mixer, medium-power and high-power amplifiers, high-power isolator, coupler, power detector, and DC supply unit.

The power budget of the transmitting channel is given in Table 4.2.

Five stages of power amplifiers amplify the input signal from 0 to 40dBm. The fifth stage is a 10 W power amplifier with high efficiency. The amplifier may transmit 10 W in high-power mode or 4 W in low-power mode. The DC bias voltage of the power amplifier is automatically controlled by the RF controller to set the power amplifier to the required mode and power level. A −30dB coupler and a power detector are used to measure the output power level. The measured power level is

TABLE 4.2 Transmitting Channel Power Budget

Component	Gain/Loss (dB)	P_{out} (dBm)
Input	0	0
VCA	−4	−4
Tr. line loss	−0.5	−4.5
Mixer	2	−2.5
Filter	−2	−4.5
Low-power amplifiers	20.5	16
Filter	−2	14
Medium-power amplifier	16	30
Power amplifier	11.5	41.5
Isolator	−0.5	41
Diplexer	−1	40
Total	40.0	40

transferred via an A/D converter to the RF controller to monitor the output power level of the transmitting channel by varying the attenuation of the voltage-controlled attenuator. This feature ensures low DC power consumption and high efficiency of the RF head. The RF controller sets the transmitting channel to on and off. A temperature sensor is used to measure the RF head temperature. The RF controller set the RF head vent to on and off according to the measured RF head temperature.

4.3.4.1 *Transmitting Channel Specifications*

Frequency range: 1626.5–1660.5 MHz

I.F. frequency range: 99.5–133.5 MHz

L.O. frequency: 1760 MHz

Input power: 0dBm

Output power (high mode): 38–40dBm

Output power (low mode): 34–36dBm

Power consumption: 42 W

4.3.4.2 *Diplexer Specifications*   A very compact and lightweight diplexer connects the receiving and transmitting channels to the antenna. The diplexer simple structure and easy manufacturability ensure lower costs in production than similar diplexers.

Transmit Filter

Passband frequency range: 1626.5–1660.5 MHz

Passband insertion loss: 0.7dB

Passband VSWR < 1.3 : 1

Rejection >54dB at 1525–1559 MHz

Receive Filter

Passband frequency range: 1525–1559 MHz

Passband insertion loss < 1.3dB

Pass Band VSWR < 1.3 : 1

Rejection >65dB at 1626.5–1660.5 MHz

Size: $86 \times 36 \times 25$ mm

4.3.5   Transmitting Channel Fabrication

A photo of the RF head prototype for Inmarsat-M ground terminal is shown in Figure 4.5. A major parameter in the transmitting channel design was to achieve a low target price in the fabrication of hundreds of units. The components were selected to meet the target price given to each component and the electrical requirements. Low production cost is achieved by using SMT technology to manufacture the transmitting channel. A quick trimming procedure is required in the fabrication of the transmitting channel to achieve the required output power and efficiency. The output power and spurious level of the transmitting channel are tested in the fabrication procedure of each RF head. Around 300 transmitting channels have been manufactured during the first production cycle.

A photo of the RF head modules for Inmarsat-M ground terminal is shown in Figure 4.6. The RF head is separated to five sections: receiving and transmitting channels, diplexer, synthesizers, RF controller, and a DC supply unit. A metallic fence and cover separate between the transmitting and receiving channels.

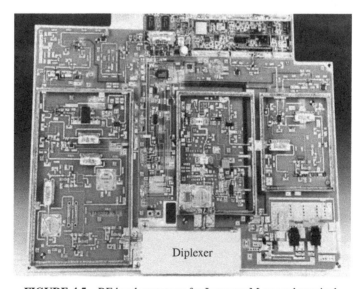

FIGURE 4.5   RF head prototype for Inmarsat-M ground terminal.

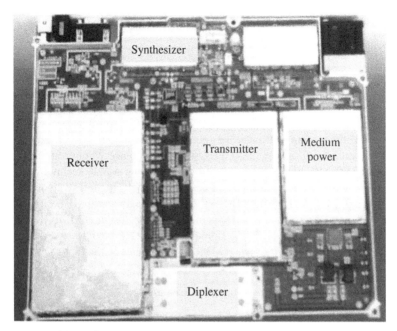

FIGURE 4.6   RF head modules for Inmarsat-M ground terminal.

4.3.6   RF Controller

The RF controller is based on an 87C51 microcontroller. The RF controller communicates with the system controller via a full-duplex serial bus. The communication is based on message transfer. The RF controller sets the transmitting channel to on and off by controlling the DC voltage switching unit. The RF controller monitors the output power level of the transmitting channel by varying the attenuation of the voltage-controlled attenuator in the transmitting channel. The RF controller sets the transmitting channel to burst or scpc modes with high- or low-power level. The RF controller produces the clock data and enable signals for the Rx and Tx synthesizers.

4.3.7   Concluding Remarks

A compact and low-cost RF head for Inmarsat-M applications is presented in this section. The RF head is part of a portable satellite communication ground terminal, "Caryphone," which supplies phone and fax services to the customer.

The RF controller automatically monitors the output power level to ensure low DC power consumption. A DC to DC converter supplies to the power amplifier a controlled DC bias voltage to set the power amplifier to high-power level mode, 10 W or 4 W.

The receiving channel noise figure is less than 2.2dB. The total gain of the receiving channel is 76dB with gain stability of 1dB as function of temperature.

The RF head size is $30 \times 20 \times 2.5$ cm and weighs less than 1 kg.

The "Caryphone" terminal has been approved by Inmarsat on August 1995. Up to date 300 terminals has been manufactured.

4.4   MONOLITHIC MICROWAVE INTEGRATED CIRCUITS

4.4.1   Introduction

Communication and wireless systems in microwave and mm-wave frequencies is currently in continuous growth. Radio Frequency modules such as front end, filters, power amplifiers, antennas, passive components, and limiters are important modules in radar and communication links, see Refs. [1–9]. The electrical performance of the modules determines if the system will meet the required specifications. Moreover, in several cases the modules performance limits the system performance. Minimization of the size and weight of the RF modules is achieved by employing MMIC technology. Design parameters that may be neglected at low frequencies in MIC modules cannot be ignored in the design of wideband integrated MMIC RF modules. Powerful RF design software, such as ADS and HFSS, are required to achieve accurate design of MMIC RF modules in mm-wave frequencies. Accurate design of mm-wave RF modules is crucial. It is an impossible mission to tune mm-wave MMIC RF modules in the fabrication process.

4.4.2   Monolithic Microwave Integrated Circuits

MMIC are circuits in which active and passive elements are formed on the same dielectric substrate, as presented in Figure 4.7, by using a deposition scheme as epitaxy, ion implantation, sputtering, evaporation, and diffusion. In Figure 4.7 the MMIC chip consists passive elements such as resistors, capacitors, inductors, and a field-effect transistor (FET).

4.4.2.1   MMIC Design Facts

MMIC components can't be tuned. Accurate design is crucial in the design of MMIC circuits. Accurate design may be achieved by using 3D electromagnetic software.

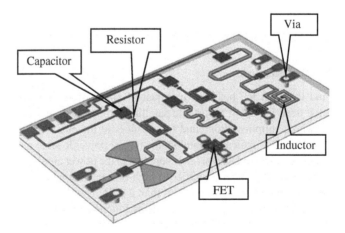

FIGURE 4.7   MMIC basic components.

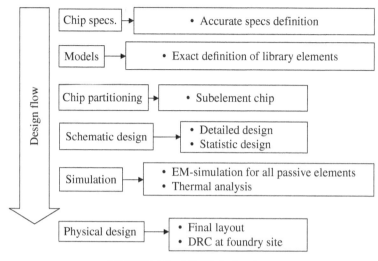

FIGURE 4.8 MMIC design flow.

Materials employed in the design of MMIC circuits are GaAs, InP, GaN, and silicon–germanium (SiGe).

Large statistic scattering of all electrical parameters cause sensitivity of the design.

FAB runs are expensive, around $200,000 per run. Miniaturization of components yields lower cost of the MMIC circuits. Figure 4.8 presents MMIC design flow.

The designer goal is to comply with customer specifications in one design iteration.

4.4.2.2 MMIC Technologies Features

- 0.25 μm GaAs PHEMT for power applications to Ku Band
- 0.15 μm GaAs PHEMT for applications to high Ka Band
- GaAs PIN process for low-loss power switching applications
- Future new process: InP HBT, SiGe, GaN,RFCMOS, RFMEMS

Figure 4.9 presents a GaAs wafer layout. Wafer size may be 3″, 5″, or 6″.

4.4.2.3 Types of Components Designed

- Amplifiers: LNA, general, power amplifiers, wideband power amplifiers, distributed TWA
- Mixers: balanced, star, subharmonic
- Switches: PIN, PHEMT, T/R matrix
- Frequency multipliers: active, passive

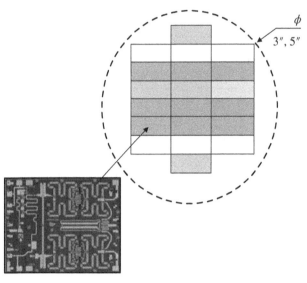

FIGURE 4.9   GaAs wafer layout.

- Modulators: QPSK, QAM (PIN, PHEMT)
- Multifunction: Rx chip, Tx chip, switched amp chip, LO chain

FET: Field-effect transistor
BJT: Bipolar junction transistor
HEMT: High-electron mobility transistor
PHEMT: Pseudomorphic HEMT
MHEMT: Metamorphic HEMT
DHBT: Double heterostructure bipolar transistor
CMOS: Complementary metal-oxide semiconductor

Table 4.3 presents types of devices fabricated by using MMIC Technology.

TABLE 4.3   MMIC Technology

Material	FET	BJT	Diode
III–V based	PHEMT GaAs	HBT GaAs	Schottky GaAs
	HEMT InP	DHBT InP	
	MHEMT GaAs		
	HEMT GaN		
Silicon	CMOS	HBT SiGe	

4.4.3 Advantages of GaAs versus Silicon

MMICs are originally fabricated by using GaAs, a III–V compound semiconductor. MMICs are dimensionally small (from around 1 to 10 mm^2) and can be mass produced. GaAs has some electronic properties which are better than those of silicon. It has a higher saturated electron velocity and higher electron mobility, allowing transistors made from GaAs to function at frequencies higher than 250 GHz. Unlike silicon junctions, GaAs devices are relatively insensitive to heat due to their higher bandgap. Also, GaAs devices tend to have less noise than silicon devices especially at high frequencies which is a result of higher carrier mobility and lower resistive device parasitic. These properties recommend GaAs circuitry in mobile phones, satellite communications, microwave point-to-point links, and higher-frequency radar systems. It is used in the fabrication of Gunn diodes to generate microwave. GaAs has a direct bandgap, which means that it can be used to emit light efficiently. Silicon has an indirect bandgap and so is very poor at emitting light. Nonetheless, recent advances may make silicon LEDs and lasers possible. Due to its lower bandgap though, Si LEDs cannot emit visible light and rather work in IR range while GaAs LEDs function in visible red light. As a wide direct bandgap material and resulting resistance to radiation damage, GaAs is an excellent material for space electronics and optical windows in high-power applications.

Silicon has three major advantages over GaAs for integrated circuit manufacturer. First, silicon is a cheap material. In addition, a Si crystal has an extremely stable structure mechanically, and it can be grown to very large diameter boules and can be processed with very high yields. It is also a decent thermal conductor thus enables very dense packing of transistors, all very attractive for design and manufacturing of very large ICs. The second major advantage of Si is the existence of silicon dioxide—one of the best insulators. Silicon dioxide can easily be incorporated onto silicon circuits, and such layers are adherent to the underlying Si. GaAs does not easily form such a stable adherent insulating layer and does not have stable oxide either. The third, and perhaps most important, advantage of silicon is that it possesses a much higher hole mobility. This high mobility allows the fabrication of higher-speed P-channel FET, which are required for CMOS logic. Because they lack a fast CMOS structure, GaAs logic circuits have much higher power consumption, which has made them unable to compete with silicon logic circuits. The primary advantage of Si technology is its lower fabrication cost compared with GaAs. Silicon wafer diameters are larger. Typically 8″ or 12″ compared with 4″ or 6″ for GaAs. Si wafer costs are much lower than GaAs wafer costs, contributing to a less expensive Si IC.

Other III–V technologies, such as indium phosphide (InP), offer better performance than GaAs in terms of gain, higher cutoff frequency, and low noise. However they are more expensive due to smaller wafer sizes and increased material fragility.

SiGe is a Si-based compound semiconductor technology offering higher-speed transistors than conventional Si devices but with similar cost advantages.

Gallium Nitride (GaN) is also an option for MMICs. Because GaN transistors can operate at much higher temperatures and work at much higher voltages than GaAs transistors, they make ideal power amplifiers at microwave frequencies. In Table 4.4 properties of the material used in MMIC technology are compared.

TABLE 4.4   Comparison of Material Properties

Property	Si	Si or Sapphire	GaAs	InP
Dielectric constant	11.7	11.6	12.9	14
Resistivity (Ω/cm)	10^3–10^5	$>10^{14}$	10^7–10^9	10^7
Mobility (cm^2/v-s)	700	700	4300	3000
Density (g/cm^3)	2.3	3.9	5.3	4.8
Saturation velocity (cm/s)	9×10^6	9×10^6	1.3×10^7	1.9×10^7

TABLE 4.5   Summary of Semiconductor Technology

	Si CMOS	SiGe HBT	InP HBT	InP HEMT	GaN HEMT
Cutoff frequency (GHz)	>200	>200	>400	>600	>200
Published MMICs (GHz)	170	245	325	670	200
Output power	Low	Medium	Medium	Medium	High
Gain	Low	High	High	Low	Low
RF noise	High	High	High	Low	Low
Yield	High	High	Medium	Low	Low
Mixed signal	Yes	Yes	Yes	No	No
$1/f$ noise	High	Low	Low	High	High
Breakdown voltage (V)	−1	−2	−4	−2	>20

4.4.4   Semiconductor Technology

Cutoff frequency of Si CMOS MMIC devices is lower than 200 GHz. Si CMOS MMIC devices are usually low-power and low-cost devices. Cutoff frequency of SiGe MMIC devices is lower than 200 GHz. SiGe MMIC devices are used as medium-power high-gain devices. Cutoff frequency of InP HBT devices is lower than 400 GHz. InP HBT devices are used as medium-power high-gain devices. Cutoff frequency of InP HEMT devices is lower than 600 GHz. InP HEMT devices are used as medium-power high-gain devices. In Table 4.5 properties of MMIC technologies are compared. Figure 4.10 presents a 0.15 μm PHEMT on GaAs substrate.

4.4.5   MMIC Fabrication Process

MMIC fabrication process consists of several controlled processes in a semiconductor FAB. The process is listed in the next paragraph.

In Figure 4.11 a MESFET cross section on GaAs substrate is shown.

MMIC fabrication process list

Wafer fabrication—Preparing the wafer for fabrication.

Wet cleans—Wafer cleaning by wet process.

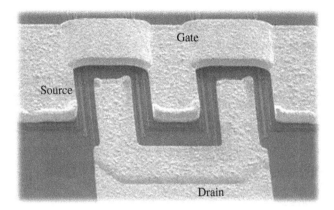

FIGURE 4.10   0.15 μm PHEMT on GaAs substrate.

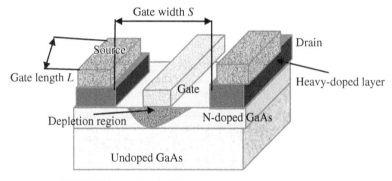

FIGURE 4.11   MESFET cross section on GaAs substrate.

Ion implantation—Dopants are embedded to create regions of increased or decreased conductivity. Selectively implant impurities create p- or n-type semiconductor regions.

Dry etching—Selectively remove materials.

Wet etching—Selectively remove materials chemical process.

Plasma etching—Selectively remove materials.

Thermal treatment—High-temperature process to remove stress.

Rapid thermal anneal—High-temperature process to remove stress.

Furnace anneal—After ion implantation, thermal annealing is required. Furnace annealing may take minutes and causes too much diffusion of dopants for some applications.

Oxidation—Substrate oxidation, for example, dry oxidation—$Si + O_2 \rightarrow SiO_2$; wet oxidation $Si + 2H_2O \rightarrow SiO_2 + 2H_2$.

Chemical vapor deposition (CVD)—Chemical vapor deposited on the wafer. Pattern defined by photoresist.

Physical vapor deposition (PVD)—Vapor produced by evaporation or sputtering deposited on the wafer. Pattern defined by photoresist.

Molecular Beam Epitaxy (MBE)—A beam of atoms or molecules produced in high vacuum. Selectively grow layers of materials. Pattern defined by photoresist.

Electroplating—Electromechanical process used to add metal.

Chemical mechanical polish (CMP)

Wafer testing—Electrical test of the wafer.

Wafer backgrinding

Die preparation

Wafer mounting

Die cutting

Lithography—Lithography is the process of transferring a pattern onto the wafer by selectively exposing and developing photoresist. Photolithography consists of four steps, the order depends on whether we are etching or lifting off the unwanted material.

Contact lithography—A glass plate is used that contains the pattern for the entire wafer. It is literally led against the wafer during exposure of the photoresist. In this case the entire wafer is patterned in one shot.

Electron-beam lithography is a form of direct-write lithography. Using E-beam lithography you can write directly to the wafer without a mask. Because an electron beam is used, rather than light, much smaller features can be resolved.

Exposure can be done with light, UV light, or electron beam, depending on the accuracy needed. E-beam provides much higher resolution than light, because the particles are bigger (greater momentum), the wavelength is shorter.

4.4.5.1 *Etching versus Lift-off Removal Processes*   There are two principal means of removing material, etching and lift-off.

The steps for an etch-off process are:

1. Deposit material
2. Deposit photoresist
3. Pattern (expose and develop)
4. Remove material where it is not wanted by etching

Etching can be isotropic (etching wherever we can find the material we like to etch) or anisotropic (directional, etching only where the mask allows). Etches can be dry (reactive ion etching or RIE) or wet (chemical). Etches can be very selective (only etching what we intend to etch) or nonselective (attacking a mask to the substrate).

In a lift-off process, the photoresist *forms a mold* into which the desired material is deposited. The desired features are completed when photoresist B under unwanted areas is dissolved, and unwanted material is "lifted off."

Lift-off processes are:

* Deposit photoresist
* Pattern
* Deposit material conductor or insulator
* Remove material where it is not wanted by *lifting off*

In Figure 4.12 a MESFET cross section on GaAs substrate is shown. In Figure 4.13 a MMIC resistor cross section is shown.

In Figure 4.14 a MMIC capacitor cross section is shown.

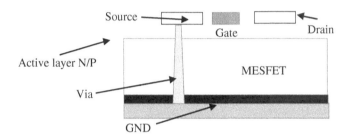

FIGURE 4.12　MESFET cross section.

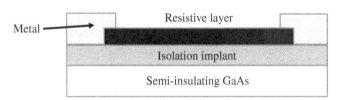

FIGURE 4.13　Resistor cross section.

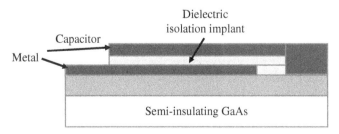

FIGURE 4.14　Capacitor cross section.

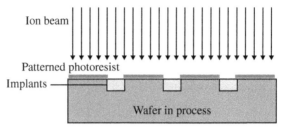

FIGURE 4.15   Ion implantation.

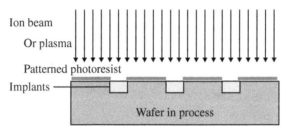

FIGURE 4.16   Ion etch.

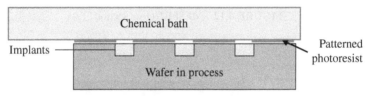

FIGURE 4.17   Wet etch.

Figure 4.15 presents the ion implantation process. Figure 4.16 presents the ion etch process. Figure 4.17 presents the wet etch process.

4.4.6   Generation of Microwave Signals in Microwave and mm Wave

Microwaves signals can be generated by solid-state devices and vacuum tube-based devices. Solid-state microwave devices are based on semiconductors such as silicon or GaAs and include FETs, BJTs, Gunn diodes, and IMPATT diodes. Microwave variations of BJTs include the heterojunction bipolar transistor (HBT) and microwave variants of FETs include the MESFET, the HEMT (also known as HFET), and LDMOS transistor. Microwaves can be generated and processed using integrated circuits, MMIC. They are usually manufactured using GaAs wafers, though SiGe and heavy-dope silicon are increasingly used. Vacuum tube-based devices operate on the ballistic motion of electrons in a vacuum under the influence of controlling electric or magnetic fields and include the magnetron, klystron, traveling-wave tube (TWT), and gyrotron. These devices work in the density-modulated mode, rather than the current-modulated mode. This means that they work on the basis of clumps of electrons flying ballistically through them, rather than using a continuous stream.

4.4.7 MMIC Circuit Examples and Applications

Figure 4.18 presents a wideband mm-wave power amplifier. The input power is divided by using a power divider. The RF signal is amplified by power amplifiers and combined by a power combiner to get the desired power at the device output.

Figure 4.19 presents a wideband mm-wave upconverter. Figure 4.20 presents a Ka band pin diode nonreflective SPDT. MMIC process cost is listed in Table 4.6.

MMIC applications

- Ka band satellite communication
- 60 GHz wireless communication
- Automotive radars
- Imaging in security
- Gbit WLAN

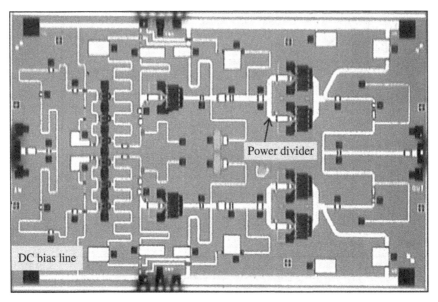

FIGURE 4.18 Wideband power amplifier.

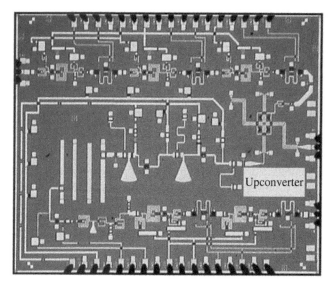

FIGURE 4.19   Ka band upconverter.

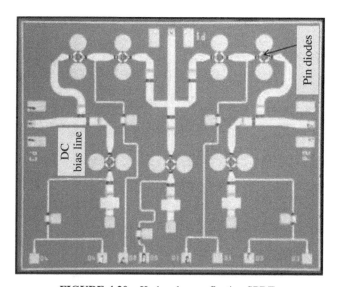

FIGURE 4.20   Ka band nonreflective SPDT.

TABLE 4.6   MMIC Cost

	Si CMOS	SiGe HBT	GaAs HEMT	InP HEMT
Chip cost ($/mm^2)	0.01	0.1–0.5	1–2	10
Mask cost (M$/mask set)	1.35	0.135	0.0135	0.0135

4.5 CONCLUSIONS

The electrical performance of the modules determines if the system will meet the required specifications. In several cases the modules performance limits the system performance. Minimization of the size and weight of the RF modules is achieved by employing MMIC and MIC technology. However, integration of MIC and MMIC components and modules raise several technical challenges. Design parameters that may be neglected at low frequencies cannot be ignored in the design of wideband integrated RF modules. Powerful RF design software, such as ADS and HFSS, are used to achieve accurate design of RF modules in mm-wave frequencies. This section describes the design, Performance, and fabrication of a compact and low-cost MIC RF head for Inmarsat-M applications. Surface-mount MIC technology is employed to fabricate a compact low-cost RF head for satellite communication application.

REFERENCES

[1] Rogers J, Plett C. *Radio Frequency Integrated Circuit Design*. Boston: Artech House; 2003.

[2] Maluf N, Williams K. *An Introduction to Microelectromechanical System Engineering*. Boston: Artech House; 2004.

[3] Sabban A. Microstrip Antenna Arrays. In: Nasimuddin N, editor. *Microstrip Antennas*. Croatia: InTech; 2011. p. 361–384.

[4] Sabban A. Applications of mm wave microstrip antenna arrays. International Symposium on Signals, Systems and Electronics, 2007. ISSSE '07 Conference; 2007 July 30 to August 2; Montreal, QC.

[5] Gauthier GP et al. A 94GHz micro-machined aperture-coupled microstrip antenna. *IEEE Trans Antenna Propag* 1999;47 (12):1761–1766.

[6] Milkov MM. Millimeter-wave imaging system based on antenna-coupled bolometer [M. Sc. thesis]. UCLA; 2000.

[7] de Lange G, Konistis K, Hu Q. A 3∗3 mm-wave micro machined imaging array with sis mixers. *Appl Phys Lett* 1999;75 (6):868–870.

[8] Rahman A, de Lange G, Hu Q. Micromachined room temperature microbolometers for mm-wave detection. *Appl Phys Lett* 1996;68 (14):2020–2022.

[9] Maas SA. *Nonlinear Microwave and RF Circuits*. Boston: Artech House; 1997.

[10] Sabban A. *Low Visibility Antennas for Communication Systems*. Boca Raton: Taylor & Francis Group; 2015.

5

PRINTED ANTENNAS FOR WIRELESS COMMUNICATION SYSTEMS

Printed antennas are the perfect antenna solution for wireless and medical communication systems. Printed antennas possess attractive features such as low profile, lightweight, small volume, and low production cost. These features are crucial for wireless compact wearable communication systems. In addition, the benefit of a compact low-cost feed network is attained by integrating the feed structure with the radiating elements on the same substrate. Printed antennas are used in communication systems that employ MIC and MMIC technologies.

5.1   PRINTED ANTENNAS

Printed antennas possess attractive features such as low profile, flexible, lightweight, small volume, and low production cost. Printed antennas are widely presented in books and papers in the last decade [1–18]. The most popular type of printed antennas are microstrip antennas. However, PIFA, slot, and dipole printed are widely used in communication systems. Printed antennas may be employed in communication links, seekers, and in biomedical systems.

5.1.1   Introduction to Microstrip Antennas

Microstrip antennas are printed on a on a dielectric substrate with low dielectric losses. Cross section of the microstrip antenna is shown in Figure 5.1. Microstrip antennas are

---

Wideband RF Technologies and Antennas in Microwave Frequencies, First Edition. Dr. Albert Sabban.
© 2016 John Wiley & Sons, Inc. Published 2016 by John Wiley & Sons, Inc.

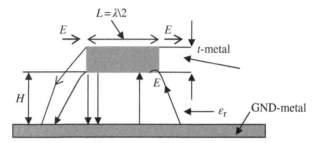

FIGURE 5.1 Microstrip antenna cross section.

thin patches etched on a dielectric substrate ε_r, see Refs. [1–7]. The substrate thickness, H, is less than 0.1λ.

Advantages of microstrip antennas

- Low cost to fabricate.
- Conformal structures are possible.
- Easy to form a large uniform array with half-wavelength spacing.
- Lightweight and low volume.

Disadvantages of microstrip antennas

- Limited bandwidth (usually 1–5%, but much more is possible with increased complexity)
- Low power handling

The electric field along the radiating edges is shown in Figure 5.2. The magnetic field is perpendicular to the E-field according to Maxwell's equations. At the edge of the strip ($X/L = 0$ and $X/L = 1$), the H-field drops to zero, because there is no conductor to carry the RF current; it is maximum in the center. The E-field intensity is at maximum magnitude (and opposite polarity) at the edges ($X/L = 0$ and $X/L = 1$) and zero at the center. The ratio of E- to H-field is proportional to the impedance that we see when we feed the patch. Microstrip antennas may be fed by a microstrip line or by a coaxial line or probe feed. By adjusting the location of the feed point between the center and the edge, we can get any impedance, including $50\,\Omega$. Microstrip e antenna shape may be square, rectangular, triangle, circle, or any arbitrary shape as shown in Figure 5.3.

The dielectric constant that controls the resonance of the antenna is the effective dielectric constant of the microstrip line. The antenna dimension W is given by Equation 5.1:

$$W = \frac{c}{2f\sqrt{\epsilon_{\text{eff}}}} \tag{5.1}$$

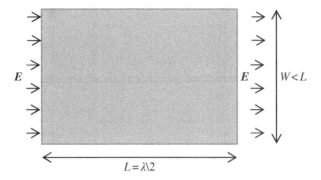

FIGURE 5.2 Rectangular microstrip antenna.

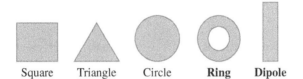

FIGURE 5.3 Microstrip antenna shapes.

The antenna bandwidth is given in Equation 5.2:

$$BW = \frac{H}{\sqrt{\epsilon_{\text{eff}}}}. \tag{5.2}$$

The gain of microstrip antenna is between 0 and 7dBi. The microstrip antenna gain is function of the antenna dimensions and configuration. We may increase printed antenna gain by using microstrip antenna array configuration. In a microstrip antenna array, the benefit of a compact low-cost feed network is attained by integrating the RF feed network with the radiating elements on the same substrate. Microstrip antenna feed networks is presented in Figure 5.4. Figure 5.4a presents a parallel feed network. Figure 5.4b presents a parallel–series feed network.

5.1.2 Transmission Line Model of Microstrip Antennas

In the transmission line model, TLM, of patch microstrip antennas, the antenna is represented as two slots connected by a transmission line. TLM is presented in Figure 5.5. TLM is not an accurate model. However it gives a good physical under-standing of patch microstrip antennas. The electric field along and underneath the patch depend on the z coordinate as given in Equation 5.3.

$$E_x \sim \cos\left(\frac{\pi z}{L_{\text{eff}}}\right) \tag{5.3}$$

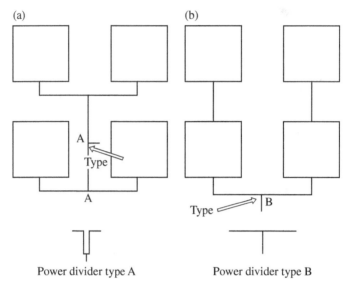

FIGURE 5.4   Configuration of microstrip antenna array. (a) Parallel feed network and (b) parallel–series feed network.

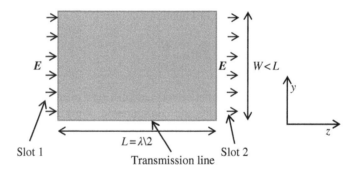

FIGURE 5.5   Transmission line model of patch microstrip antennas.

At $z=0$ and $z=L_{\text{eff}}$ the electric field is maximum. At $z=\dfrac{L_{\text{eff}}}{2}$ the electric field equal zero.

For $\dfrac{H}{\lambda_0}<0.1$ the electric field distribution along the x-axis is assumed to be uniform. The slot admittance may be written as given in Equations 5.4 and 5.5:

$$G=\frac{W}{120\lambda_0}\left[1-\frac{1}{24}\left(\frac{2\pi H}{\lambda_0}\right)^2\right] \quad \text{for} \ \frac{H}{\lambda_0}<0.1 \qquad (5.4)$$

$$B = \frac{W}{120\lambda_0}\left[1 - 0.636 \ \ln\left(\frac{2\pi H}{\lambda_0}\right)^2\right] \quad \text{for} \quad \frac{H}{\lambda_0} < 0.1 \qquad (5.5)$$

B represents the capacitive nature of the slot. In $G = 1/R$, R represents the radiation losses. When the antenna is resonant, the susceptances of both slots cancel out at the feed point for any position of the feed point along the patch. However, the patch admittance depends on the feed point position along the z-axis. At the feed point the slot admittance is transformed by the equivalent length of the transmission line. The width W of the microstrip antenna controls the input impedance. Larger widths can increase the bandwidth. For a square patch antenna fed by a microstrip line, the input impedance is around $300\,\Omega$. By increasing the width, the impedance can be reduced. The antenna admittance is given in Equation 5.6.

$$Y(l_1) = Z_0 \frac{1 + j(Z_L/Z_0)\tan\beta l_1}{(Z_L/Z_0) + j\tan\beta l_1} = Y_1$$

$$Y_{\text{in}} = Y_1 + Y_2 \qquad (5.6)$$

5.1.3 Higher-Order Transmission Modes in Microstrip Antennas

In order to prevent higher-order transmission modes, we should limit the thickness of the microstrip substrate to 10% of a wavelength. The cutoff frequency of the higher-order mode is given in Equation 5.7:

$$f_c = \frac{c}{4H\sqrt{\varepsilon - 1}} \qquad (5.7)$$

5.1.4 Effective Dielectric Constant

Part of the fields in the microstrip antenna structure exists in air and the other part of the fields exists in the dielectric substrate. The effective dielectric constant is somewhat less than the substrate's dielectric constant.

The effective dielectric constant of the microstrip line may be calculated by Equation 5.8a and 5.8b as function of W/H:

$$\text{For} \left(\frac{W}{H}\right) < 1 \qquad (5.8a)$$

$$\varepsilon_e = \frac{\varepsilon_r + 1}{2} + \frac{\varepsilon_r - 1}{2}\left[\left(1 + 12\left(\frac{H}{W}\right)\right)^{-0.5} + 0.04\left(1 - \left(\frac{W}{H}\right)\right)^2\right]$$

$$\text{For} \left(\frac{W}{H}\right) \geq 1 \qquad (5.8b)$$

$$\varepsilon_e = \frac{\varepsilon_r + 1}{2} + \frac{\varepsilon_r - 1}{2} \left[\left(1 + 12\left(\frac{H}{W}\right)\right)^{-0.5} \right]$$

This calculation ignores strip thickness and frequency dispersion, but their effects are negligible.

5.1.5 Losses in Microstrip Antennas

Losses in microstrip line are due to conductor loss, radiation loss, and dielectric loss.

5.1.5.1 Conductor Loss   Conductor loss may be calculated by using Equation 5.9:

$$\alpha_c = 8.686 \log\left(\frac{R_S}{(2WZ_0)}\right) \quad \text{dB/length}$$

(5.9)

$$R_S = \sqrt{\pi f \mu \rho} \quad \text{skin resistance}$$

Conductor losses may also be calculated by defining an equivalent loss tangent δ_c, given by $\delta_c = \delta_s/h$, where $\delta_s = \sqrt{2/\omega\mu\sigma}$. Where σ is the strip conductivity, h is the substrate height, and μ is the free space permeability.

5.1.5.2 Dielectric Loss   Dielectric Loss may be calculated by using Equation 5.10:

$$\alpha_d = 27.3 \frac{\varepsilon_r}{\sqrt{\varepsilon_{eff}}} \frac{\varepsilon_{eff} - 1}{\varepsilon_r - 1} \frac{\text{tg}\delta}{\lambda_0} \quad \text{dB/cm}$$

(5.10)

$$\text{tg}\delta = \text{dielectric loss coefficent}$$

5.1.6 Patch Radiation Pattern

The patch width, W, controls the antenna radiation pattern. The coordinate system is shown in Figure 5.6. The normalized radiation pattern is approximately given by Equations 5.11 and 5.12

$$E_\theta = \frac{\sin((k_0W/2)\sin\theta\sin\varphi)}{(k_0W/2)\sin\theta\sin\varphi} \cos\left(\frac{k_0L}{2}\sin\theta\cos\varphi\right)\cos\varphi$$

(5.11)

$$k_0 = \frac{2\pi}{\lambda}$$

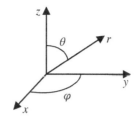

FIGURE 5.6 Coordinate system.

$$E_\varphi = \frac{\sin\left((k_0 W/2)\sin\theta\sin\varphi\right)}{(k_0 W/2)\sin\theta\sin\varphi}\cos\left(\frac{k_0 L}{2}\sin\theta\cos\varphi\right)\cos\theta\sin\varphi$$

$$k_0 = \frac{2\pi}{\lambda}$$

(5.12)

The magnitude of the fields is given by Equation 5.13

$$f(\theta,\varphi) = \sqrt{E_\theta^2 + E_\varphi^2}$$

(5.13)

5.2 TWO LAYERS STACKED MICROSTRIP ANTENNAS

Two layers microstrip antennas was presented first in Refs. [1, 5–8]. The major disadvantage of single-layer microstrip antennas is narrow bandwidth. By designing a double-layer microstrip antenna, we can get a wider bandwidth.

In the first layer the antenna feed network and a resonator are printed. In the second layer the radiating element is printed. The electromagnetic field is coupled from the resonator to the radiating element. The resonator and the radiating element shape may be square, rectangular, triangle, circle, or any arbitrary shape. The distance between the layers is optimized to get maximum bandwidth with the best antenna efficiency. The spacing between the layers may be air or a foam with low dielectric losses.

A circular polarization double-layer antenna was designed at 2.2 GHz. The resonator and the feed network was printed on a substrate with relative dielectric constant of 2.5 with thickness of 1.6 mm. The resonator is a square microstrip resonator with dimensions $W = L = 45$ mm. The radiating element was printed on a substrate with relative dielectric constant of 2.2 with thickness of 1.6 mm. The radiating element is a square patch with dimensions $W = L = 48$ mm. The patch was design as circular polarized antenna by connecting a 3dB 90° branch coupler to antenna feed lines, as shown in Figure 5.7. The antenna bandwidth is 10% for VSWR better than 2 : 1. The measured antenna beam width is around 72°. The measured antenna gain is 7.5dBi.

Measured results of stacked microstrip antennas are listed in Table 5.1.

Results in Table 5.1 indicated that bandwidth of two layers microstrip antennas may be around 9–15% for VSWR better than 2 : 1. In Figure 5.13 a stacked microstrip

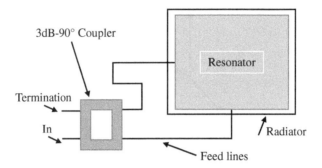

FIGURE 5.7   Circular polarized microstrip stacked patch antenna.

TABLE 5.1   Measured Results of Stacked Microstrip Antennas

Antenna	F (GHz)	Bandwidth (%)	Beamwidth (%)	Gain (dBi)	Side Lobe (dB)	Polarization
Square	2.2	10	72	7.5	−22	Circular
Circular	2.2	15	72	7.9	−22	Linear
Annular disc	2.2	11.5	78	6.6	−14	Linear
Rectangular	2.0	9	72	7.4	−25	Linear
Circular	2.4	9	72	7	−22	Linear
Circular	2.4	10	72	7.5	−22	Circular
Circular	10	15	72	7.5	−25	Circular

FIGURE 5.8   A microstrip stacked patch antenna.

antenna is shown. The antenna feed network is printed on FR4 dielectric substrate with dielectric constant of 4 and 1.6 mm thick. The radiator is printed on RT/duroid 5880 dielectric substrate with dielectric constant of 2.2 and 1.6 mm thick. The antenna electrical parameters was calculated and optimized by using ADS software. The dimensions of the microstrip stacked patch antenna shown in Figure 5.8 are 33 × 20 × 3.2 mm. The computed S_{11} parameters are presented in Figure 5.9. Radiation

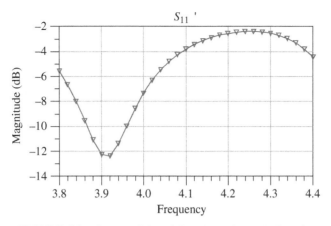

FIGURE 5.9   Computed S_{11} of the microstrip stacked patch.

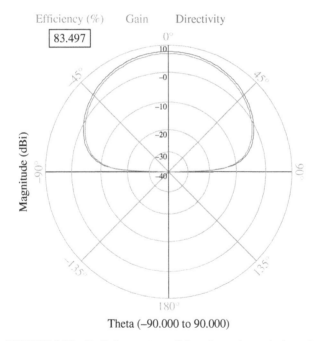

FIGURE 5.10   Radiation pattern of the microstrip stacked patch.

pattern of the microstrip stacked patch is shown in Figure 5.10. The antenna bandwidth is around 5% for VSWR better than 2.5 : 1. The antenna bandwidth is improved to 10% for VSWR better than 2.0 : 1 by adding 8 mm air spacing between the layers. The antenna beamwidth is around 72°. The antenna gain is around 7dBi.

5.3   STACKED MONOPULSE Ku BAND PATCH ANTENNA

A monopulse double-layer antenna was designed at 15 GHz. The monopulse double-layer antenna consists of four circular patch antenna as shown in Figure 5.11. The resonator and the feed network were printed on a substrate with relative dielectric constant of 2.5 with thickness of 0.8 mm. The resonator is a circular microstrip resonator with diameter, $a = 4.2$ mm. The radiating element was printed on a substrate with relative dielectric constant of 2.2 with thickness of 0.8 mm. The radiating element is a circular microstrip patch with diameter, $a = 4.5$ mm. The four circular patch antennas are connected to three 3dB 180° rat-race couplers via the antenna feed lines, as shown in Figure 5.11. The comparator consists of three strip-line 3dB 180° rat-race couplers printed on a substrate with relative dielectric constant of 2.2 with thickness of 0.8 mm. The comparator has four output ports: a sum port $\sum$, difference port Δ, elevation difference port ΔEl, and azimuth difference port ΔAz as shown in Figure 5.16. The antenna bandwidth is 10% for VSWR better than $2:1$. The antenna beamwidth is around 36°. The measured antenna gain is around 10dBi. The comparator losses around 0.7dB.

5.3.1   Rat-Race Coupler

A rat-race coupler is shown in Figure 5.12. The rat-race circumference is 1.5 wavelengths. The distance from A to Δ port is $3\lambda\backslash4$. The distance from A to $\sum$ port is $\lambda\backslash4$. For an equal-split rat-race coupler, the impedance of the entire ring is fixed at

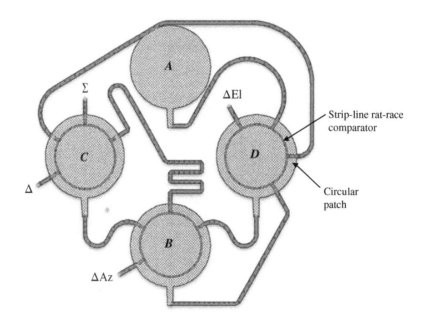

FIGURE 5.11   A microstrip stacked monopulse antenna.

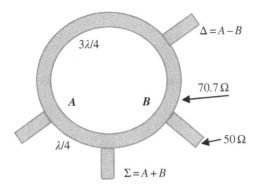

FIGURE 5.12 Rat-race coupler.

$1.41 \times Z_0$ or 70.7 Ω for $Z_0 = 50$ Ω. For an input signal V, the outputs at ports 2 and 4 are equal in magnitude but 180° out of phase.

5.4 LOOP ANTENNAS

Loop antennas are used as receive antennas in communication and medical systems. Loop antennas may be printed on a dielectric substrate or manufactured as a wired antenna. In this paragraph several loop antennas are presented.

5.4.1 Small Loop Antenna

The small loop antenna is shown in Figure 5.13. The shape of the loop antenna may be circular or rectangular. These antennas have low radiation resistance and high reactance. It is difficult to match the antenna to a transmitter. Loop Antennas are most often used as receive antennas, where impedance mismatch loss can be accepted. Small loop antennas are used as field strength probes, in pagers, and in wireless measurements.

The loop lies in the x–y plane. The radius a of the small loop antenna is smaller than a wavelength ($a \ll \lambda$). Loop antenna electric field is given Equation 5.14. Loop antenna magnetic field is given in Equation 5.15:

$$E_\phi = \eta \frac{(ak)^2 I_0 \sin\theta}{4r} e^{j(\omega t - \beta r)} \tag{5.14}$$

$$H_\theta = -\frac{(ak)^2 I_0 \sin\theta}{4r} e^{j(\omega t - \beta r)} \tag{5.15}$$

The variation of the radiation pattern with direction is $\sin\theta$, the same as dipole antenna. The fields of a small loop have the E- and H-fields switched relative to that of a short dipole. The E-field is horizontally polarized in the x–y plane.

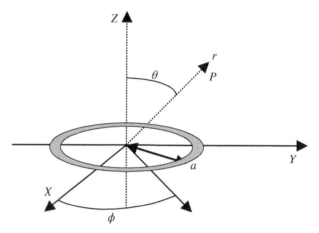

FIGURE 5.13   Small loop antenna.

The small loop is often referred to as the dual of the dipole antenna, because if a small dipole had magnetic current flowing (as opposed to electric current as in a regular dipole), the fields would resemble that of a small loop. The short dipole has a capacitive impedance (imaginary part of the impedance is negative). The impedance of a small loop is inductive (positive imaginary part). The radiation resistance (and ohmic loss resistance) can be increased by adding more turns to the loop. If there are N turns of a small loop antenna, each with a surface area S, the radiation resistance for small loops can be approximated as given in Equation 5.16:

$$R_{rad} = \frac{31,329 N^2 S^2}{\lambda^4} \tag{5.16}$$

For a small loop, the reactive component of the impedance can be determined by finding the inductance of the loop. For a circular loop with radius a, and wire radius r, the reactive component of the impedance is given by Equation 5.17.

$$X = 2\pi a f \mu \left(\ln \left(\frac{8a}{r} \right) - 1.75 \right) \tag{5.17}$$

Loop antennas behave better in vicinity of the human body than dipole antennas. The reason is that the electric near fields in dipole antenna are very strong. For $r \ll 1$, the dominant component of the field vary as $1/r^3$. These fields are the dipole near fields. In this case the waves are standing waves and the energy oscillate in the antenna near zone and are not radiated to the open space. The real part of the pointing vector equal to zero. Near the body the electric fields decays rapidly. However, the magnetic fields are not affected near the body. The magnetic fields are strong in the near field of the loop antenna. These magnetic fields give rise to the loop antenna radiation. The loop antenna radiation near the human body is stronger than the dipole radiation near the human body. Several loop antennas are used as "wearable antennas."

5.4.2 Printed Loop Antenna

The diameter of a printed loop antenna is around half wavelength. Loop antenna is dual to a half-wavelength dipole. Several loop antennas was designed for medical systems at frequency range between 400 and 500 MHz. In Figure 5.14 a printed loop antenna is presented. The antenna was printed on FR4 with 0.5 mm thickness. The loop diameter is 45 mm. The loop antenna VSWR is around 4 : 1. The printed loop antenna radiation pattern at 435 MHz is shown in Figure 5.15. The loop antenna gain is around 1.8 dBi. The antenna with a tuning capacitor is shown in Figure 5.16. The loop antenna VSWR without the tuning capacitor was 4 : 1. This loop antenna may be

FIGURE 5.14 Printed loop antenna.

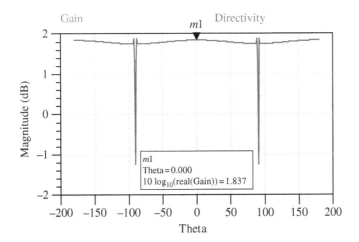

FIGURE 5.15 Printed loop antenna radiation pattern at 435 MHz.

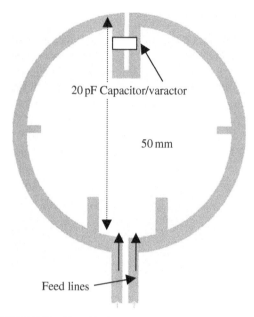

FIGURE 5.16   Tunable loop antenna without ground plane.

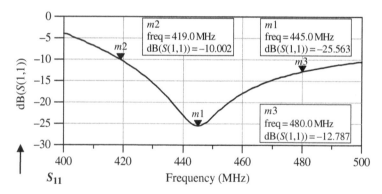

FIGURE 5.17   Computed S_{11} of loop antenna, without ground plane, with a tuning capacitor.

tuned by adding a capacitor or varactor as shown in Figure 5.17. Matching stubs are employed to tune the antenna to the resonant frequency. Tuning the antenna allows us to work in a wider bandwidth as shown in Figure 5.18. Loop antennas are used as receive antennas in medical systems. The loop antenna radiation pattern on human body is shown Figure 5.19.

The computed 3D radiation pattern and the coordinate used in this chapter are shown in Figure 5.19.

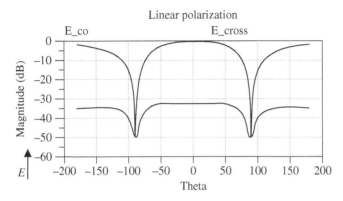

FIGURE 5.18 Radiation pattern of loop antenna on human body.

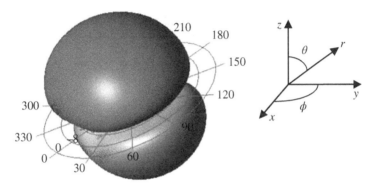

FIGURE 5.19 Loop antenna 3D radiation pattern.

5.4.3 RFID Loop Antennas

RFID loop antennas are widely used. Several RFID loop antennas are presented in Ref. [10]. RFID loop antennas have low efficiency and narrow bandwidth. As an example the measured impedance of a square four-turn loop at 13.5 MHz is $0.47 + j107.5 \, \Omega$. A matching network is used to match the antenna to $50 \, \Omega$. The matching network consists of a 56 pF shunt capacitor, $1 \, \text{k}\Omega$ shunt resistor, and another 56 pF capacitor. This matching network has narrow bandwidth. The antenna is printed on a FR4 substrate. The antenna dimensions are $32 \times 52.4 \times 0.25$ mm. The antenna layout is shown in Figure 5.20. A photo of the RFID antenna is shown in Figure 5.21. S_{11} results of the printed loop antenna are shown in Figure 5.22. The antenna S_{11} parameter is better than −9.5dB without an external matching network. The computed radiation pattern is shown in Figure 5.23. The RFID antenna beamwidth is around 160°.

5.4.4 New Loop Antenna with Ground Plane

A new loop antenna with ground plane has been designed on Kapton substrates with relative dielectric constant of 3.5 and thickness of 0.25 mm. The antenna is presented

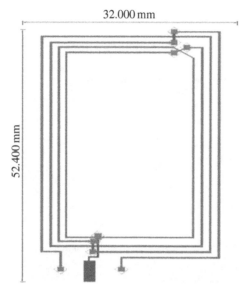

FIGURE 5.20 A square four-turn loop antenna.

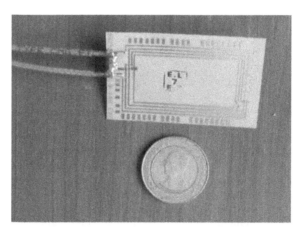

FIGURE 5.21 Four-turn 13.5 MHz loop rectangular antenna.

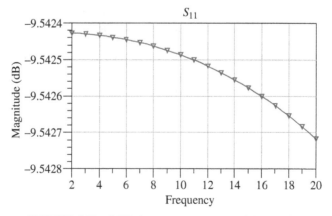

FIGURE 5.22 RFID loop antenna computed S_{11} results.

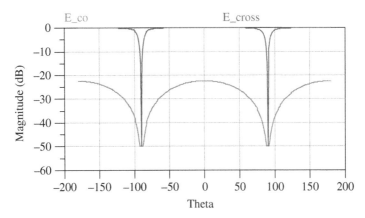

FIGURE 5.23 RFID loop antenna radiation pattern.

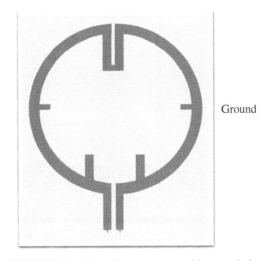

Ground

FIGURE 5.24 Printed loop antenna with ground plane.

in Figure 5.24. Matching stubs are employed to tune the antenna to the resonant frequency. The loop antenna with ground plane diameter is 45 mm. The antenna was
designed by employing ADS software. The antenna center frequency is 427 MHz.
The antenna bandwidth for VSWR better than 2 : 1 is around 12% as shown in
Figure 5.25. The printed loop antenna radiation pattern at 435 MHz is shown in
Figure 5.26. The loop antenna gain is around 1.8dBi. The loop antenna with ground
plane beamwidth is around 100°.

Printed loop antenna with ground plane with shorter tuning stubs is presented
in Figure 5.27. The loop antenna with ground with shorter tuning stubs plane

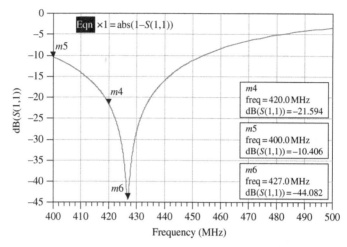

FIGURE 5.25 Computed S_{11} of loop antenna with ground plane.

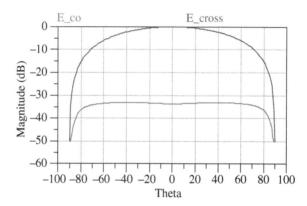

FIGURE 5.26 Radiation pattern of loop antenna with ground plane.

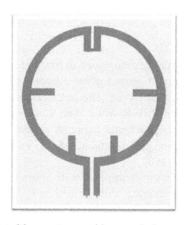

FIGURE 5.27 Printed loop antenna with ground plane and short tuning stubs.

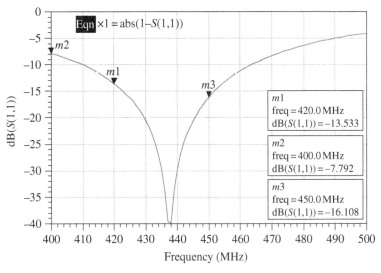

FIGURE 5.28   Radiation pattern of loop antenna with ground plane.

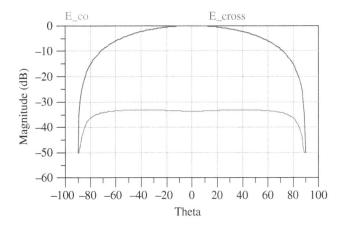

FIGURE 5.29   Radiation pattern of loop antenna with ground plane.

diameter is 45 mm. The antenna center frequency is 438 MHz. The antenna bandwidth for VSWR better than $2:1$ is around 12% as shown in Figure 5.28. The printed loop antenna with shorter tuning stubs radiation pattern at 435 MHz is shown in Figure 5.34. The loop antenna gain is around 1.8 dBi. The loop antenna with ground plane beamwidth is around $100°$. The printed loop antenna with shorter tuning stubs 3D radiation pattern at 430 MHz is shown in Figures 5.29 and 5.30.

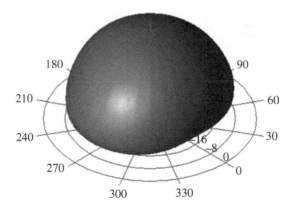

FIGURE 5.30   3D radiation pattern of loop antenna with ground plane.

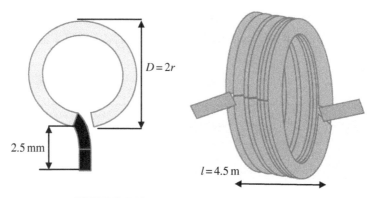

FIGURE 5.31   Wire seven-turn loop antenna.

5.5   WIRED LOOP ANTENNA

A wire seven-turn loop antenna is shown in Figure 5.31. The antenna length $l = 4.5$ mm. The loop diameter is 3.5 mm. Equation 5.18 is an approximation to calculate the inductance value for air coil loop antenna with N turns, diameter r, and length l:

$$L(nH) = \frac{3.94 r_{mm} N^2}{0.9\frac{\ell}{r} + 1} = \frac{0.1 r_{mil} N^2}{0.9\frac{\ell}{r} + 1} \qquad (5.18)$$

An approximation to calculate the inductance value for air coil loop antenna with N turns, diameter r, and length l is given in Equation 5.19:

$$L(nH) \approx \frac{r_{mm}^2 N^2}{2l + r} \qquad (5.19)$$

FIGURE 5.32   Wire loop antenna on PCB board.

For length l = 4.5 mm, wire diameter 0.6 mm, loop diameter 3.5 mm, and N = 7, the inductance is around L = 52 nH. The quality factor of this seven-turn wire loop is around 100.

For length l = 1.7 mm, wire diameter 0.6 mm, loop diameter 6.5 mm, and N = 2, the inductance is around L = 45 nH.

For length l = 2 mm, wire diameter 0.6 mm, loop diameter 7.62 mm, and N = 2, the inductance is around L = 42.1 nH.

For length l = 0.5 mm, wire diameter 0.6 mm, loop diameter 7 mm, and N = 2, the inductance is around L = 87 nH.

For length l = 2.5 mm, wire diameter 0.6 mm, loop diameter 5 mm, and N = 2, the inductance is around L = 20.7 nH.

The wire loop antenna has very low radiation efficiency. The amount of power radiated is a small fraction of the input power. The antenna efficiency is around 0.01%, −41dB. The ratio between the antenna's dimension and wavelength is around 1 : 100. Small antennas are characterized by radiation resistance R_r. The radiated power is given by $I^2 R_r$, where I is the current through the coil. Remember that the current through the coil is Q times the current through the antenna. For example, for current of 2 mA and Q of 20, if $R_r = 10^{-3}$ Ω, the radiated power is around −32dBm.

Figure 5.32 presents a wire loop antenna with 2.5 turns on a PCB board. Figure 5.33 presents a wire loop antenna with seven turns on a PCB board.

5.6   RADIATION PATTERN OF A LOOP ANTENNA NEAR A METAL SHEET

E and H plane 3D radiation pattern of a wire loop antenna in free space is shown in Figure 5.34.

E and H plane 3D radiation pattern of a wire loop antenna near a metal sheet as shown in Figure 5.35 was computed by employing HFSS software.

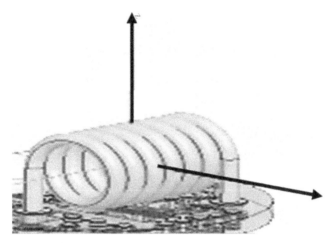

FIGURE 5.33   Seven turns loop antenna on PCB board.

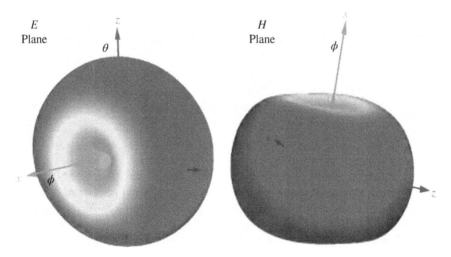

FIGURE 5.34   *E* and *H* plane radiation pattern of loop antenna in free space.

Figure 5.36 presents the *E* and *H* plane radiation pattern of loop antenna for distance of 30 cm from a metal sheet. We can see that a metal sheet in the vicinity of the antenna split the main beam and creates holes of around 20dB in the radiation pattern.

Figure 5.37 presents the *E* and *H* plane radiation pattern of loop antenna located 10 cm from a metal sheet as presented in Figure 5.38. We can see that a metal sheet in the vicinity of the antenna split the main beam and creates holes up to 15dB in the radiation pattern.

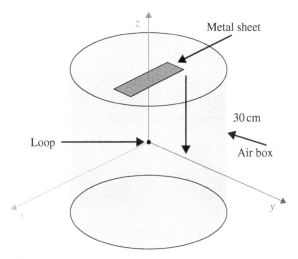

FIGURE 5.35 Loop antenna located near a metal sheet.

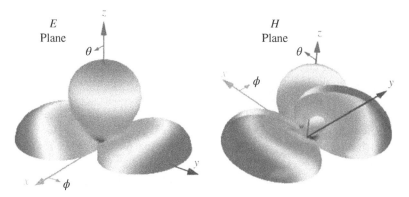

FIGURE 5.36 E and H plane radiation pattern of a loop antenna for distance of 30 cm from a metal sheet.

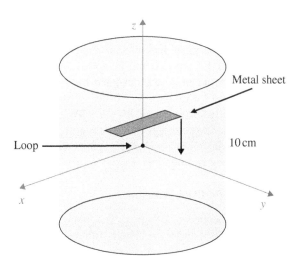

FIGURE 5.37 Loop antenna located 10 cm from a metal sheet.

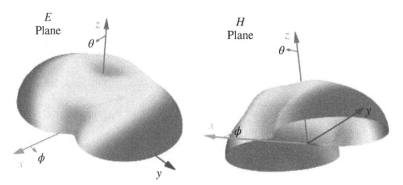

FIGURE 5.38   *E* and *H* plane radiation pattern of a loop antenna for distance of 10 cm from a metal sheet.

5.7   PLANAR INVERTED-F ANTENNA

The planar inverted-F antenna (PIFA) possess attractive features such as low profile, small size, and low fabrication costs [16–18]. PIFA antenna bandwidth is higher than the bandwidth of the conventional patch antenna, because the PIFA antenna thickness is higher than the thickness of patch antennas. The conventional PIFA antenna is a grounded quarter wavelength patch antenna. The antenna consists of ground plane, a top plate radiating element, feed wire, and a shorting plate or via holes from the radiating element to the ground plane as shown in Figure 5.39.

The patch is shorted at the end; the fringing fields that are responsible for radiation are shorted on the far end, so only the fields nearest the transmission line radiate. Consequently, the gain is reduced, but the patch antenna maintains the same basic properties as a half-wavelength patch. However the antenna length is reduced by 50%. The feed location may be placed between the open and shorted end. The feed location controls the antenna input impedance.

5.7.1   Grounded Quarter Wavelength Patch Antenna

A grounded quarter wavelength patch antenna was designed on FR4 substrate with relative dielectric constant of 4.5 and 1.6 mm thickness at 3.85 GHz. The antenna is shown in Figure 5.40. The antenna was designed by using ADS software. The antenna dimensions are $34 \times 17 \times 1.6$ mm. S_{11} results of the antenna are shown in Figure 5.41. The antenna bandwidth is around 6% for VSWR better than 3 : 1 without a matching network. The radiation pattern is shown in Figure 5.42. The antenna beamwidth is around 76°. The grounded quarter wavelength patch antenna gain is around 6.7dBi. The antenna efficiency is around 92%.

5.7.2   A New Double Layers PIFA Antenna

A new double layers PIFA antenna was designed. The first layer is a grounded quarter wavelength patch antenna printed on FR4 substrate with relative dielectric constant of 4.5 and 1.6 mm thickness. The second layer is a rectangular patch antenna printed on

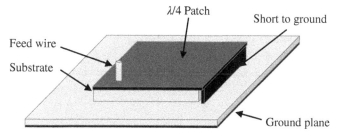

FIGURE 5.39   Conventional PIFA antenna.

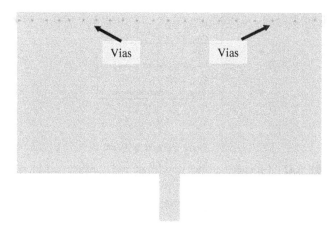

FIGURE 5.40   Grounded quarter wavelength patch antenna.

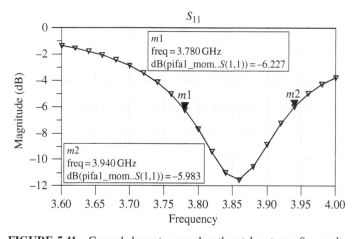

FIGURE 5.41   Grounded quarter wavelength patch antenna S_{11} results.

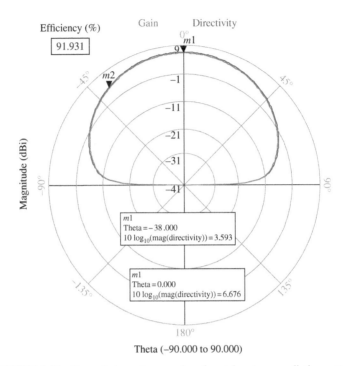

FIGURE 5.42   Grounded quarter wavelength patch antenna radiation pattern.

FIGURE 5.43   Double layers PIFA antenna.

Duroid substrate with relative dielectric constant of 2.2 and 1.6 mm thickness. The antenna is shown in Figure 5.43. The antenna was designed by employing ADS software. The antenna dimensions are $34 \times 17 \times 3.2$ mm. S_{11} results of the antenna are shown in Figure 5.44. The antenna is a dual-band antenna. The first resonant

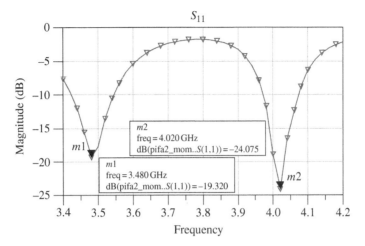

FIGURE 5.44 Double layers PIFA antenna S_{11} results.

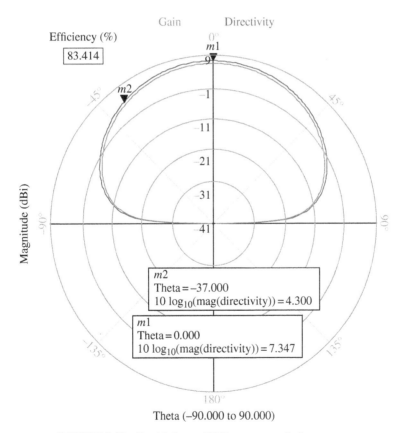

FIGURE 5.45 Double layers PIFA antenna radiation pattern.

frequency is 3.48 GHz. The second resonant frequency is 4.02 GHz. The radiation pattern is shown in Figure 5.45. The antenna beamwidth is around 74°. The antenna gain is around 7.4dBi. The antenna efficiency is around 83.4%.

REFERENCES

[1] James JR, Hall PS, Wood C. *Microstrip Antenna Theory and Design*. London: Peter Per-egrinus; 1981.

[2] Sabban A, Gupta KC. Characterization of radiation loss from microstrip discontinuities using a multiport network modeling approach. IEEE Trans Microwave Theory Tech 1991;39 (4):705–712.

[3] Sabban A. Multiport network model for evaluating radiation loss and coupling among discontinuities in microstrip circuits [Ph.D. thesis]. University of Colorado at Boulder; 1991.

[4] Kathei PB, Alexopoulos NG. Frequency-dependent characteristic of microstrip, MTT-33, discontinuities in millimeter-wave integrated circuits. IEEE Trans Microwave Theory Tech 1985;MTT-33:1029–1035.

[5] Sabban A. A new wideband stacked microstrip antenna. IEEE Antenna and Propagation Symposium; June 1983; Houston, TX.

[6] Sabban A. Wideband microstrip antenna arrays. IEEE Antenna and Propagation Symposium; June 1981; MELCOM, Tel-Aviv.

[7] Sabban A. Microstrip antennas. IEEE Symposium; October 1979; Tel-Aviv.

[8] Sabban A, Navon E. A MM-waves microstrip antenna array. IEEE Symposium; March 1983; Tel-Aviv.

[9] de Lange G, Konistis K, Hu Q. A 3∗3 mm-wave micro machined imaging array with sis mixers. Appl Phys Lett 1999;75 (6):868–870.

[10] Rahman A, de Lange G, Hu Q. Micromachined room temperature microbolometers for mm-wave detection. Appl Phys Lett 1996;68 (14):2020–2022.

[11] Milkov MM. Millimeter-wave imaging system based on antenna-coupled bolometer [M. Sc. thesis]. UCLA; 2000.

[12] Jack MD, Ray M, Varesi J, Grinberg J, Fetterman H, Dolezal FA. Architecture and method of coupling electromagnetic energy to thermal detectors. US patent 6,329,655. November 2001.

[13] Sinclair GN et al. Passive millimeter wave imaging in security scanning. Proc SPIE 2000;4032:40–45.

[14] Kompa G, Mehran R. Planar waveguide model for computing microstrip components. Electron Lett 1975;11 (9):459–460.

[15] Luukanen A., Sipila H, Viitanen V-P. Imaging system functioning on submillimeter waves. US patent 6,242,740. June 2001.

[16] Wang F et al. Enhanced-bandwidth PIFA with T-shaped ground plane. Electron Lett 2004;40 (23):1504–1505.

[17] Kim B, Hoon J, Choi H. Small wideband PIFA for mobile phones at 1800 MHz. IEEE Veh Technol Conf 2004;1:27–29.

[18] Kim B, Park J, Choi H. Tapered type PIFA design for mobile phones at 1800 MHz. IEEE Veh Technol Conf 2005;2:1012–1014.

6

MIC AND MMIC MILLIMETER-WAVE RECEIVING CHANNEL MODULES

6.1   18–40 GHz COMPACT RF MODULES

Compact wideband RF modules are crucial in mm-wave direction finding, radars, seekers, and communication systems. Compact 18–40 GHz direction finding system is shown in Figure 6.1. Figures 6.2 and 6.3 present the block diagram of an 18–40 GHz direction finding system. The direction finding system consists of 14 direction finding front-end modules, two omnidirectional front-end modules, switch filter bank (SFB) modules, Frequency Source Unit (FSU), and downconverter units. RF components and modules are presented in books and papers (see Refs. [1–10]).

Figure 6.4 presents the block diagram of a 6–18 GHz up-/downconverter unit. The up-/downconverter unit consists of low noise amplifiers, 6–18 GHz switch filter bank modules, FSU, and downconverter units.

6.2   18–40 GHz FRONT END

Development and design considerations of a compact wideband 18–40 GHz front end are described in this section. The RF modules and the system were designed by using ADS software and momentum RF software. There is a good agreement between computed and measured results.

Wideband RF Technologies and Antennas in Microwave Frequencies, First Edition. Dr. Albert Sabban.
© 2016 John Wiley & Sons, Inc. Published 2016 by John Wiley & Sons, Inc.

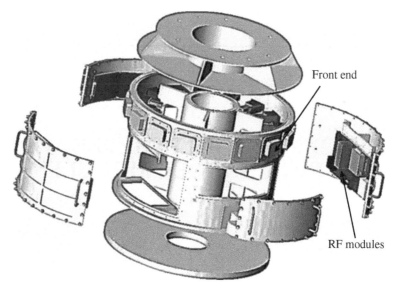

FIGURE 6.1   Direction finding system.

6.2.1   18–40 GHz Front-End Requirements

The front-end electrical specifications are listed in Table 6.1. The front-end design presented in this section meets the front-end electrical specifications.

6.2.1.1   *Physical Characteristics: Interfaces and Connectors*

Interface	Type
RF input	Waveguide WRD180 (double ridge)
RF output	K connector
DC supply	D type
Control	D type

Dimensions: $60 \times 40 \times 20$ mm

6.2.2   Front-End Design

Front-end block diagram is shown in Figure 6.5. The front-end module consists of a limiter and a wideband 18–40 GHz, filtronic, low noise amplifier LMA406. The LMA406 gain is around 12dB with 4.5dB noise figure and 14dBm saturated output power. The LNA dimensions are 1.44×1.1 mm. We used a wideband PHEMT MMIC SPDT manufactured by Agilent, AMMC-2008. The SPDT insertion loss is lower than 2dB. The isolation between the SPDT input port and the output ports is better than 25dB. The SPDT 1dBc compression point is around 14dBm. The SPDT dimensions

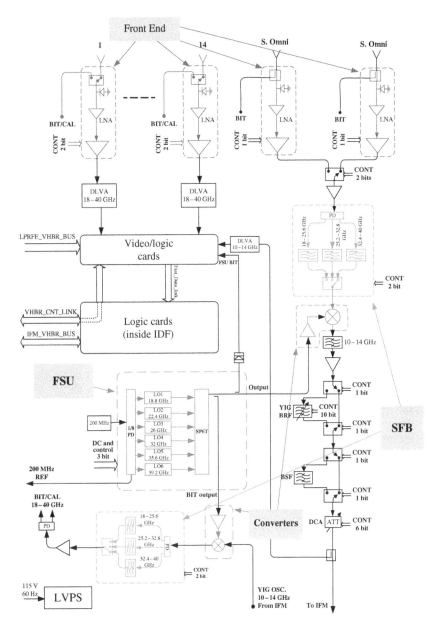

FIGURE 6.2   Block diagram of an 18–40 GHz direction finding system.

are $1 \times 0.7 \times 0.1$ mm. The front-end electrical characteristics were evaluated by using Agilent ADS software and by using SYSCAL software. Figure 6.6 presents the front-end module noise figure and gain for LNA noise figure of 6dB. The overall computed module noise figure is 9.46dB. The module gain is 21dB. Figure 6.7 presents the front-end module noise figure and gain for LNA noise figure of 5.5dB. The overall computed module noise figure is 9.25dB. The module gain is 21dB.

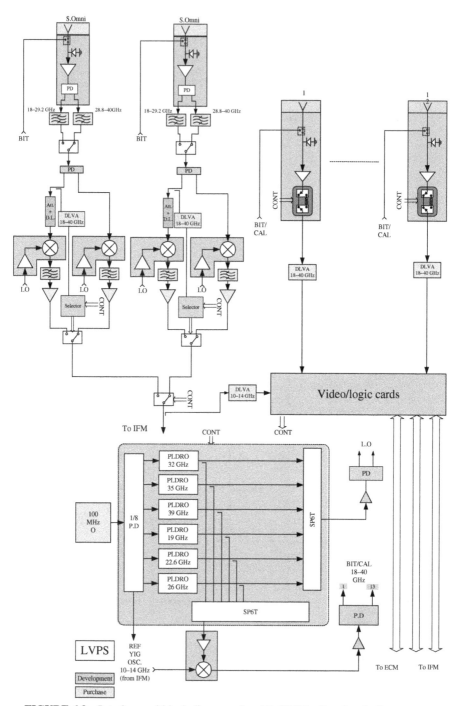

FIGURE 6.3   Interface and block diagram of an 18–40 GHz direction finding system.

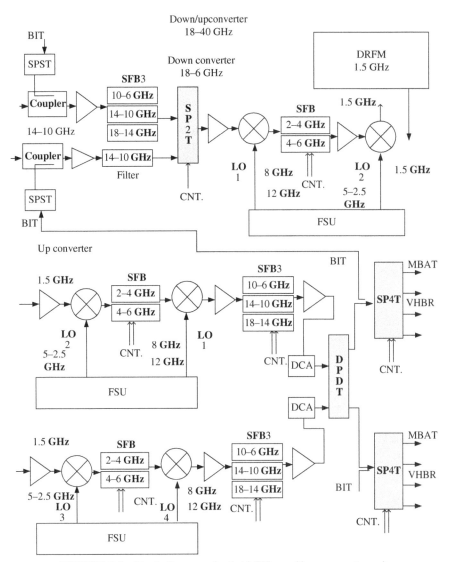

FIGURE 6.4 Block diagram of a 6–18 GHz up-/downconverter unit.

The MMIC amplifiers and the SPDT are glued to the surface of the mechanical box. The MMIC chips are assembled on covar carriers. During development it was found that the spacing between the front-end carriers should be less than 0.03 mm in order to achieve flatness requirements and VSWR better than 2 : 1.

The front-end voltage and current consumption are listed in Table 6.2. The front-end module has high gain and low gain channels. The gain difference between high gain and low gain channels is around 15–20dB. Measured front-end gain is presented in Figure 6.8. The front-end gain is around 20 ± 4dB for the 18–40 GHz frequency range.

TABLE 6.1   Front-End Electrical Specifications

Parameter	Requirements	Performance
Frequency range	18–40 GHz	Comply
Gain	24/3dB typical, switched by external control. (−40dB or lower for OFF state); gain response versus frequencies will be defined by designer as soon as possible (TBD)	Comply
Gain flatness	±0.5dB max. for any 0.5 GHz BW in 18–40 GHz. ±2dB max. for any 4 GHz BW in 18–40 GHz. ±3dB max. for the whole range 18–40 GHz	Comply
Noise figure (high gain)	10dB max. for 40°C baseplate temperature; 11dB over temperature	Comply
Input power range	−60 to 10dBm	Comply
Output power range	−39 to 11dBm not saturated; 13dBm saturated	Comply
Linearity	Output 1dB compression point at 12dBm min. Third intercept point (Ip3) at 21dBm single tone second harmonic power −25dBc max. for 10dBm output	Comply
VSWR	2 : 1	Comply
Power input protection	No damage at +30dBm CW and +47dBm pulses (for average power higher than 30dBm) input power at 0.1–40 GHz. Test for pulses: PW = 1 μs, PRF = 1 kHz	Comply
Power supply voltages	±5 V, ±15 V	Comply
Control logic	LVTTL standard "0" = 0–0.8 V; "1" = 2.0–3.3 V	Comply
Switching time	Less then 100 ns	Comply
Nonharmonic spurious (output)	−50dBm max. (when it isn't correlative with the input signals)	Comply
Video leakage	Video leakage signals will be below the RF output level for terminated input	Comply
Dimension	60 × 40 × 20 mm	Comply

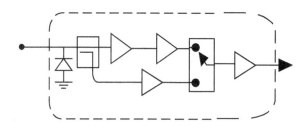

FIGURE 6.5   Front-end block diagram.

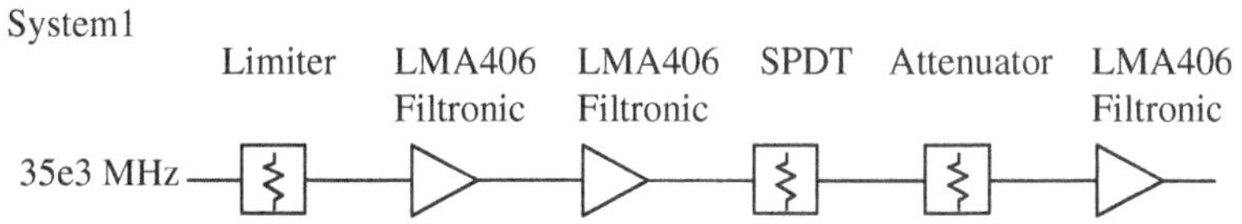

System1

	Limiter	LMA406 Filtronic	LMA406 Filtronic	SPDT	Attenuator	LMA406 Filtronic	Total

35e3 MHz

							Total
NF (dB)	3.00	6.00	6.00	3.00	3.00	6.00	9.46
Gain (dB)	−3.00	10.00	10.00	−3.00	−3.00	10.00	21.00
OIP3 (dBm)	30.00	25.00	25.00	30.00	30.00	25.00	23.14

Input pwr (dBm)	−60.00	System temp (K)	290.00	IM offset (MHz)	0.025	
OIP2 (dBm)	23.14	OIP3 (dBm)	23.14	Output P1dB (dBm)	11.37	
IIP2 (dBm)	2.14	IIP3 (dBm)	2.14	Input P1dB (dBm)	−8.63	
OIM2 (dBm)	−101.14	OIM3 (dBm)	−163.28	Compressed (dB)	0.00	
ORR2 (dB)	62.14	ORR3 (dB)	124.28	Gain, actual (dB)	21.00	
IRR2 (dB)	31.07	IRR3 (dB)	41.43			
SFDR2 (dB)	61.34	SFDR3 (dB)	81.79	Gain, linear (dB)	21.00	
AGC controlled range:		Min input (dBm)	N/A	Max input (dBm)	N/A	

FIGURE 6.6   Front-end module design for LNA NF = 6dB.

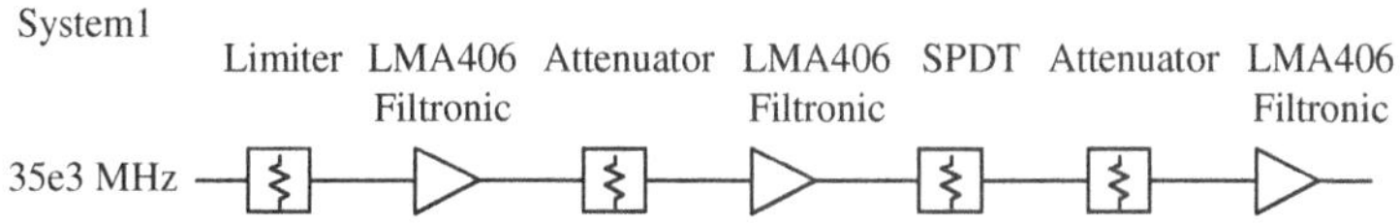

System1

	Limiter	LMA406 Filtronic	Attenuator	LMA406 Filtronic	SPDT	Attenuator	LMA406 Filtronic	Total

35e3 MHz

								Total
NF (dB)	3.00	5.50	3.00	5.50	3.00	1.00	5.50	9.25
Gain (dB)	−3.00	10.50	−3.00	10.50	−3.00	−1.00	10.00	21.00
OIP3 (dBm)	30.00	25.00	30.00	25.00	30.00	30.00	25.00	23.61

Input pwr (dBm)	−60.00	System temp (K)	290.00	IM offset (MHz)	0.025	
OIP2 (dBm)	24.04	OIP3 (dBm)	23.61	Output P1dB (dBm)	11.74	
IIP2 (dBm)	3.04	IIP3 (dBm)	2.61	Input P1dB (dBm)	−8.26	
OIM2 (dBm)	−102.04	OIM3 (dBm)	−164.23	Compressed (dB)	0.00	
ORR2 (dB)	63.04	ORR3 (dB)	125.23	Gain, actual (dB)	21.00	
IRR2 (dB)	31.52	IRR3 (dB)	41.74			
SFDR2 (dB)	61.90	SFDR3 (dB)	82.24	Gain, linear (dB)	21.00	
AGC controlled range:		Min input (dBm)	N/A	Max input (dBm)	N/A	

FIGURE 6.7   Front-end module design for LNA NF = 5.5dB.

TABLE 6.2   Front-End Module Voltage and Current Consumption

Voltage (V)	3	5	−12	−5	5 digital
Current (A)	0.25	0.15	0.1	0.1	0.1

6.2.3   High Gain Front-End Module

To achieve a high gain front-end module, a medium power Hittite MMIC amplifier, HMC283, was added to the front-end module presented in Figure 6.9. The HMC283 gain is around 21dB with 10dB noise figure and 21dBm saturated output power. The

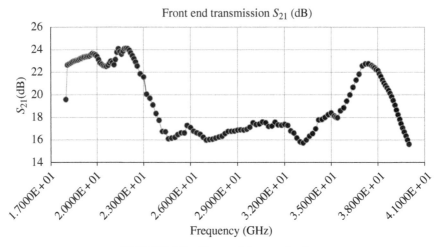

FIGURE 6.8   Measured front-end gain.

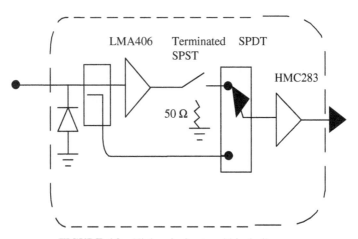

FIGURE 6.9   High gain front-end block diagram.

amplifier dimensions are 1.72×0.9 mm. The high gain front-end module block diagram is shown in Figures 6.9 and 6.10. The front-end module has high gain and low gain channels. The gain difference between high gain and low gain channels presented in Figure 6.9 is around 15–20dB. The gain difference between high gain and low gain channels presented in Figure 6.10 is around 10–15dB. Detailed block diagram of high gain module is shown in Figure 6.11.

6.2.4   High Gain Front-End Design

The high front-end electrical characteristics were evaluated by using Agilent ADS software and by using SYSCAL software. Figure 6.12 presents the front-end module

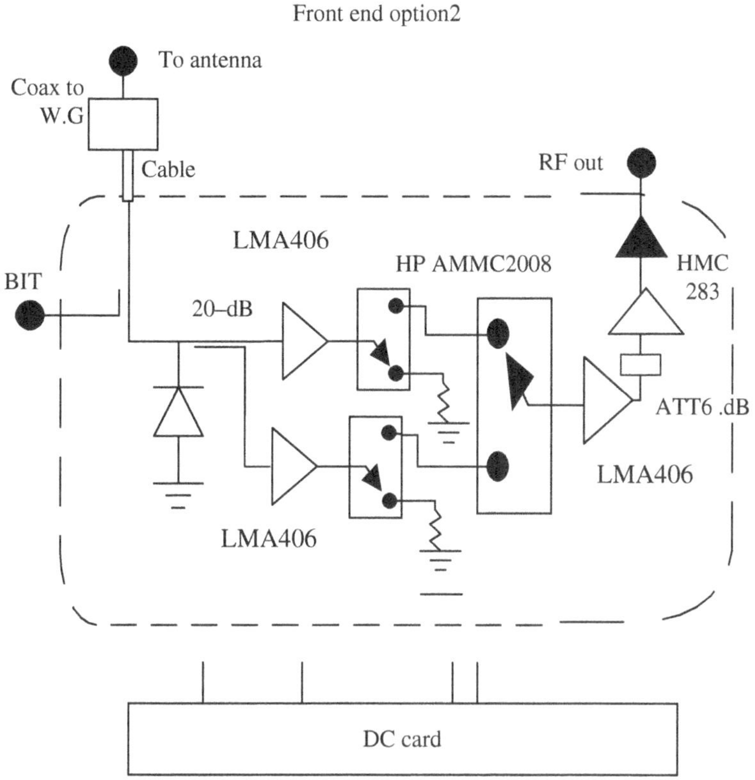

Front end option2

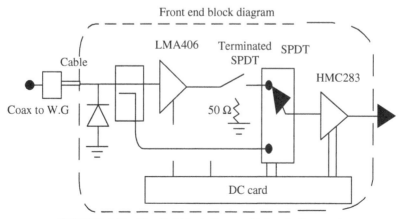

FIGURE 6.10 High gain front-end block diagram with amplifier in the low gain channel.

FIGURE 6.11 Detailed block diagram high gain front end.

System1

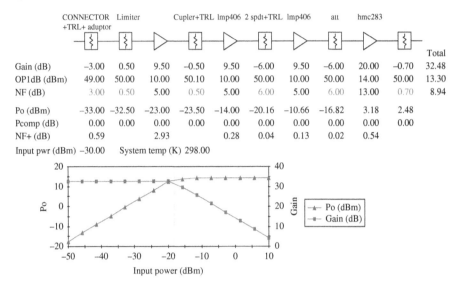

	CONNECTOR +TRL+ aduptor	Limiter	Cupler+TRL	lmp406	2 spdt+TRL	lmp406	att	hmc283		Total
Gain (dB)	−3.00	0.50	9.50	−0.50	9.50	−6.00	9.50	−6.00	20.00	−0.70	32.48
OP1dB (dBm)	49.00	50.00	10.00	50.10	10.00	50.00	10.00	50.00	14.00	50.00	13.30
NF (dB)	3.00	0.50	5.00	0.50	5.00	6.00	5.00	6.00	13.00	0.70	8.94
Po (dBm)	−33.00	−32.50	−23.00	−23.50	−14.00	−20.16	−10.66	−16.82	3.18	2.48	
Pcomp (dB)	0.00	0.00	0.00	0.00	0.00	0.00	0.00	0.00	0.00	0.00	
NF+ (dB)	0.59		2.93		0.28	0.04	0.13	0.02	0.54		

Input pwr (dBm) −30.00   System temp (K) 298.00

FIGURE 6.12   Front-end module design for LNA NF = 9.5dB.

							Total
NF (dB)	3.50	5.00	5.00	5.00	6.00	9.00	10.02
Gain (dB)	−3.50	11.00	−5.00	11.00	−6.00	22.00	29.50
OIP3 (dBm)		20.00		20.00		20.00	19.87
NF+ (dB)		2.85		0.56		0.57	
Po (dBm)	−63.50	−52.50	−57.50	−46.50	−52.50	−30.50	

Input pwr (dBm) −60.00   System temp (K)   290.00

FIGURE 6.13   Front-end module design for LNA NF = 5dB.

noise figure and gain for LNA noise figure of 9.5dB. The overall computed module noise figure is 13.3dB. The module gain is 32.48dB. Figure 6.13 presents the front-end module noise figure and gain for LNA noise figure of 5dB. The overall computed module noise figure is 10dB. The module gain is 29.5dB.

Measured results of front-end modules are listed in Table 6.3. HMC283 assembly is shown in Figure 6.14. A photo of the front end is shown in Figure 6.15. Performance and cost of available mm-wave RF components are listed in Table 6.4. Costs of available MMIC and MIC mm-wave RF components are listed in Table 6.5. There is a good agreement between computed and measured results.

A photo of the compact wideband 18–40 GHz RF modules is shown in Figure 6.16.

TABLE 6.3   Measured Results of Front-End Modules

Parameter	DF6	DF4	DF3	DF2	DF1	OMNI02	OMNI01
High gain max.	31	32	32.5	32.5	31	31.5	32
High gain min.	26	26	27.5	27	26	28.5	28
High gain avg.	29	29	29	29	29	30	30
Amp. bal.	5	6	5	5	4	3	4
S_{11} (dB)	4.5	5	5	5	5	5	4
S_{22} (dB)	7.5	6	5	6	5	7	6
Isolation (dB)	9	9	10	10	6.5	21.5	22.5
Low gain max.	19	18	17	17	17	16.5	18
Low gain min.	13	10	7.5	12	12	10.5	11
Low gain avg.	15	14	12	14	14	13.5	14.5
Amp. bal.	6	8	9.5	5	5	6	7
P1Db 30 GHz	11.6	11.93	11.7	11.4	10.9	14	15.96
P1Db 40 GHz	13.96	14.5	15.58	15.28	14	14.48	16.8
NF 30 GHz	8.68	9.48	8.65	8.45	10.5	8.14	8.75
NF 40 GHz	9.28	10.1	8.64	9.17	10.24		8.75

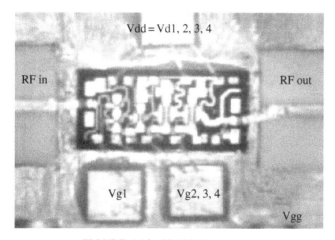

FIGURE 6.14   HMC283 assembly.

FIGURE 6.15   18–40 GHz front-end module.

TABLE 6.4 Performance and Cost of Available Millimeter-Wave RF Components

Component	Model#	Freq. Range (GHz)	NF/IL (dB)	Gain (dB)	Power Out 1dBc (dBm)	Coupling (dB)	Isolation (dB)	VSWR max.	Unit Cost (K$)
Switch general microwave	F9022	18-40	4	na		na	40	2.30 : 1	2.1
Coupler KRYTAR	184020	18-40	1.3	na		20		1.70 : 1	0.8
LNA space labs	SLKKa-30-6	18-40	6	30	17	na		1.80 : 1	2.2
WG/K adapter	R45	18-40	0.75					1.25 : 1	0.5
MICROTECH	240130	18-40						1.6 : 1	

TABLE 6.5 Cost of Available mm-Wave RF Components

Application	Component	Model#	Type	Unit Cost ($)
RF front end	Switch	F9022	Drop in	1800
	General microwave			
RF front end	LNA filtronic	LMA406	MMIC	24
RF front end	LNA/VGA UMS	CHA2097A	MMIC	42
RF front end	Limiter diode	GC4800	Diode	100
SFB/FSU	Switch	HMMC2027 HP	MMIC	80
SFB/FSU	Switch	TLCSWY04	MMIC	80
FSU	PLDRO	RADIATEK 18.8G RPLO-B	Drop in	1600
FSU	PLDRO	RADIATEK 22.4G RPLO-B	Drop in	1600
FSU	PLDRO	RADIATEK RPLO-B26-G	Drop in	2300
FSU	PLDRO	RADIATEK 32.G RPLO-B	Drop in	2400
FSU	PLDRO	RADIATEK 35.6G RPLO-B	Drop in	2400
FSU	PLDRO	RADIATEK 39.2G RPLO-B	Drop in	2400
FSU/up-/down converters	Amplifier	HMMC5040 HP	MMIC	50
FSU/up-/down converters	Amplifier	RMDA1840	MMIC	100
		RAYTHEON		
Up-/down converter	MIXER	MARKI	Drop in	800
		M2-0243		
RF front end	WG/K adapter	Wrd180	Adapter	500
	MDC	22180		

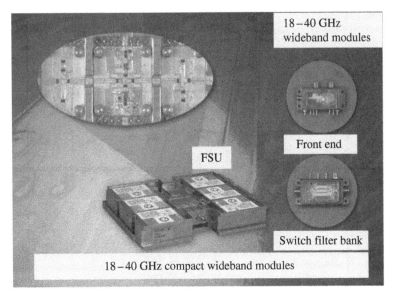

FIGURE 6.16   Photo of 18–40 GHz compact modules.

6.3   18–40 GHz INTEGRATED COMPACT SWITCHED FILTER BANK MODULE

6.3.1   Introduction

Filter bank modules are an important unit in radar and communication links. The electrical performance of the filter bank determines if the system will meet the required specifications. Moreover in several cases the filter bank performance limits the system performance. This section describes the design and development of an integrated low-cost 18–40 GHz wideband compact filter bank module. Design and fabrication considerations are presented. The switched filter bank consists of three channels. The filter consists of nine side coupled sections. Wideband MMIC switches are employed to select the required channel. The passband bandwidth of each channel is around 8 GHz. The insertion loss is around 12dB with ±1dB flatness. The rejection at ±7 GHz from center frequency is 40dB. The rejection at ±11 GHz from center frequency is around 60dB. The filter bank dimensions are $20 \times 50 \times 10$ mm.

6.3.2   Description of the Filter Bank

The block diagram of the filter bank module is shown in Figure 6.17. The filter bank module consists of three wideband filters. The filter consists of nine side coupled sections printed on a 5 mil alumina substrate. The filter input and output ports are connected to wideband MMIC SPDT switches via 1–2dB attenuators. The attenuators are employed to adjust each channel losses to the required value. The module losses are adjusted to be higher in the low frequencies and lower in the high frequencies. This

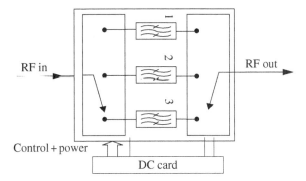

FIGURE 6.17   Block diagram of the filter bank module.

feature improves the system flatness over the frequency range. The filters are glued to the surface of the mechanical box. The input and output switches are assembled on a covar carrier. During development it was found that the spacing between the filters and the carrier should be less than 0.03 mm in order to achieve ±1dB flatness and VSWR better than 2 : 1.

6.3.3   Switch Filter Bank Specifications

Parameter	Requirements
Frequency range	18–40.1 GHz
Number of channels	3
Channel passband	8 GHz
Rejection	40dB min. at F0±8 GHz
	50dB min. at F0±11GHz
Flatness	±1.2dB max. for 4 GHz
I.L	
CH-1	12–14.5dB
CH-2	10.5–13.5dB
CH-3	9–12dB
Switching time	100 ns
Input power	25dBm max.
VSWR	2.5 : 1 max.
Control	2 LVTTL lines
Power supply voltages	±5 VDC heat dissipation, 1 W max.
Dimension	$50 \times 20 \times 10$ mm

6.3.4   Filter Design

The filters have been designed by employing AWR and ADS software. Single filter response requirements are presented in Figure 6.18. The SFB expected frequency response is listed in Table 6.6 and presented in Figure 6.19. Comparison between discrete and integrated designs is presented in Table 6.7. The filter consists of side coupled sections printed on a 5 mil alumina substrate. We applied optimization tools to determine the number of sections and filter configuration needed to meet the specifications. The computed results of the filters are shown in Figures 6.20, 6.21, 6.22,

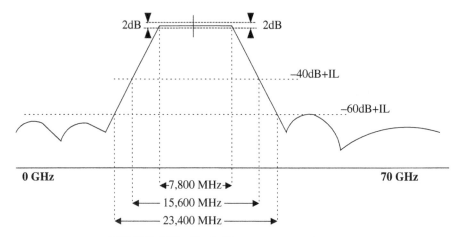

FIGURE 6.18   Single filter response requirements.

TABLE 6.6   Switch Filter Bank Requirements

CH (GHz)	Rejection −60dB	Rejection −40dB	Passband −3dB	Passband −3dB	Rejection −40dB	Rejection −60dB
CH-1	10.1	14	17.9	25.7	29.6	33.5
CH-2	17.3	21.2	25.1	32.9	36.8	40.7
CH-3	24.5	28.4	32.3	40.1	44	47.9

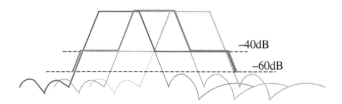

FIGURE 6.19   SFB expected frequency response.

TABLE 6.7   Comparison between Discrete and Integrated Designs

Parameter Design	Dimension (cm)	Weight (kg)	Price (K$)
Integrated	5.5 × 2.5 × 1.5	0.05	2.2
Discrete	12 × 6 × 3	1	10

CH-1 FILTER

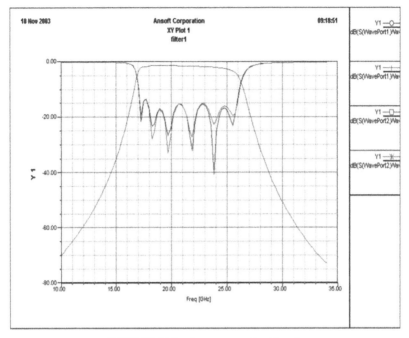

FIGURE 6.20 *S* parameters for filter 1.

CH-2 FILTER

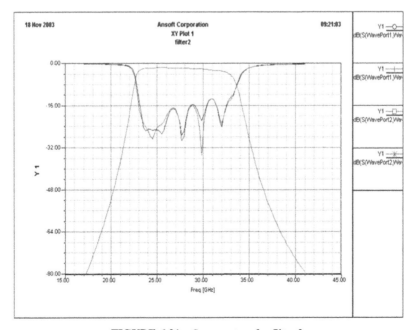

FIGURE 6.21 *S* parameters for filter 2.

CH–3 FILTER

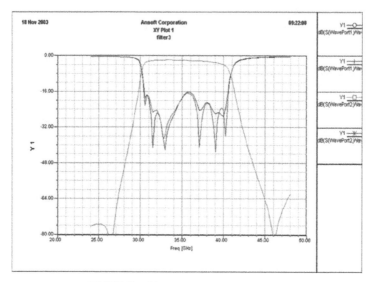

FIGURE 6.22   *S* parameters for filter 3.

System1

	SPDT	Filter bank	Attenuator	SPDT	Attenuator	
35e3 MHz		/6\				
		Bessel				Total
NF (dB)	2.00	Fo: 30,000	1.00	2.00	1.00	9.01
Gain (dB)	−2.00	BW: 8000	−1.00	−2.00	−1.00	−9.01
OIP3 (dBm)	30.00	Gain: −3.01	30.00	30.00	30.00	20.32

Input pwr (dBm)	−60.00	System temp (K)	290.00	IM offset (MHz)	0.025	
OIP2 (dBm)	16.42	OIP3 (dBm)	20.32	Output P1dB (dBm)	5.47	
IIP2 (dBm)	25.43	IIP3 (dBm)	29.34	Input P1dB (dBm)	15.48	
OIM2 (dBm)	−154.44	OIM3 (dBm)	−247.69	Compressed (dB)	0.00	
ORR2 (dB)	85.43	ORR3 (dB)	178.67	Gain, actual (dB)	−9.01	
IRR2 (dB)	42.72	IRR3 (dB)	59.56			
SFDR2 (dB)	45.68	SFDR3 (dB)	63.52	Gain, linear (dB)	−9.01	
AGC controlled range:		Min input (dBm)	N/A	Max input (dBm)	N/A	

FIGURE 6.23   Switch filter bank analysis results.

and 6.23. The sensitivity of the design to substrate tolerances such as height and dielectric constant has been optimized. We fabricated the filter configuration that was less sensitive to production tolerances (Fig. 6.26).

Figure 6.24 presents the filter bank computed S_{11} and S_{21} parameters by using ADS software. Figure 6.25 presents the expanded S_{12} computed results of the filter bank.

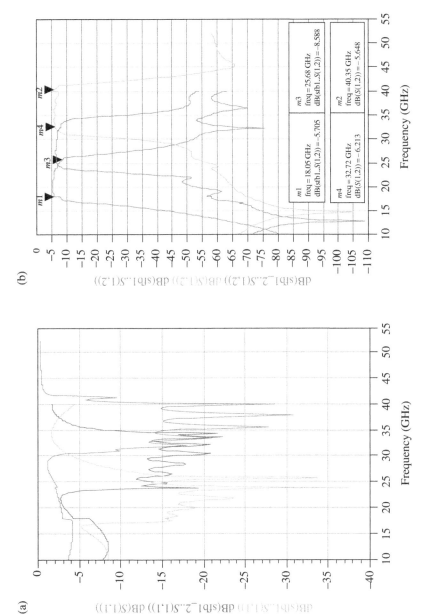

FIGURE 6.24 (a) SFB S_{11} computed results. (b) S_{21} SFB computed results. $S_{12} = S_{21}$.

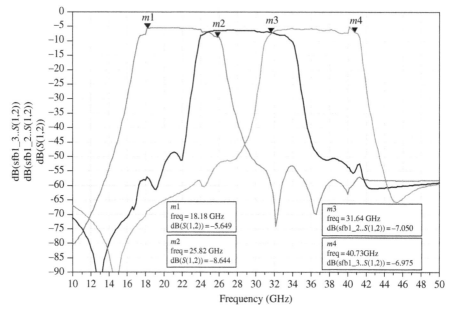

FIGURE 6.25   Switch filter bank S_{12} computed results.

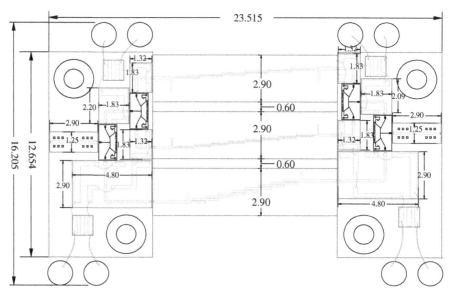

FIGURE 6.26   SFB layout.

FIGURE 6.27 SFB picture.

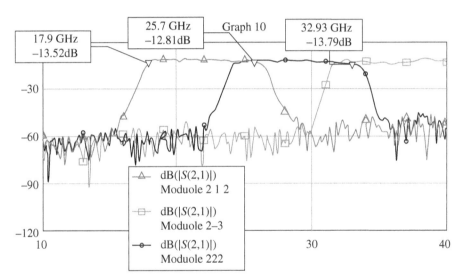

FIGURE 6.28 Switch filter bank measured results of unit number 1.

Figure 6.26 presents the filter bank layout. Figure 6.27 shows a picture of the switched filter bank. We received a good agreement between the computed and measured results. Figure 6.28 presents the switch filter bank measured S_{21} results of the first unit measured during the production process.

Figure 6.29 presents the measured S_{12} parameter of filter 2 during the production process. Figure 6.30 presents the measured S_{12} parameter of the switch filter bank as measured in the production line. Figure 6.31 presents the measured S_{11} parameter of the switch filter bank as measured in the production line. The switch filter bank losses at low frequencies are around 12dB, and at the high frequencies they are around

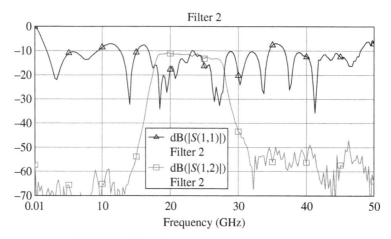

FIGURE 6.29 Measured *S* parameters for filter 2.

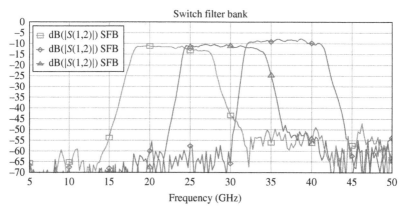

FIGURE 6.30 Switch filter bank measured S_{12} results.

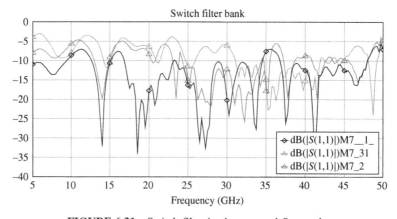

FIGURE 6.31 Switch filter bank measured S_{11} results.

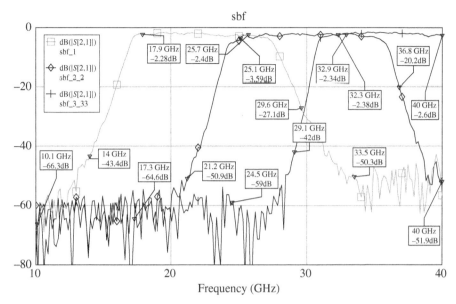

FIGURE 6.32   Switch filter bank detail of measured S_{12} results.

9dB. Figure 6.32 presents the detailed measured S_{11} parameter of the switch filter bank.

In this section we presented the design and performance of a new compact and low-cost switched filter bank module. The filter passband bandwidth is around 8 GHz. The switched filter bank insertion loss is around 9.5dB in the high-frequency range and 12dB in the low-frequency range. The filter flatness is better than ±1dB. The rejection at ±7 GHz from center frequency is 40dB. The rejection at ±11 GHz from center frequency is better than 55dB. More details and measured results will be presented during the conference.

6.4   FSU PERFORMANCE

The FSU consists of six fixed frequency source in mm-wave 18–40 GHz. Frequency is selected by an external control unit.

6.4.1   FSU Requirements

The FSU electrical specifications are listed in Table 6.8. The FSU design and fabrication process meet all the specifications.

TABLE 6.8   Electrical Requirements

Parameter	Requirements
Output discrete frequencies	
2 mm-wave outputs required, first LO output and second BIT output	LO1: 18800 MHz
	LO2: 22400 MHz
	LO3: 26000 MHz
Freq. selected by external control	LO4: 32000 MHz
	LO5: 35600 MHz
	LO6: 39200 MHz
Leakage of other frequencies (when LO output is active)	−55dBc min., −45dBc min. else. For BIT output not active
Output power for LO output	
LO1: 18800MHz	+15.0dBm ± 1dB
LO2: 22400MHz	+15.2dBm ± 1dB
LO3: 26000MHz	+15.4dBm ± 1dB
LO4: 32000MHz	+15.6dBm ± 1dB
LO5: 35600MHz	+15.8dBm ± 1dB
LO6: 39200MHz	+16.0dBm ± 1dB
Output power for BIT output	
LO1: 18800MHz	+5.0dBm ± 1dB
LO2: 22400MHz	+5.2dBm ± 1dB
LO3: 26000MHz	+5.4dBm ± 1dB
LO4: 32000MHz	+5.6dBm ± 1dB
LO5: 35600MHz	+5.8dBm ± 1dB
LO6: 39200MHz	+6.0dBm ± 1dB
Output power stability/ripple	AC component of amplitude ripple in frequency range 100 Hz to 100 MHz shall be less than −60dBm
Isolation between 2 mm-wave active outputs	35dB min.
Input power reference (200 MHz)	−1dBm ± 1dB
FSU BIT output voltage (video)	On status 0.2–0.6 mV/μW (TBD)
	Off status (noise) 0.1 mV max.
Load VSWR; impedance 50 Ω	2.0 : 1 max.
Switching speed	200 ns max.
Spurious	−60dBc (max.) for frequency range 9–15 GHz. Outside this band can be −50dBc
Harmonics	−30dBc (max.)
Subharmonics	−60dBc (max.) for subharmonics freq. range 10–14 GHz
Phase noise LO1	100 Hz offset: reference noise + 20log(N) +6 (dBc/Hz) max.
	1 kHz offset: reference noise + 20log(N) +6 (dBc/Hz) max.
	10 kHz offset: reference noise + 20log(N) +6 (dBc/Hz) max.
	100 kHz offset: −110dBc/Hz max.
	1 MHz offset: −130dBc/Hz max.
	∗N = (LO freq.)/(ref freq.)

TABLE 6.8 (*Continued*)

Parameter	Requirements
Phase noise LO2	Same as LO1
Phase noise LO3	Same as LO1
Phase noise LO4	100 Hz offset: reference noise + 20log(N) +6 (dBc/Hz) max.
	1 kHz offset: reference noise + 20log(N) +6 (dBc/Hz) max.
	10 kHz offset: reference noise + 20log(N) +6 (dBc/Hz) max.
	100 kHz offset: −100dBc/Hz max.
	1 MHz offset: −120dBc/Hz max.
Phase noise LO5	Same as LO4
Phase noise LO6	Same as LO4
Reference signal (for information only)	External 200 MHz
	Temperature stability 0.1 ppm
	Aging 0.1 ppm/year
	Phase noise
	Offset 100 Hz: −110dBc/Hz min.
	Offset 1 kHz: −125dBc/Hz min.
	Offset 50 kHz: −130dBc/Hz min.
Load VSWR—no damage	The unit shall withstand without damage any load VSWR at any phase including short and open
Control logic; logic states	**LVTTL standard**
	"0" = 0–0.8 V, "1" = 2.0–3.3 V
	5 bits for control
	0–3 bits (LSBs): state selection input
	4–9 bits: PLDRO locked indication
Power supply input ripple	Up to 300 mVp-p at 0.05 ÷ 400 MHz (for ±12 VDC)
	Up to 100 mVp-p at 0.05 ÷ 400 MHz (for ±5 VDC)
Heat dissipation	25 W max.

6.4.1.1 Electrical Specifications The FSU block diagram is presented in Figure 6.33. The FSU consists of 6 PLDROs, dielectric resonators, switching matrix, logic unit, and amplifiers.

The FSU conceptual and interface block diagram is shown in Figure 6.34.

6.4.2 Electrical Interfaces and Connectors

The FSU electrical interfaces are listed in Table 6.9.

6.4.3 Environmental Conditions

FSU environmental conditions are listed in Table 6.10.

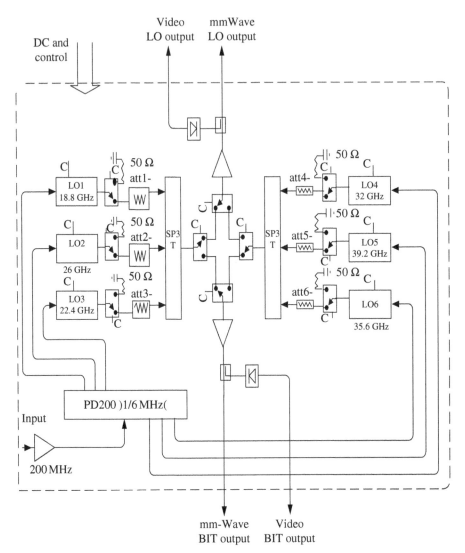

FIGURE 6.33   FSU principle block diagram.

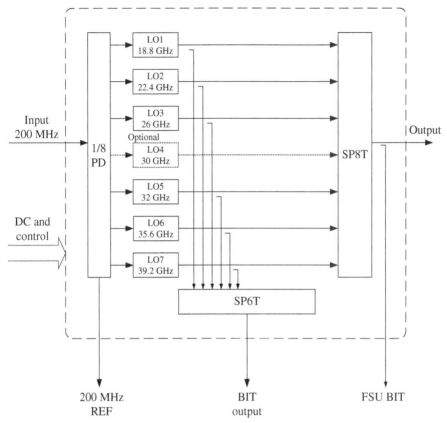

FIGURE 6.34   FSU conceptual and interface block diagram.

TABLE 6.9   FSU: Connectors and Interfaces

Interface	Type
RF input (reference 200 MHz)	SMA connector
RF output mm-wave 2	K connector
Video outputs	D type
DC supply, control/status indication	

TABLE 6.10   FSU Environmental Conditions

Environmental Event	Exposed Equipment ESM Antennas, MBAT Antennas, VHBT
Temperature high	Operating: +85°C
	MIL-STD-810D method: 501.2; procedure: II
	Nonoperating: +85°C
	MIL-STD-810D method: 501.2; procedure: I
Temperature low	Operating: −6°C
	MIL-STD-810D method: 502.2; procedure: II
	Nonoperating: −20°C
	MIL-STD-810D method: 502.2; procedure: I
Humidity	95% RH; 25–55°C noncondensing
	MIL-STD-810D method: 507.2; procedure: I
Shock	25 G 11 ms half sine for vertical
	12 G 11 ms half sine for lateral
	MIL-STD-810D method: 516.3
Vibration	MIL-STD-810F method: random vibrations 514.5
	Procedure: category 21; Figure 514.5C-15
	Test level: 0.3 g; 1–100 Hz 1 h each axis

6.4.4   Input DC Voltages and Currents

The FSU input DC terminals shall withstand the ripple levels listed in Table 6.11. ±5 V to input: ±50 mVptp. ±12 V input: ±100 mVptp.

TABLE 6.11   FSU: Voltage and Current Consumption: Preliminary

Voltage (V)	+5 V Digital	+12 V Analog	−12 V Analog	+5 V Analog	−5 V Analog
Current (mA)	300	2000	300	1000	300
Power (W)	1.5	24	3.6	5	1.5

6.4.5   Electronic Compatibility

The component shall be qualified to stand electronic compatibility requirements and shall meet the specification limits of the tests listed in Table 6.12.

6.4.6   FSU Isolation

FSU isolation is listed in Table 6.13.

6.4.7   Built-in Test

The FSU has the following support built-in test (BIT) facilities. PLDRO locked indication (6 bits) is an indication for proper operation of the 6 LOs—"1" when the specific LO is locked. For RF output for BIT, see Figure 6.34. For video detection of the RF output, see Figure 6.34.

TABLE 6.12   EMI Test Requirements

Test	Description
CE01	Conducted emissions, power leads (30 Hz to 15 kHz)
CE03	Conducted emissions, power leads (15 kHz to 50 MHz)
CS01	Conducted susceptibility, power leads (30 Hz to 50 kHz)
CS02	Conducted susceptibility, power leads (30 kHz to 400 MHz)
CS06	Conducted emissions, power leads, spikes
RE02	Radiated emissions, electric field (14 kHz to 10 GHz)
RS01	Radiated susceptibility, magnetic field (30 Hz to 50 kHz)
RS02	Radiated susceptibility, magnetic field, electric fields, spikes (400 V/5 μs), and power frequency
RS03	Radiated susceptibility, electric field (14 kHz to 18 GHz), 10 V/m

TABLE 6.13   Isolation Requirement Matrix

For LO Outputs Operating	Isolation of −55dB (Design Goal) for Other LO Outputs
LO4 or LO5	LO1, LO2, LO3
LO1 or LO6	LO3, LO4
LO2 or LO3	LO4, LO5, LO6

TABLE 6.14   FSU: Preliminary Mechanical Characteristics

No.	Parameter	Requirements
1	Dimension	$200 \times 200 \times 50$ mm max.
2	Weight	TBD

6.4.8   FSU Physical Characteristics

FSU weight and dimensions are listed in Table 6.14.

Cooling Requirements—The FSU will be mounted on cooled baseplate. The baseplate temperature will be −6 to +85°C.

Painting/Coating—The FSU will be coated with hard sulfuric anodize coating to prevent metal corrosion.

Grounding and Bonding—The FSU shall have external provisions for low-impedance grounding to system chassis, for example, grounding bolts and conductive surfaces.

Shielding—The FSU enclosure shall provide adequate shielding from electric and magnetic fields.

Filtering—Adequate filtering shall be provided at the input power to the FSU. All DC input and return wires shall be filtered.

6.4.9   Logic Requirements

The logic matrix is listed in Table 6.15.

TABLE 6.15 Logic Matrix

0–3 bits	0001	0010	0011	0100	0101	0110	1001	1010	1011	1100	1101	1110
Command word	State 1	State 2	State 3	State 4	State 5	State 6	State 7	Stare 8	State 9	State 10	State 11	State 12
LO output frequency	LO4	LO5	LO6	LO1	LO2	LO3	LO4	LO5	LO6	LO1	LO2	LO3
BIT output frequency	OFF	OFF	OFF	OFF	OFF	OFF	LO1	LO2	LO3	LO4	LO5	LO6

For states (0000, 0111, 1000, 1111) the 2 RF outputs shall be OFF.

6.5 FSU DESIGN AND ANALYSIS

At the first stage of the design process, we compared the implementation of the FSU by discrete components or as integrated super component.

6.5.1 Comparison of FSU Implementation by Discrete Components or as Super Component

The dimensions and cost of FSU implemented by discrete components are listed in Table 6.16. The dimensions and cost of integrated super component FSU are listed in Table 6.17. It is risky to design and manufacture an integrated FSU comparable to a discrete FSU. The design cost of an integrated FSU is almost twice the design cost of a discrete FSU. However, the weight of the integrated FSU is 1 kg, while the weight of the discrete FSU is around 1.5 kg. The manufacturing cost of an

TABLE 6.16 FSU Implemented by Discrete Component Dimensions and Cost

Component	W/Diameter (mm)	L (mm)	H (mm)	Quantity	Estimated Cost ($/Unit)	Total ($)
200 MHz amplifier				1	200	200
1/8 PD				1	400	400
PLDRO	50	60	20	6	2200	13,200
Attenuator	16	50	—	6	100	600
SP6T	50	60	15	1	4000	4,000
mm-wave amplifier	25	45	8	1	2000	2,000
Coupler				1	500	500
mm-wave PD				1	1000	1,000
Cables				20	50	1,000
Total						22,900

TABLE 6.17 FSU Super Component Dimensions and Cost

Component	Ref.	W/Diameter (mm)	L (mm)	H (mm)	Quantity	Estimated Cost ($/Unit)	Total ($)
200 MHz amplifier		Drop in			1	100	100
1/8 PD		Print			1	100	100
PLDRO	Drop in	50	50	20	6	2,200	13,200
SP6T	Drop in	50	50	15	1	4,000	4,000
mm-wave amplifier		10	20	5	2	100	100
Extra cost					1	100	100
Total							17,600

integrated FSU is almost half of the manufacturing cost of a discrete FSU. The dimensions and volume of an integrated FSU are almost half of the dimensions and volume of a discrete FSU. Electrical performance of an integrated FSU is almost the same as the performance of a discrete FSU.

6.5.2 FSU Analysis

The integrated FSU was designed by employing SYSCAL and ADS software.

FSU component arrangement block diagram is presented in Figure 6.35. The FSU consists of 6 PLDROs, dielectric resonators, switching matrix, logic unit, and amplifiers. We applied optimization tools to determine the best selection of the FSU components and configuration needed to meet the FSU specifications. The computed results of the FSU are shown in Figures 6.36 and 6.37. The sensitivity of the design

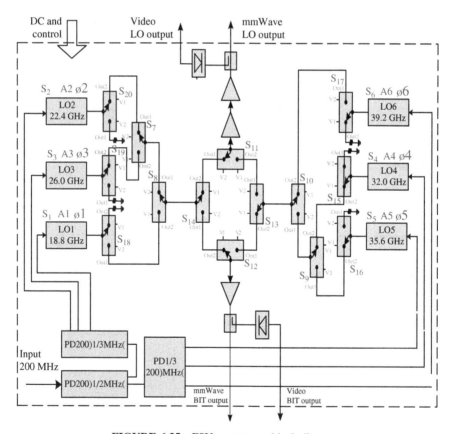

FIGURE 6.35   FSU component block diagram.

LO channel

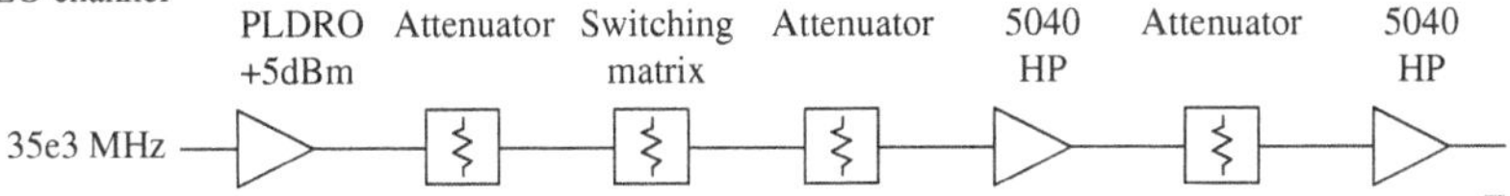

	PLDRO +5dBm	Attenuator	Switching matrix	Attenuator	5040 HP	Attenuator	5040 HP	Total
NF (dB)	5.00	4.00	15.00	2.00	9.00	5.00	9.00	19.29
Gain (dB)	11.00	−4.00	−15.00	−2.00	20.00	−5.00	19.44	24.43*
OIP3 (dBm)	23.00	35.00	35.00	35.00	28.00	38.00	30.00	28.82
OP1dB (dBm)	17.00	30.00	30.00	30.00	20.00	30.00	20.00	
Po (dBm)	5.00	1.00	−14.00	−16.00	4.00	−1.00	18.43*	18.43*

Input pwr (dBm) –6.00	System temp (K) 290.00	IM offset (MHz) 0.025

OIP2 (dBm)	28.51	OIP3 (dBm) 28.82	Output P1dB (dBm) 19.97
IIP2 (dBm)	3.51	IIP3 (dBm) 3.82	Input P1dB (dBm) −4.03
OIM2 (dBm)	8.35	OIM3 (dBm) −2.34	Compressed (dB) 0.57
ORR2 (dB)	10.08	ORR3 (dB) 20.78	Gain, actual (dB) 24.43
IRR2 (dB)	5.04	IRR3 (dB) 6.93	Inp backoff (dBm) 0.00
SFDR2 (dB)	−10.90	SFDR3 (dB) −14.33	Gain, linear (dB) 25.00
AGC controlled range:		Min input (dBm) N/A	Max input (dBm) N/A

FIGURE 6.36 FSU analysis results.

BIT channel

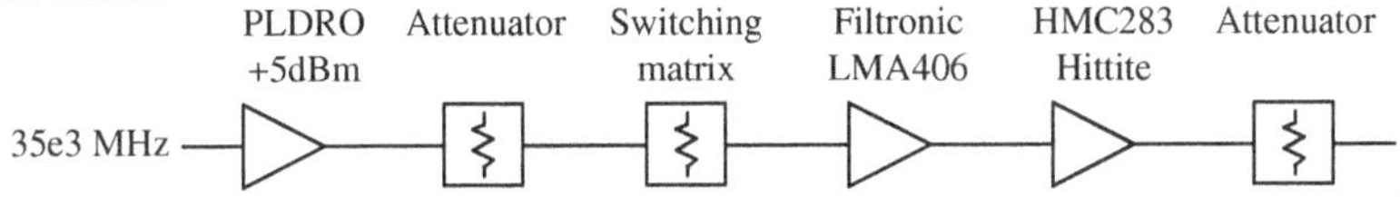

	PLDRO +5dBm	Attenuator	Switching matrix	Filtronic LMA406	HMC283 Hittite	Attenuator	Total
NF (dB)	5.00	4.00	15.00	5.00	9.00	4.00	14.38
Gain (dB)	11.00	−4.00	−15.00	9.96	16.88	−4.00	14.84
OIP3 (dBm)	23.00	35.00	35.00	27.00	27.00	38.00	21.27
OP1dB (dBm)	17.00	28.00	26.00	12.00	18.00	30.00	
Po (dBm)	5.00	1.00	−14.00	−4.04	12.84	8.84	8.84

Input pwr (dBm) –6.00	System temp (K) 290.00	IM offset (MHz) 0.025

OIP2 (dBm)	22.05	OIP3 (dBm) 21.27	Output P1dB (dBm) 13.62
IIP2 (dBm)	7.05	IIP3 (dBm) 6.27	Input P1dB (dBm) −0.38
OIM2 (dBm)	−4.38	OIM3 (dBm) −16.02	Compressed (dB) 0.16
ORR2 (dB)	13.21	ORR3 (dB) 24.86	Gain, actual (dB) 14.84
IRR2 (dB)	6.61	IRR3 (dB) 8.29	Inp backoff (dBm) 0.00
SFDR2 (dB)	−6.68	SFDR3 (dB) −9.42	Gain, linear (dB) 15.00
AGC controlled range:		Min input (dBm) N/A	Max input (dBm) N/A

FIGURE 6.37 FSU BIT channel analysis results.

to alumina substrate tolerances such as height and dielectric constant has been optimized. There is a good agreement between computed and measured results.

The FSU channel analysis results, gain and noise figure, are presented in Figure 6.36. The FSU gain is around 24dB. The FSU noise figure is around 19dB.

The FSU BIT channel analysis results are presented in Figure 6.37. The FSU BIT channel gain is around 14dB. The FSU BIT channel noise figure is around 15dB.

6.5.3   FSU Thermal Analysis

Thermal analysis is based on the following data:

Environmental temperature 55°C

Wind velocity 3 m/s

FSU power consumption 25 W

FSU voltages and current consumption are listed in Table 6.18.

Figure 6.38 presents FSU component thermal analysis, for white package colure.

TABLE 6.18   FSU Voltages and Current Consumption

Voltage (V)	Current (A)
12	1.3
5	1
5 digital	0.3
−12	0.3
−5	0.3

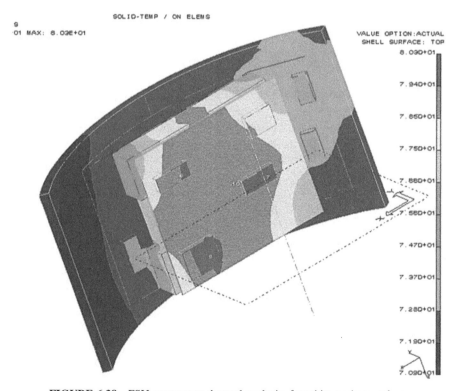

FIGURE 6.38   FSU component thermal analysis, for white package colure.

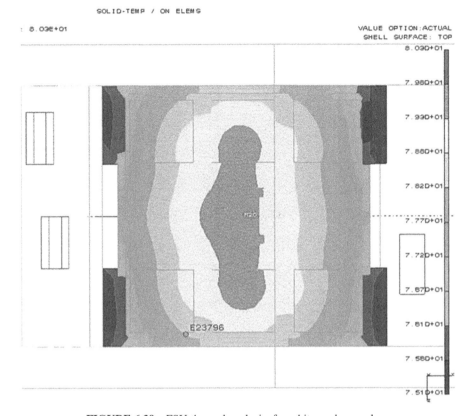

SOLID-TEMP / ON ELEMS

: 8.03E+01

VALUE OPTION:ACTUAL
SHELL SURFACE: TOP

FIGURE 6.39   FSU thermal analysis, for white package colure.

Figure 6.39 presents FSU thermal analysis, for white package colure.

Figure 6.40 presents FSU thermal analysis, for gray package colure.

For white colure package the temperature on the component base is around 78°C. For black colure package the temperature on the component base is around 86°C.

6.5.4   FSU Interfaces and Layout

The FSU module is controlled by the FSU logic unit.

6.5.4.1   FSU Logic Unit Requirements   The FSU logic unit is implemented on the FSU card. It is placed in the Omni downconverter section of the direction finding system. Table 6.19 summarizes the demands from the logic unit.

The FSU logic unit primary functions are:

1. To control RF switches located on the FSU card
2. To maintain an interface with the direction finding system controller card
3. To translate commands coming from the direction finding system card to desired states of the FSU

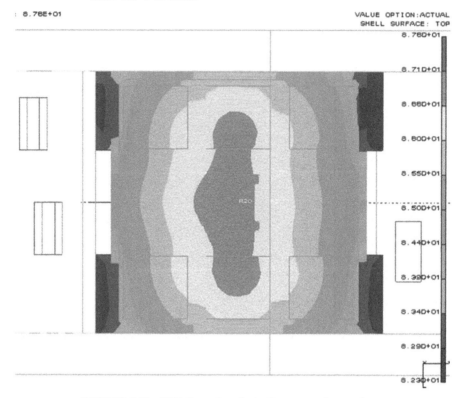

FIGURE 6.40   FSU thermal analysis, for gray package colure.

TABLE 6.19   FSU Logic Unit Requirements

No.	Requirement	Description
1.	Power supply	P5V ± 5% 0.5 A
2.	Power consumption	Power dissipation in steady state <0.5 W
3.	Dimensions	50 mm × 50 mm
4.	Temperature range	Reference document is raf#92912—logic card's section
5.	Logic interface	The logic unit shall maintain a 12-bit interface with the VHBR_TC
6.	JTAG	The access, to the XPLD JTAG TAP, should be available from a maintenance connector on the FSU card
7.	Test signals	The logic unit should provide test signals from the XPLD to a test connector for debugging and maintenance
9.	Timing constraints	The logic unit should be able maintain a minimum of 50 ns switching rate
10	Performance	The logic unit shall perform the state machine presented in Table 6.4

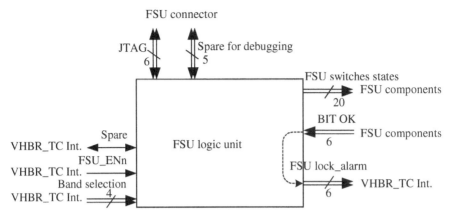

FIGURE 6.41   FSU logic unit's general block diagram.

Logic Unit Definition   Figure 6.41 describes the FSU logic unit. The FSU logic unit contains an XPLD device, which defines a truth table. According to four control lines (band selection), a unique combination is made for 20 output signals. For 20 different switches, the truth table is defined in Table 6.22. In addition this XPLD buffers 6 "OK" indication bits toward the direction finding system.

The XPLD is LC5512MV-75F256I: This device is manufactured by Lattice.

This device has its own E^2rom memory for configuring itself on power-up.

It has JTAG compliance, up to 193 I/O pin, 2 internal PLLs, and 512 macrocells (logic blocks). In practice 30 I/O pins are used.

FSU Module Interface   Table 6.20 defines the FSU logic unit's ICD.

The FSU logic interface is divided into three sections:

1. Direction finding system digital interface—for FSU control and indication

2. Direction finding system analog interface—for FSU detector monitoring

3. Local interface for JTAG configuration and debugging

6.5.4.2   Direction Finding System Digital Interface   The interface contains 12-bit interface for full control performance. The digital interface function is described in Table 6.21.

6.5.4.3   Direction Finding System Analog Interface   The analog interface contains two analog signals and one reference signal, which are monitored in the direction finding system ADC device.

The three analog signals are:

1. FSU_2_TC_RF_Sense (1:0)—two signals—yielded from the detectors located in the FSU

2. FSU_2_TC_RF_Sense_RET—one signal—the analog ground that is used as a reference for the detector output monitoring

TABLE 6.20 Direction Finding System FSU Logic Unit Interface

Section	Pin # (Conn)	Line Connection	Line Description	Current (mA)	Technology	Dir	Max. Toggle Rate (MHz)
1	13	TC to FSU	FSU band select (1)	10	LVTTL	I	5
	14	TC to FSU	FSU band select (2)	10	LVTTL	I	5
	31	TC to FSU	FSU band select (3)	10	LVTTL	I	5
	29	TC to FSU	FSU band select (4)	10	LVTTL	I	5
	30	TC to FSU	FSU enable	10	LVTTL	I	5
	48	TC to FSU	Spare output/input	10	LVTTL	I/O	5
	27	FSU to TC	FSU OK—lock alarm (1)	10	LVTTL	O	20
	10	FSU to TC	FSU OK—lock alarm (2)	10	LVTTL	O	20
	43	FSU to TC	FSU OK—lock alarm (3)	10	LVTTL	O	20
	45	FSU to TC	FSU OK—lock alarm (4)	10	LVTTL	O	20
	42	FSU to TC	FSU OK—lock alarm (5)	10	LVTTL	O	5
	44	FSU to TC	FSU OK—lock alarm (6)	10	LVTTL	O	5
	9	TC to FSU	Return digital ground	—	GND	O	—
3	11	Local	Spare output/input	10	LVTTL	I/O	5
	12	Local	Spare output/input	10	LVTTL	I/O	5
	46	Local	Spare output/input	10	LVTTL	I/O	5
	28	Local	Spare output/input	10	LVTTL	I/O	5
	47	Local	Spare output/input	10	LVTTL	I/O	5
	19	Local	JTAG data	10	LVTTL	I	5
	21	Local	JTAG data	10	LVTTL	I	5
	20	Local	JTAG data	10	LVTTL	I	5
	22	Local	JTAG data	10	LVTTL	O	5
	1	Local	Positive 3.3 V DC	—	Power 3.3 V	O	—
	3	Local	Return digital ground	—	GND	O	—
2	6	FSU to TC	Detector LO	—	Analog	O	—
	5	FSU to TC	Detector BIT	—	Analog	O	—
	25	FSU to TC	Return analog ground	—	Analog Ret	O	—
Supply	2	TC to FSU	Positive 5 V digital DC	600	Pos. 5 V ± 3%	I	—
Supply	4	FSU to TC	Return digital ground	—	GND	O	—

TABLE 6.21 FSU: Direction Finding System Digital Interface

Controlled Element Name (Direction Finding System *Dedicated*)	#bits for Control	States/Comments	Logic State	I/O	Type	Required Switching Time
FSU (*TC_2_FSU_CTRL(12:1)*)	12 FSU(1–3) band selection bits	Band 1, BIT OFF	0001	O	LVTTL	1 μs max.
		Band 2, BIT OFF	0010			
		Band 3, BIT OFF	0011			
		Band 4, BIT OFF	0100			
		Band 5, BIT OFF	0101			
		Band 6, BIT OFF	0110			
		Band 1, BIT ON	1001			
		Band 2, BIT ON	1010			
		Band 3, BIT ON	1011			
		Band 4, BIT ON	1100			
		Band 5, BIT ON	1101			
		Band 6, BIT ON	1110			
	FSU(4) indication	FSU OK	6 bits	I	LVTTL	100 ms

TABLE 6.22 FSU Logic Unit's State Machine

0–3 bits	0001	0010	0011	0100	0101	0110	1001	1010	1011	1100	1101	1110	0000[a]
Command word	State 1	State 2	State 3	State 4	State 5	State 6	State 7	State 8	State 9	State 10	State 11	State 12	State 13
LO output frequency	LO4	LO5	LO6	LO1	LO2	LO3	LO4	LO5	LO6	LO1	LO2	LO3	LO OFF
BIT output frequency	OFF	OFF	OFF	OFF	OFF	OFF	LO1	LO2	LO3	LO4	LO5	LO6	BIT OFF
S_1	0	0	0	1	0	0	1	0	0	1	0	0	0
S_2	0	0	0	0	1	0	0	1	0	0	1	0	0
S_3	0	0	0	0	0	1	0	0	1	0	0	1	0
S_4	1	1	0	0	0	0	1	0	0	1	0	0	0
S_5	0	0	1	0	0	0	0	1	0	0	1	0	0
S_6	0	0	0	0	0	1	0	0	1	0	0	1	0
S_7	0	0	1	1	0	0	1	0	1	1	0	1	0
S_8	1	1	0	1	0	0	1	0	0	1	0	0	0
S_9	1	0	1	0	0	1	1	0	1	1	0	1	0
S_{10}	0	0	1	1	0	0	0	1	1	0	0	0	0
S_{11}	1	1	1	0	1	1	1	1	1	1	0	0	0
S_{12}	1	1	0	1	1	0	1	1	0	1	1	1	1
S_{13}	0	0	0	1	0	1	0	0	0	0	1	1	0
S_{14}	0	0	0	0	0	0	0	0	0	0	1	1	0
S_{15}	1	1	1	0	1	0	1	1	1	1	0	1	0
S_{16}	0	1	0	1	1	1	1	1	0	0	1	0	0
S_{17}	1	1	1	1	1	1	1	1	1	1	1	1	1
S_{18}	1	1	1	0	1	1	1	1	1	0	1	1	1
S_{19}	0	0	0	0	0	1	0	0	1	0	0	1	0
S_{20}	0	0	0	0	1	0	0	1	0	0	1	0	0

[a]For state #13 LO signal is OFF.

Local Interface   The local interface is being used for:

1. **Configuration**: JTAG—configuration is being used when there is a need to change the truth table or when a debugging version is needed.
2. **Debugging**: In debugging mode, any signal can be routed toward five dedicated pins to be monitored.

Table 6.22 describes the states that shall be accomplished in the FSU module.

6.6   FSU FABRICATION

Figure 6.42 presents the mechanical fabrication drawing of the FSU module. The FSU module dimensions are $16.55 \times 21.4 \times 3.83$ cm. The FSU consists of six PLDROs, dielectric resonators, switching matrix, logic unit, and amplifiers. The PLDROs are connected to the switching matrix by low-loss semirigid cables as shown in Figure 6.42. Figure 6.43 presents the FSU module component placing and arrangement. Figure 6.44 presents the FSU module mechanical package. Figure 6.45 presents

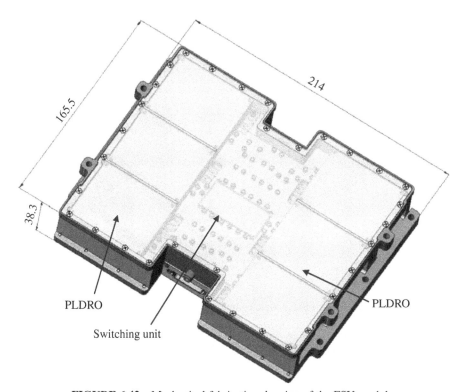

FIGURE 6.42   Mechanical fabrication drawing of the FSU module.

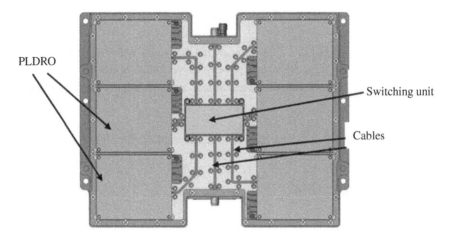

PLDRO

Switching unit

Cables

FIGURE 6.43   Top view drawing of the FSU module.

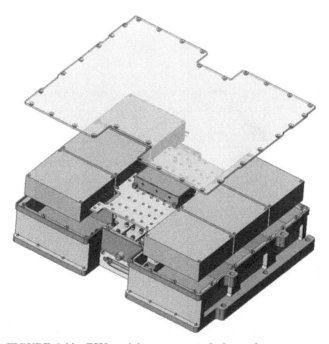

FIGURE 6.44   FSU module component placing and arrangement.

the FSU switching matrix unit top view. The FSU switching matrix consists of 14 SPDT switches. The SPDT switches are mounted on three covar carriers as presented in Figure 6.46. In Figure 6.47, FSU presents the switching matrix unit dimensions and mechanical package. The FSU switching matrix dimensions are

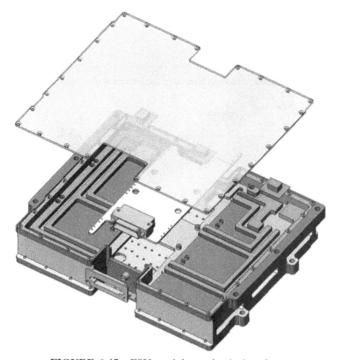

FIGURE 6.45 FSU module mechanical package.

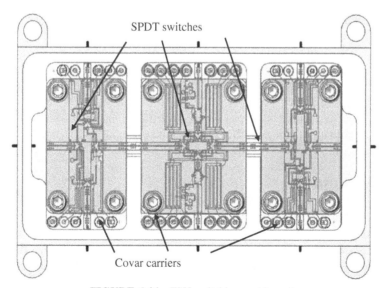

FIGURE 6.46 FSU switching matrix unit.

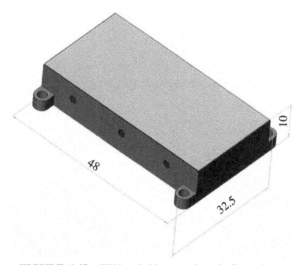

FIGURE 6.47   FSU switching matrix unit dimensions.

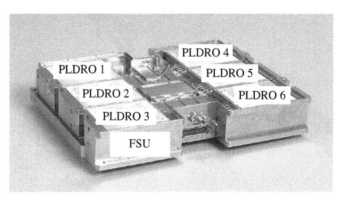

FIGURE 6.48   Fabricated FSU module photo.

4.8 × 3.25 × 1 cm. Figure 6.48 presents a fabricated FSU module photo. Figure 6.49 presents a photo of a fabricated FSU switching matrix unit.

6.7   CONCLUSIONS

Compact 18–40 GHz wideband RF modules for direction finding, seekers, and communication systems were presents in this chapter. The direction finding system consists of 14 direction finding front-end modules, two omnidirectional front-end modules, switch filter bank (SFB) modules, FSU, and downconverter units. The RF modules were designed by using ADS momentum software.

There is a good agreement between computed and measured results.

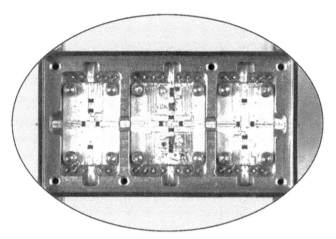

FIGURE 6.49 Photo of a fabricated FSU switching matrix unit.

REFERENCES

[1] Rogers J, Plett C. *Radio Frequency Integrated Circuit Design*. Boston: Artech House; 2003.

[2] Maluf N, Williams K. *An Introduction to Microelectromechanical System Engineering*. Boston: Artech House; 2004.

[3] Sabban A. Microstrip Antenna Arrays. In Nasimuddin N, editor. *Microstrip Antennas*. Rijeka: InTech; 2011. p. 361–384.

[4] Sabban A. Applications of MM wave microstrip antenna arrays. International Symposium on Signals, Systems and Electronics, 2007. ISSSE'07 Conference; July 30 to August 2, 2007; Montreal, QC.

[5] Gauthier GP *et al*. A 94GHz micro-machined aperture-coupled microstrip antenna. IEEE Trans Antenna Propag 1999;47 (12):1761–1766.

[6] Milkov MM. Millimeter-wave imaging system based on antenna-coupled bolometer [M. Sc. thesis]. UCLA; 2000.

[7] de Lange G, Konistis K, Hu Q. A 3*3 mm-wave micro machined imaging array with sis mixers. Appl Phys Lett 1999;75 (6):868–870.

[8] Rahman A, de Lange G, Hu Q. Micromachined room temperature microbolometers for mm-wave detection. Appl Phys Lett 1996;68 (14):2020–2022.

[9] Maas SA. *Nonlinear Microwave and RF Circuits*. Boston: Artech House; 1997.

[10] Sabban A. *Low Visibility Antennas for Communication Systems*. Boca Raton: Taylor & Francis Group; 2015.

7

INTEGRATED OUTDOOR UNIT FOR MILLIMETER-WAVE SATELLITE COMMUNICATION APPLICATIONS

The block diagram of an integrated outdoor unit (ODU) for mm-wave satellite communication applications is presented in Figure 7.1. The ODU consists of a receiving and a transmitting channel.

7.1   THE ODU DESCRIPTION

The outdoor system is divided into three main parts: the transmitter, the "ODU;" the antenna assembly; and the receiver front end, the "LNB." The antenna assembly consists of two main parts: the dish and mounting structure and the dual channel feed horn. The ODU module design is presented in this section. RF component design and theory are presented in books and in articles (see Refs. [1–10]).

At the low-frequency I/O terminal, four types of input and output signals are fed: transmitting (Tx) input IF signal (in frequency band of 2500–3000 GHz), DC input power supply (28 V ± 10%), output reference (10 MHz), and monitor and control at (22 kHz) IN/OUT signal. The DC input is fed through the multiplexer to the power supply from which regulated voltages are supplied to the different parts of the ODU. The output 10 MHz reference signal is obtained from a 100 MHz internal crystal oscillator by 1 : 10 frequency divider and is routed to the I/O connector through the multiplexer. The monitor and control (22 kHz) signal is separated from the other signals by the multiplexer between the I/O connector and the controller. The control command ON/OFF is processed by the controller and applied to the Solid State Power

Wideband RF Technologies and Antennas in Microwave Frequencies, First Edition. Dr. Albert Sabban.
© 2016 John Wiley & Sons, Inc. Published 2016 by John Wiley & Sons, Inc.

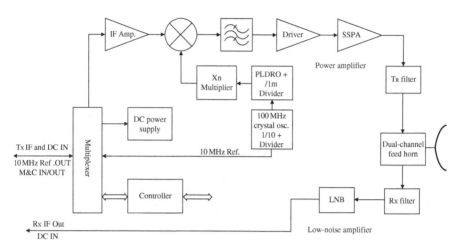

FIGURE 7.1 ODU basic block diagram.

Amplifier (SSPA) supply ON/OFF switch. The monitoring signals, originating from different parts of the ODU, are processed by the controller and installed in the 22 kHz serial link. The PLDRO lock/unlock monitoring signal is used also (via the controller) for switching the SSPA supply ON/OFF accordingly (automatic shutdown). Additional alarm monitoring signal will be provided by switching the 10 MHz signal from ON to OFF state (by the controller) if any of the alarm signal (PLDRO unlocked, SSPA supply in the OFF state, or power supply alarm) appears at the controller inputs. The input IF signal (in frequency band 2500–3000 GHz) is separated from the other inputs by the multiplexer and is fed to the IF amplifier. The IF signal is amplified to an appropriated level (the IF amplifiers amplify the input IF signal from the minimum input level to the nominal level at the mixer input) and is fed to the mixer input.

The frequency band of the input signal is converted to the transmitted frequency band by the mixer and the LO signal. The band-pass filter, following the mixer output, attenuates the LO signal leaking (at 27 GHz), image signal, and other spurious products. The filtered Tx signal is amplified by the driver and the power amplifier (PA) to the appropriate output power level. The Tx band pass following the power amplifier attenuates all the spurious signal levels below the relevant specified levels and also the output noise, in the receiving (Rx) frequency band, below the thermal noise at the LNB (Rx link) input. The LO source consists of a PLDRO, multiplier, and a band-pass filter. The PLDRO is locked to a 100 MHz (internal) crystal oscillator. The output signal (at 9 or 13.5 GHz) is multiplied by the frequency multiplier to the specified frequency (27 GHz) and filtered to eliminate the spurious signals. All the components shown in Figure 7.1 are installed in a mechanical enclosure, which protects them from the outside environment and dissipates the heat, by convection, to the air outside the enclosure. Two connectors are mounted on the enclosure (box): input I/O (type F connector) and output (ISO PBR 320 for WR28 waveguide). The ODU box will be mounted directly to the dual channel feed horn through the waveguide flange. The ODU, the dual channel feed horn, and the LNB are assembled on a special mounting cradle, which is mounted on the antenna boom and allows the rotation of the ODU, the LNB, and the antenna feed around the feed axis for polarization adjustment. The ODU specifications are listed in Table 7.1.

TABLE 7.1 ODU Specifications

Description	Specification
Transmit frequency range	29.5–30.0 GHz
Receive frequency range, in two bands (10.7–12.75 GHz)	
Low	10.7–11.7 GHz
High	11.7–12.75 GHz
Transmit on one linear (horizontal (*H*) or vertical (*V*)) polarization plane. Receive dual linear orthogonal polarized signal by switching between *H* and *V* polarization	
Selecting the polarization orientation. Continuous adjustment of the polarization orientation	*H* or *V*
Identical antenna tilt angle for	±45° Ka and Ku signals
Min. EIRP and 1dB gain compression	
Max. antenna diameter 0.75 m	40dBW
Max. antenna diameter 0.95 m	45dBW
Max. antenna diameter 1.30 m	50dBW
Off-axis EIRP density, max.	10 W/m^2
EIRP stability over temperature	±2dB
G/T figure of merit	>14dB/K
Spurious emission below the total EIRP	60dB max.
In 29.5–30.0 GHz band	
Outside 29.5–30.0 GHz band	
Noise emission max. in 29.5–30.0 GHz band	
SSPA ON	−75dBW/Hz of EIRP
SSPA OFF	−105dBW/Hz of EIRP
Phase noise, max. : freq. offset	
(For the transmitter) 10 Hz	−32dBC/Hz
100 Hz	−62dBC/Hz
1 kHz	−72dBC/Hz
10 kHz	−82dBC/Hz
>100 kHz	−92dBC/Hz
Spurious phase noise at AC line frequency. Level of the sum of all phase noise components in frequency range between the AC line freq. and 1 MHz	−32dBC max. −38dBC max.
Amplitude variation (transmitter only) in any 200 kHz band	0.2dB max.
2 MHz band	0.4dB max.
40 MHz band	1.0dB max.
500 MHz band	2.5dB max.
Group delay variations (transmitter only) in any 2 MHz band	2 ns max. ptp
40 MHz band	4 ns max. ptp
Radiation pattern (transmitter only)	
$1.8° < \Phi < 7°$	$29 \div 25\log\Phi$ dBi max.
$7° < \Phi < 9.2°$	8dBi max.

(Continued)

TABLE 7.1   *(Continued)*

Description	Specification
$9.2° < \Phi < 48°$	$32 \div 25\log \Phi$ dBi max.
$48° < \Phi < 180°$	0dBi max.
Tx cross-polar gain: $1.8° < \Phi < 7°$	$19 \div 25\log\Phi$ dBi max.
$7° < \Phi < 9.2°$	−2dBi max.
Tx cross-polar isolation within 1/10dB beam contour	−25/−22dB max.
Antenna pointing, manual adjustment	
Elevation	10–50°
Azimuth	0–360°
Pointing accuracy (mechanical coarse and fine adjustment)	10% of the 3dB beam width
Tx power consumption: SIT I	20 W
SIT II	30 W
SIT III	50 W
Safety SSPA switch OFF	Automatic shutdown
Avoiding interference to other users	
Monitoring functions	Lock alarm
	SSPA ON/OFF status
	Power supply alarm
	Presence det. alarm
Control	SSPA ON/OFF
SSPA ON/OFF indication on ODU	Green/red LED
Operation environment	
Temperature	−30° to +50°
Solar radiation	500 W/m² max.
Humidity	0–100% (condensing)
Rain	40 mm/h max.
Wind	45 km/h max.
Survival conditions: Temperature	−40° to +60°
Solar radiation	1000 W/m²
Humidity	0–100% (condensing)
Precipitation: Rain	100 mm/h max.
Freezing rain	12 mm/h max.
Snowfall	50 mm/h max.
Static load	25 mm of ice on all surfaces
Wind	120 km/h max.
Storage and transportation	
Temperature	−40° to +70°
Shock and vibration	As required for commercial freights
IDU and ODU, DC power supply	28 V ± 10%
Tx IF IN (2.5–3 GHz)	−40dBm
M&C: 10 MHz ref. output power	22 kHz PWK
Frequency stability	−40dBm ± 20 ppm
Connector: Tx output flange	Type F
(Tx to antenna feed)	ISO PBR 320

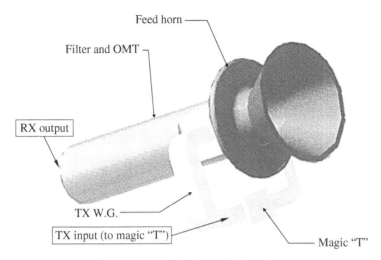

Feed horn

Filter and OMT

RX output

TX W.G.

TX input (to magic "T")

Magic "T"

FIGURE 7.2   Outdoor unit drawing.

7.2   THE LOW NOISE UNIT: LNB

The LNB is of universal standard and one with a circular C-120 flange at the input. The LNB is mounted directly to the dual channel feed horn through the C-120 flange, see Figure 7.2. This LNB is produced by several manufacturers and is available on the commercial market. The full description and specification of the LNB are listed in Table 7.1 with an exception that the maximum noise figure should be 0.9dB for both receiving bands. The LNB output may be connected through the ODU to enable the use of only one coaxial cable between the indoor unit, the IDU, and the outdoor assembly. For that purpose, a special power supply must be added into the ODU enclosure, an appropriate protocol must be specified, and the multiplexer must be redesigned to allow the filtering of the different signals at the ODU's I/O terminal.

7.3   SSPA OUTPUT POWER REQUIREMENTS

Given: a. Dual band feed horn's insertion loss			0.5dB max.
b. Transmitter filter and WG28 WG loss			0.2dB max.
d. Mismatch loss (WSWR < 1.35)			0.1dB max.
Total loss			0.8dB max.
	SIT I	SIT II	SIT III
Antenna gain at Ka-band (dBi)	45.0	47.0	49.5
Tx to feed output loss (dB)	0.8	0.8	0.8
Net. gain (dBi)	44.2	46.2	48.7
Specified EIRP (dBW)	40.0	45.0	50.0
Minimum SSPA output (dBW)	−4.2	−1.2	+1.3

7.4 ISOLATION BETWEEN RECEIVING AND TRANSMITTING CHANNELS

Since the Tx and the Rx signal may appear at the feed with the same polarization, the isolation between the transmitter (Tx) port and the receiver (Rx) port is mainly achieved by the Rx input filter. The main objectives of the Rx input filter is to prevent the gain degradation of the LNB and to eliminate spurious response (mixing with the harmonics of the LOs in the LNB) and to reject signals at the image frequencies.

The isolation may be estimated for the following data:

Max. Tx signal level at the Tx port	+33dBm
Output 1dB compression point of the LNB	+5dBm
LNB gain (max.)	60dB

According to these assumptions the input 1dB compression point of the LNB is around −55dBm, and the isolation must be at least 88dB.

The LNA inside the LNB is tuned to the 10.7–12.75 frequency band. Following the LNA, a band-pass filter is used to reject the image signal.

7.5 SSPA

7.5.1 Specifications

1. Input power: −15dBm.
2. The output power requirements result from the system parameters listed in Table 7.2.

7.5.2 SSPA General Description

The module is an integral MIC assembly, which includes the upconverter, SSPA, output power detector, and transition to the antenna. The key element for the realization of the ODU SSPA is the output stage basic MMIC power amplifier.

TABLE 7.2 Output Power Requirements for Three SIT Situations

	EIRP Spec. (dBW)	Required PA Output *(1) dBm (dBW)	PA to Feed Output Loss	Antenna Gain (dBi)
SIT1	40	25.8 (−4.2dBW)	0.8	45
SIT2	45	28.8 (−1.2dBW)	0.8	47
SIT3	50	31.3 (+1.3dBW)	0.8	49.5

*(1) Output power required at P1dB.

7.5.3   SSPA Electrical Design

1. Input power: −15dBm.
2. The output power requirements result from the following system parameters: The SSPA block diagram is shown in Figure 7.3.

Power amplifier power and gain budget:

1. Input power: −15dBm.
2. The output power requirements result from the system parameters given in Tables 7.3, 7.4, and 7.5.

Power supply requirement are listed in Table 7.6.

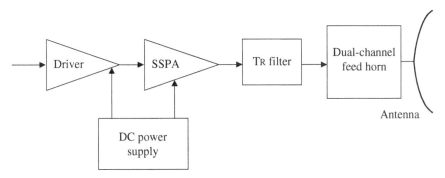

FIGURE 7.3   Block diagram of the SSPA unit.

TABLE 7.3   Situation 1

	P_{in}	Low PA	Medium PA	High PA
Power (dBm)	−15	−5	13.5	26
Gain (dB)		10	18.5	12.5
P1dB (dBm)		8	19	27

Power supply: ±5 V, 7 W.

TABLE 7.4   Situation 2

	P_{in}	Low PA	Medium PA	High PA
Power (dBm)	−15	−3	16	28.5
Gain (dB)		12	19	12.5
P1dB (dBm)		8	19	29

Power supply: ±5 V, 10 W.

TABLE 7.5   Situation 3

	P_{in}	Low PA	Medium PA	High PA
Power (dBm)	−15	−1	18	31
Gain (dB)		14	19	13
P1dB (dBm)		8	19	31.5

Power supply: ±5 V, 15 W.

TABLE 7.6   Supplies

	Drain Voltage (V)	Gate Voltage (V)	DC Power (W)	Drain Current (A)
SIT1	5	−5	7	1.4
SIT2	5	−5	10	2
SIT3	5	−5	15	3

7.5.4   Advanced Packaging Techniques for High-Power 30 GHz Transmit Channels

Utilizing advanced packaging techniques will boost the high-power 30 GHz transmit chains. Current MIC technology suffers from significant problems at millimeter waves in several aspects.

Performance—Poor input/output and interconnect quality due to transitions and wire parasitic effects. Tolerances are not tight enough, and therefore it is hard to predict parasitic effects. Grounding problems cause unpredictable behavior and degraded performance.

Other considerations: heat spread density problems, in/out connection to the package, and the yield of the MMIC chips and its assembly.

Cost—Much delicate human labor is required to produce millimeter-wave MIC module. This labor includes obviously assembly and the critical step of test and tune to compensate for parasitic effects. These limitations are overcome by addressing several issues. New types of substrate material with reduced dielectric constant lower loss than common substrates, enabling better realization of MIC passive elements.

Multilayer ceramic and polymer material substrate will allow DC/RF crossover and reduce the number of wire bonds required. An excellent heat conductive material must be used for the carrier. Embedded elements will reduce the overall part count and assembly complexity. Higher thermal conductivity materials will allow higher power device utilization. Automated assembly will increase the reproducibility, which is critical in media transitions.

7.6   THE ODU MECHANICAL PACKAGE

The ODU package combines all feed RF components and driving electronics (Fig. 7.4). The RF chain consists of a horn feed, a transducer, a filter, and the LNB. The ODU package is designed to ensure full compliance with the SOW

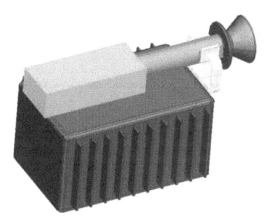

FIGURE 7.4 ODU package.

mechanical requirements, most notable harsh atmospheric conditions of precipitation and temperature. The package houses the SSPA, upconverter, PLDRO, controller, and a power supply. The ODU is mounted on the boom of the antenna using mechanics, which allow adjustment of the polarization angle, while maintaining accurate position of the horn on the antenna's focal point. The ODU's package is designed to cost-effectiveness and to large volume production, while it protects its inner components from precipitation, dissipates the heat generated within, and is coated and painted to absorb minimum solar radiation.

7.7 LOW NOISE AND LOW-COST K-BAND COMPACT RECEIVING CHANNEL FOR VSAT SATELLITE COMMUNICATION GROUND TERMINAL

7.7.1 Introduction

An increasing demand for wide bandwidth in communication links makes the K-/Ka-band attractive for future commercial systems. The frequency allocations for the Ka-/K-band VSAT system are 17.7–21.2 GHz for the receiving channel and 27.5–31 GHz for the transmitting channel. Communication industry is currently in continuous growth. In particular the very small aperture terminal (VSAT) networks have gained wide use for business and private applications. Private organizations and banks are using VSAT networks to communicate between their various sites. VSAT applications cover a wide range such as telephony, message distribution, lottery, credit card approval, and inventory management. Commercial VSAT systems operate in C-band and Ku-band; however there are many advantages to develop wideband K-/Ka-band communication systems. However, only some commercial low-cost power amplifiers, low noise amplifiers, mixers, and DROs are published in commercial catalogs. Moreover, development of low-cost RF components is crucial for K-/Ka-band satellite communication industry. This section describes the design and performance of a compact and low-cost K-band receiving channel.

7.7.2   Receiving Channel Design

The major objectives in the design of the receiving channel were electrical specifications and cost. The target price in the production of thousands was around $80.

7.7.3   Receiving Channel Specifications

The receiving channel specification was listed in Table 7.1.

7.7.3.1   Specifications

RF frequency range	18.8–19.3 GHz, 19.7–20.2 GHz
IF frequency range	0.95–1.45 GHz
Gain	50dB
Noise figure	2dB
Input VSWR	2 : 1
Output VSWR	2 : 1
Spurious level	−40dBc
Frequency stability versus temperature	±2 MHz
Supply voltage	±5 V
Connectors	K connectors
Operating temperature	−40 to 60°C
Storage temperature	−50 to 80°C
Humidity	95%

7.7.4   Description of the Receiving Channel

A block diagram of the receiving channel is shown in Figure 7.5.

The receiving channel consists of an RF side coupled band-pass filter, low noise amplifier, mixer, DRO, IF filter, and MMIC downconverter block. Noise figure, gain, and IP$_3$ budget are given in Figure 7.6.

A receiving channel with improved NF (0.95dB) and gain (77dB) are obtained by adding a gain block after the LNA. However, from cost consideration we decided to realize the first configuration shown in Figure 7.6. The major objectives in the design of the receiving channel were specifications and cost. The target price in the production of thousands was around $80.

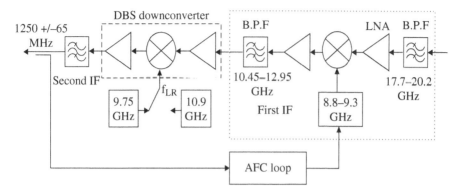

FIGURE 7.5 Block diagram of the receiving channel.

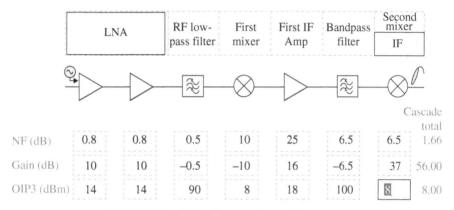

	LNA	RF low-pass filter	First mixer	First IF Amp	Bandpass filter	Second mixer IF	Cascade total
NF (dB)	0.8	0.8	0.5	10	25	6.5	6.5	1.66
Gain (dB)	10	10	−0.5	−10	16	−6.5	37	56.00
OIP3 (dBm)	14	14	90	8	18	100	8	8.00

FIGURE 7.6 Noise figure, gain, and IP₃ budget.

7.7.5 Development of the Receiving Channel

A MIC and MMIC LNA were developed for K-band VSAT applications. The dimensions of the MMIC LNA are much smaller than the MIC LNA. However, the NF of the MIC LNA is around 1.2–1.5dB, and the NF of the MMIC LNA is around 1.7–2dB. Components of the receiving channel were printed on 10 mil-thick RT–duroid 5880 substrate. Drawings of the MMIC LNA and of the receiving channel are shown in Figures 7.7 and 7.8. A photograph of the receiving channel with a MMIC LNA is shown in Figure 7.9.

7.7.6 Measured Test Results of the Receiving Channel

The receiving channel was tested and the measured test results are summarized in Table 7.7.

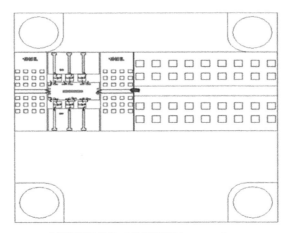

FIGURE 7.7   MMIC LNA on carrier.

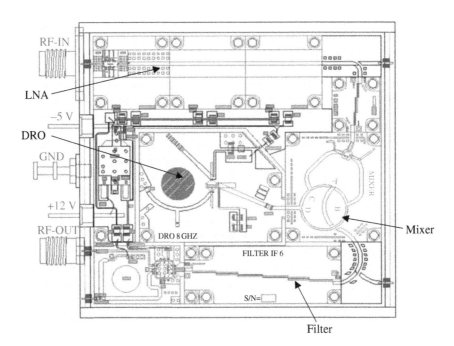

FIGURE 7.8   ODU receiving channel.

7.7.7   Conclusions

A low noise and low-cost K-band compact receiving channel for VSAT applications has been presented in this section. The receiving channel main features are NF 2dB, channel gain 50dB, frequency stability versus temperature ±2 MHz, and spurious

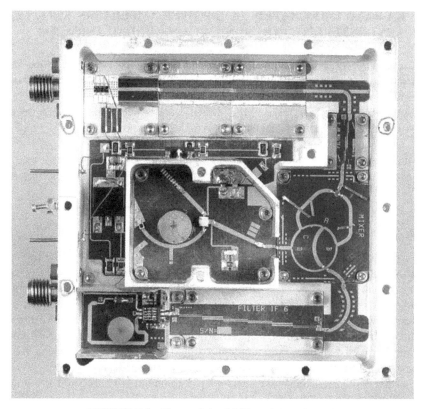

FIGURE 7.9   Photo of the ODU receiving channel.

TABLE 7.7   Receiving Channel Measured Test Results

Parameter	Measured Results
RF frequency range	18.8–19.3 GHz
IF frequency range	0.95–2.0 GHz
Gain	50 dB
Noise figure	2 dB
Input VSWR	2 : 1
Output VSWR	2 : 1
Spurious level	−40 dBc
Frequency stability versus temperature	±2 MHz

level lower than 40 dB. The major objectives in the design of the receiving channel were specifications and cost. The target price in the production of thousands is around $60–$80.

7.8   KA-BAND INTEGRATED HIGH POWER AMPLIFIERS, SSPA, FOR VSAT SATELLITE COMMUNICATION GROUND TERMINAL

This section describes the design and performance of the new compact and low-cost Ka-band power amplifiers. The main features of the power amplifiers are 27–35dBm minimum output power for −16dBm input power over the frequency range of 27.5–31 GHz. To reduce losses MIC, MMIC, and waveguide technologies are employed in the development and fabrication of this set of power amplifiers. Employing waveguide technology has minimized losses in the power combiner.

Three power amplifiers are described in this paper. The first amplifier is a 0.5 W power amplifier, the second is a 1.5 W power amplifier, and the third amplifier is a 3.2 W power amplifier.

7.8.1   Introduction

An increasing demand for wide bandwidth in communication links makes the Ka-band attractive for future commercial systems. However, only some commercial low-cost power amplifiers and low noise amplifiers are published in commercial amplifier catalogs. This section describes the design and performance of compact and low-cost Ka-band power amplifiers. The main features of the power amplifiers are 27–35dBm minimum output power for −16dBm input power over the frequency range of 27.5–31 GHz.

Communication industry is currently in continuous growth. In particular the VSAT networks have gained wide use for business and private applications. Private organizations and banks are using VSAT networks to communicate between their various sites. VSAT applications cover a wide range such us telephony, message distribution, lottery, credit card approval, and inventory management. Commercial VSAT systems operate in C-band and Ku-band; however there are many advantages to develop wideband Ka-band communication systems. Moreover, development of low-cost high power amplifiers is crucial for Ka-band satellite communication industry.

7.8.2   Power Amplifier Specifications

The transmitting channel specification was listed in Table 7.1.

Frequency range	27.5–31 GHz
Input power	−16 to −20dBm
Output power	27, 32, 35dBm min.
Input VSWR	2 : 1
Output VSWR	2 : 1
Spurious level	−60dBc
Supply voltage	±5 V
Connectors	K connectors
Operating temperature	−30 to 60°C
Storage temperature	−50 to 80°C
Humidity	100%

7.8.3   Description of the 0.5 and 1.5 W Power Amplifiers

The block diagram of the 0.5 and 1.5 W power amplifier is shown in Figure 7.10. The 0.5 W power amplifier consists of low power MMIC amplifier, band-pass filter, medium power MMIC amplifier, and 0.5 W power amplifier. The 1.5 W power amplifier consists of the same modules as the 0.5 W power amplifier. The 0.5 W MMIC amplifier is connected to a four-way microstrip power divider, a high power module with four MMIC power amplifiers, a four-way waveguide power combiner, and a DC supply unit. Three stages of MMIC power amplifiers amplify the input signal from −16 to 18dBm. The fourth stage, a 0.5 W MMIC power amplifier, is connected to a four-way hybrid ring microstrip power divider printed on 10 mil duroid. The output ports of the power divider are connected to four 0.5 W MMIC power amplifiers. The 0.5 W power amplifiers are combined via a four-way waveguide power combiner to yield a 32.5dBm minimum output power level. The DC bias voltages of each MMIC amplifier have been experimentally optimized to achieve the required gain and output power level.

7.8.4   Gain and Power Budget for the 0.5 and 1.5 W Amplifiers

The power amplifiers gain and power budget is listed in Table 7.8. The expected large signal gain of the 1.5 W amplifier is 49dB. The expected output power is 33dBm.

7.8.5   Description of the 3.2 W Power Amplifier

A block diagram of the 3.2 W power amplifier is shown in Figure 7.11. The 3.2 W power amplifier consists of the same modules as the 0.5 W power amplifier. The 0.5 W MMIC amplifier is connected to a two-way microstrip power divider, a high

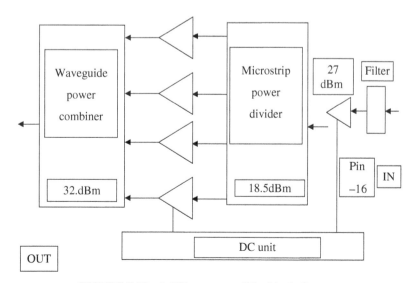

FIGURE 7.10   1.5 W power amplifier block diagram.

TABLE 7.8   Transmitter Gain and Power Budget

Component	Gain/ Loss (dB)	P_{out} (dBm) 0.5 W Amplifiers	P_{out} (dBm) 1.5 W Amplifiers
Input power	—	−16	−16
Amplifier	15	−1	−1
Filter	−1	−2	−2
Amplifier	20	18	18
Amplifier	9	27	27
Four-way power divider	−7.5	—	19.5
Amplifier	8	—	27.5
Four-way power combiner	5.5	—	33
Total	49	—	33

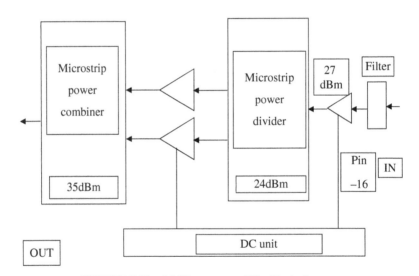

FIGURE 7.11   3.2 W power amplifier block diagram.

power module with two 2 W MMIC power amplifiers, and a two-way microstrip power combiner. Three stages of MMIC power amplifiers amplify the input signal from −16 to 18dBm. The fourth stage, a 0.5 W MMIC power amplifier, is connected to a two-way microstrip power divider printed on 5 mil alumina substrate. The output ports of the power divider are connected to two 2 W MMIC power amplifier. The 2 W output power is combined via a two-way low-loss power combiner to yield a 35dBm minimum output power level. The DC bias voltages of each MMIC amplifier have been experimentally optimized to achieve the required gain and output power level.

7.8.6 Measured Test Results

Two different 1.5 W power amplifier modules were fabricated and tested. The first power amplifier is a modular unit and consists of seven modules. The unit consists of a band-pass filter, a low power amplifier, a medium power amplifier, a four-way microstrip power divider module, four 0.5 W power amplifiers, a four-way microstrip power combiner, and a DC supply unit. The second power amplifier is an integrated power amplifier as described in Section 7.8.4. Test results of the modular and integrated power amplifier are given in Table 7.9.

The measured test results of the four-way waveguide power combiner are listed in Table 7.10.

Figure 7.12 presents the output power balance of the 0.5 W power amplifier. Figure 7.13 presents the output power and gain of the 1.5 W power amplifier. A photo of the 1.5 W power amplifier is shown in Figure 7.14. The four-way waveguide power combiner is covered by a metallic cover.

Figure 7.14 presents the low power and medium power MMIC amplifiers. A microstrip four-way power divider is used to supply the power to each of the four 0.5 W MMIC amplifiers. The output power of the 0.5 W MMIC amplifiers is combined by a four-way waveguide power combiner. A photo of the 1.5 W power amplifier with a four-way waveguide power combiner is shown in Figure 7.15.

A photo of the 1.5 W power amplifier without the four-way waveguide power combiner is shown in Figure 7.16. The first carrier in this photo is a side coupled band-pass

TABLE 7.9 Measured Test Results of the Modular and Integrated Power Amplifier

	Measured Results	
Parameter	Modular 1.5 W Amplifiers	Integrated 1.5 W Amplifiers
Frequency range (GHz)	27–31	27–31
Input power (dBm)	−16	−16
Output power (dBm)	31.5	31.5
Spurious level (dBc)	>−50	>−50
VSWR (input/output)	2 : 1	2 : 1

TABLE 7.10 Measured Test Results of the Four-Way Waveguide Power Combiner

Parameter	Measured Results
Frequency range	27–31 GHz
Insertion loss	0.5dB
Amplitude balance	0.2dB max.
Phase balance	±5° max.
VSWR (input/output)	1.5 : 1

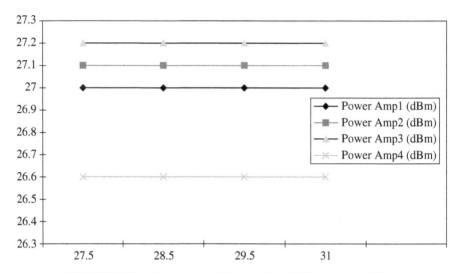

FIGURE 7.12   Output power balance of the 0.5 W power amplifier.

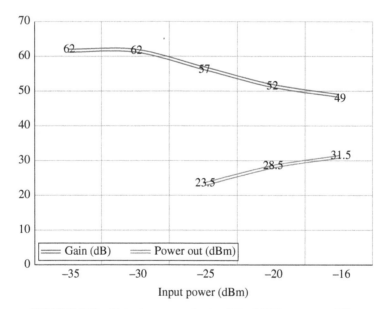

FIGURE 7.13   Output power and gain of the 1.5 W power amplifier.

filter. The second and third carrier contains medium power MMIC amplifiers. The
fourth carriers contain a 0.5 W MMIC power amplifier. The output of the 0.5 W
MMIC power amplifier is connected to a microstrip four-way power divider.

FIGURE 7.14   Photo of the 1.5 W power amplifier.

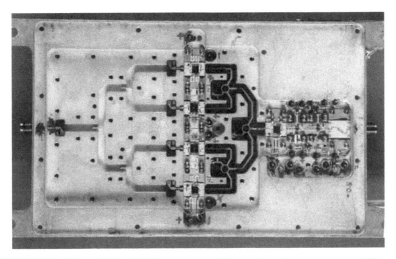

FIGURE 7.15   Photo of the 1.5 W power amplifier with a four-way waveguide power combiner.

7.9   CONCLUSIONS

In this section we presented the design and performance of compact and low-cost power amplifiers for Ka-band VSAT satellite communication ground terminal.

The main features of the amplifiers are 27–35dBm minimum output power for −16dBm input power over the frequency range of 27.5–31 GHz. The 1.5 W power amplifier size is 15.5 × 11 × 2.5 cm. To achieve low-cost design, extensive use is made of commercially available MMICs.

FIGURE 7.16   Photo of the 1.5 W power amplifier without the four-way waveguide power combiner.

MIC, MMIC, and waveguide technologies are combined in the development and fabrication of this set of power amplifiers.

REFERENCES

[1] Rogers J, Plett C. *Radio Frequency Integrated Circuit Design*. Boston: Artech House; 2003.

[2] Maluf N, Williams K. *An Introduction to Microelectromechanical System Engineering*. Boston: Artech House; 2004.

[3] Sabban A. Microstrip Antenna Arrays. In: Nasimuddin N, editor. *Microstrip Antennas*. Rijeka: InTech; 2011. p 361–384.

[4] Sabban A. Applications of MM wave microstrip antenna arrays. International Symposium on Signals, Systems and Electronics, 2007. ISSSE '07 Conference; July 30 to August 2, 2007; Montreal, QC.

[5] Gauthier GP et al. A 94 GHz micro-machined aperture-coupled microstrip antenna. IEEE Trans Antenna Propag 1999;47 (12):1761–1766.

[6] Milkov MM. Millimeter-wave imaging system based on antenna-coupled bolometer [M.Sc. thesis]. UCLA; 2000.

[7] de Lange G, Konistis K, Hu Q. A 3*3 mm-wave micro machined imaging array with sis mixers. Appl Phys Lett 1999;75 (6):868–870.

[8] Rahman A, de Lange G, Hu Q. Micromachined room temperature microbolometers for mm-wave detection. Appl Phys Lett 1996;68 (14):2020–2022.

[9] Maas SA. *Nonlinear Microwave and RF Circuits*. Boston: Artech House; 1997.

[10] Sabban A. *Low Visibility Antennas for Communication Systems*. Boca Raton: Taylor & Francis Group; 2015.

8

MIC AND MMIC INTEGRATED RF HEADS

8.1 INTEGRATED Ku-BAND AUTOMATIC TRACKING SYSTEM

The monopulse tracking principle is the most popular way to obtain accurate information about the target angular position in radar systems (see Refs. [1–9]). Tracking radars and scan radar systems use monopulse antennas in the accurate determination of angular deviation with respect to the antenna axis, off-boresight angle. Also radio communication systems that need to track a narrow beam antenna to the transmitter use monopulse tracking systems. Satellite and tracking communication systems use monopulse tracking principle to secure communication links with very narrow beam antennas. Identification military and commercial systems use this principle to obtain narrow beam information links.

Ku-band two-axis automatic tracking system may be implemented by using a four-element monopulse antenna connected to a phase comparator network. The concept is widely used in communication industry. The antenna may be a reflector antenna or an antenna array. The elements of the reflector feed may be placed at a distance of at least, $\lambda/2$, a half wavelength apart each. When the target is located along the antenna boresight, each element is equidistant from the antenna elements, and all signals are received in phase. However, when the target is off boresight, due to the difference in path length (L), a phase difference is introduced between the signals, which is used to calculate the error angle (θ). The tracking system consists of antennas, monopulse comparator, up- and downconverter, monopulse processor at baseband frequencies, and DC and control units, as shown in Figure 8.1. The development and design of

Wideband RF Technologies and Antennas in Microwave Frequencies, First Edition. Dr. Albert Sabban.
© 2016 John Wiley & Sons, Inc. Published 2016 by John Wiley & Sons, Inc.

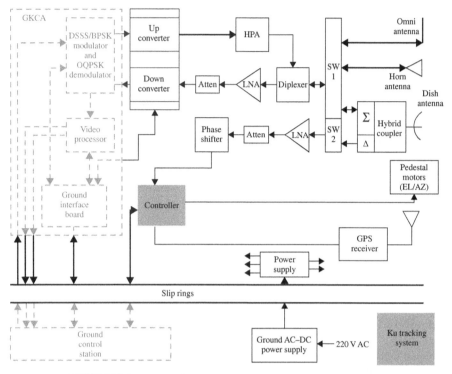

FIGURE 8.1   Ku-band automatic tracking system block diagram.

the Ku-band automatic tracking system are based on MIC, waveguide, and MMIC technology. The power divider and combiners in the power amplifier module are waveguide power combiners to minimize losses in the transmitter. Packed MMIC components are used to minimize system volume and weight.

8.1.1   Automatic Tracking System Link Budget Calculations

The downlink budget tracking system calculations are listed in Table 8.1 for three distances: 50, 100, and 200 km. The transmitting antenna gain is 20dBi.

The transmitter power is 44dBm. The receiving antenna gain is 45dBi.

The tracking system consists of three antennas: reflector, horn, and omnidirectional. The uplink tracking system budget calculations are listed in Table 8.2. The uplink budget calculations are listed in Table 8.2 for three distances: 50, 100, and 200 km. The transmitting antenna gain is 45dBi. The transmitter power is 43dBm. The receiving antenna gain is 20dBi.

The downlink budget, dish to omnidirectional channels, tracking system calculations are listed in Table 8.3. The downlink budget, dish to omnidirectional channels, calculations are listed in Table 8.3 for three distances: 70, 75, and 80 km. The transmitting antenna gain is 2.15dBi. The transmitter power is 44dBm. The receiving dish antenna gain is 45dBi.

TABLE 8.1 Downlink, DNL Link, and Budget Calculations

Parameter	Ku Band	Ku Band	Ku Band
Frequency (GHz)	15.005	15.005	15.005
Distance (km)	**50**	**100**	**200**
RX noise figure (dB)	2.0	2.0	2.0
IF bandwidth (MHz)	8	8	8
TX power (dBm)	44	44	44
TX component loss (dB)	5.5	5.5	5.5
TX antenna gain (vertical) (dBi)	20	20	20
TX pointing loss (dB)	1.5	1.5	1.5
TX random loss (dB)	0.5	0.5	0.5
TX EIRP (dBm)	56.5	56.5	56.5
Free-space loss (dB)	150	156	162
Atmospheric absorption (dB)	2.24	4.49	8.98
Precipitation absorption (dB)	0.0	0.0	0.0
Total propagation loss (dB)	152.24	160.49	170.98
RX antenna gain (dBi)	45	45	45
RX polarization loss (dB)	0.5	0.5	0.5
RX pointing loss (dB)	1.0	1.0	1.0
RX component loss (MP) (dB)	5.0	5.0	5.0
Effective carrier power at RX input (dBm)	**−57.24**	**−65.49**	**−75.98**
RX noise threshold (dBm)	103.28	103.28	103.28
Calculated C/N at RX input (dB)	46.04	37.79	27.3
Required C/N at RX input (dB)	5.2	5.2	5.2
Calculated fade margin (dB)	**40.84**	**32.59**	**22.1**

TABLE 8.2 Uplink, UPL, and Budget Calculations

Parameter	Ku Band	Ku Band	Ku Band
Frequency (GHz)	14.759	14.759	14.759
Distance (km)	**50**	**100**	**200**
RX noise figure (dB)	3.0	3.0	3.0
IF bandwidth (MHz)	16	16	16
TX power (dBm)	43	43	43
TX component loss (dB)	6.0	6.0	6.0
TX antenna gain (dBi)	45	45	45
TX pointing loss (dB)	1.5	1.5	1.5
TX random loss (dB)	0.5	0.5	0.5
TX EIRP (dBm)	80	80	80
Free-space loss (dB)	150	156	162
Atmospheric absorption (dB)	2.14	4.27	8.55
Precipitation absorption (dB)	0.0	0.0	0.0
Total propagation loss (dB)	152.14	160.77	170.55
RX antenna gain (vertical) (dBi)	20	20	20
RX polarization loss (dB)	0.5	0.5	0.5
RX pointing loss (dB)	0.5	0.5	0.5
RX component loss (dB)	8.3	8.3	8.3
RX PG (dB)	16.9	16.9	16.9
Effective carrier power at RX input (dBm)	**−44.54**	**−53.17**	**−62.95**
RX noise threshold (dBm)	99.28	99.28	99.28
Calculated C/N at RX input (dB)	54.74	46.11	36.33
Required C/N at RX input (dB)	8.5	8.5	8.5
Calculated fade margin (dB)	**46.24**	**37.61**	**27.83**

TABLE 8.3   Downlink, DNL, and Budget Calculations: Dish–Omni

Parameter	Ku Band	Ku Band	Ku Band
Frequency (GHz)	15.005	15.005	15.005
Distance (km)	**70**	**75**	**80**
RX noise figure (dB)	2.0	2.0	2.0
IF bandwidth (MHz)	8	8	8
TX power (dBm)	44	44	44
TX component loss (dB)	5.5	5.5	5.5
TX antenna gain (vertical) (dBi)	2.15	2.15	2.15
TX pointing loss (dB)	0.0	0.0	0.0
TX random loss (dB)	0.0	0.0	0.0
TX EIRP (dBm)	40.65	40.65	40.65
Free-space loss (dB)	152.9	153.5	154.0
Atmospheric absorption (dB)	2.07	2.22	2.37
Precipitation absorption (dB)	0.0	0.0	0.0
Total propagation loss (dB)	154.97	155.72	156.37
RX antenna gain (dBi)	45	45	45
RX polarization loss (dB)	0.5	0.5	0.5
RX pointing loss (dB)	1.0	1.0	1.0
RX component loss (MP) (dB)	5.0	5.0	5.0
Effective carrier power at RX Input (dBm)	**−75.82**	**−76.57**	**−77.22**
RX noise threshold (dBm)	103.28	103.28	103.28
Calculated C/N at RX input (dB)	27.46	26.71	26.06
Required C/N at RX input (dB)	5.2	5.2	5.2
Calculated fade margin (dB)	**22.26**	**21.51**	**20.86**

The downlink budget, horn to omnidirectional channels, tracking system calculations are listed in Table 8.4. The downlink budget, omnidirectional to horn antenna channels, calculations are listed in Table 8.4 for three distances: 2, 5, and 10 km. The transmitting omnidirectional antenna gain is 2.15dBi. The transmitter power is 44dBm. The receiving horn antenna gain is 14dBi.

The downlink budget, omnidirectional to omnidirectional, tracking system calculations are listed in Table 8.5. The downlink budget, omnidirectional to omnidirectional antenna channels, calculations are listed in Table 8.5 for three distances: 1, 2, and 3 km. The transmitting omnidirectional antenna gain is 2.15dBi. The transmitter power is 44dBm. The receiving omnidirectional antenna gain is 2.15dBi.

8.1.2   Ku-Band Tracking System Antennas

8.1.2.1   Monopulse Parabolic Reflector Antenna   The **parabolic reflector** antenna [1] consists of a radiating feed that is used to illuminate a reflector that is curved in the form of an accurate parabola with diameter D as presented in Figure 8.2. This shape enables a very beam to be obtained. To provide the optimum illumination of the reflecting surface, the level of the parabola illumination should be greater by 10dB in the center than that at the parabola edges. The parabolic reflector antenna gain

TABLE 8.4 Downlink, DNL Link, and Link Budget Calculations: Horn–Omni

Parameter	Ku Band	Ku Band	Ku Band
Frequency (GHz)	15.005	15.005	15.005
Distance (km)	**3**	**5**	**10**
RX noise figure (dB)	2.0	2.0	2.0
IF bandwidth (MHz)	8	8	8
TX power (dBm)	44	44	44
TX component loss (dB)	5.5	5.5	5.5
TX antenna gain (vertical) (dBi)	2.15	2.15	2.15
TX pointing loss (dB)	0.0	0.0	0.0
TX random loss (dB)	0.0	0.0	0.0
TX EIRP (dBm)	40.65	40.65	40.65
Free-space loss (dB)	125.5	129.9	136.0
Atmospheric absorption (dB)	0.13	0.22	0.45
Precipitation absorption (dB)	0.0	0.0	0.0
Total propagation loss (dB)	125.63	130.12	136.45
RX antenna gain(1) (dBi)	14	14	14
RX polarization loss (dB)	0.5	0.5	0.5
RX pointing loss (dB)	0.5	0.5	0.5
RX component loss (MP) (dB)	5.0	5.0	5.0
Effective carrier power at RX input (dBm)	**−76.98**	**−81.47**	**−87.8**
RX noise threshold (dBm)	103.28	103.28	103.28
Calculated C/N at RX input (dB)	26.3	21.81	15.48
Required C/N at RX input (dB)	5.2	5.2	5.2
Calculated fade margin (dB)	**21.1**	**16.61**	**10.28**

TABLE 8.5 Downlink and DNL Budget Calculations: Omni–Omni

Parameter	Ku Band	Ku Band	KU Band
Frequency (GHz)	15.005	15.005	15.005
Distance (km)	**1**	**2**	**3**
RX noise figure (dB)	2.0	2.0	2.0
IF bandwidth (MHz)	8	8	8
TX power (dBm)	44	44	44
TX component loss (dB)	5.5	5.5	5.5
TX antenna gain (vertical) (dBi)	2.15	2.15	2.15
TX pointing loss (dB)	0.0	0.0	0.0
TX random loss (dB)	0.0	0.0	0.0
TX EIRP (dBm)	40.65	40.65	40.65
Free-space loss (dB)	115.964	122.0	125.5
Atmospheric absorption (dB)	0.0449	0.09	0.13
Precipitation absorption (dB)	0.0	0.0	0.0
Total propagation loss (dB)	116.0	122.09	125.63
RX antenna gain (dBi)	2.15	2.15	2.15
RX polarization loss (dB)	0.5	0.5	0.5
RX pointing loss (dB)	0.0	0.0	0.0
RX component loss (MP) (dB)	5.0	5.0	5.0
Effective carrier power at RX input (dBm)	**−78.7**	**−84.79**	**−88.33**
RX noise threshold (dBm)	103.28	103.28	103.28
Calculated C/N at RX input (dB)	24.58	18.49	14.95
Required C/N at RX input (dB)	5.2	5.2	5.2
Calculated fade margin (dB)	**19.38**	**13.29**	**9.75**

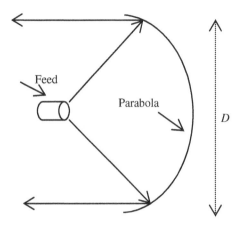

FIGURE 8.2   Parabolic antenna.

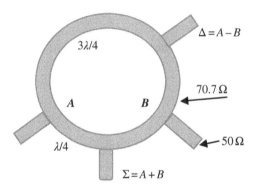

FIGURE 8.3   Rat-race coupler.

may be calculated by using Equation 8.1, where α is the parabolic reflector antenna efficiency.

Parabolic Reflector Antenna Gain

$$G \cong 10\log_{10}\left(\alpha\frac{(\pi D)^2}{\lambda^2}\right) \tag{8.1}$$

Reflector antenna specifications

Frequency: 14.5–15.3 GHz

Gain: 45dBi

Beamwidth: 0.8–0.9°

Dish diameter: 1.6 m

Monopulse Comparator: Rat-Race Coupler   A rat-race coupler is shown in Figure 8.3. The rat-race circumference is 1.5 wavelengths. The distance from A to Δ port is $3\lambda/4$. The distance from A to $\sum$ port is $\lambda/4$. For an equal-split rat-race coupler,

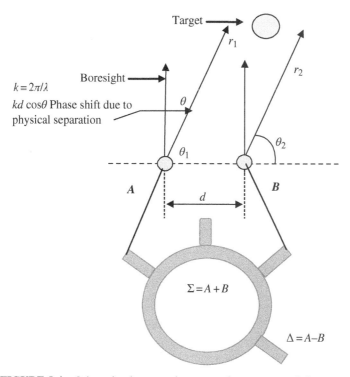

FIGURE 8.4   Orientation between the monopulse antenna and the target.

the impedance of the entire ring is fixed at $1.41 \times Z_0$ or $70.7\,\Omega$ for $Z_0 = 50\,\Omega$. For an input signal V, the outputs at ports 2 and 4 are equal in magnitude, but $180°$ out of phase.

Figure 8.4 presents the orientation between the monopulse antenna and the target. The distance between the two elements is d. A wavefront is incident at an angle θ. The phase difference between the two antennas is a phase $\Delta\Phi$. The angle θ may be calculated by using Equation 8.2:

$$\theta = \sin^{-1}\left(\frac{\lambda\Delta\Phi}{2\pi d}\right) \tag{8.2}$$

Ku-Band Monopulse Reflector Antenna Feed   A monopulse double-layer antenna was designed at 15 GHz. The monopulse double-layer antenna consists of four circular patch antenna as shown in Figure 8.5. The resonator and the feed network were printed on a substrate with relative dielectric constant of 2.5 with thickness of 0.8 mm. The resonator is a circular microstrip resonator with diameter, $a = 4.2$ mm. The radiating element was printed on a substrate with relative dielectric constant of 2.2 with thickness of 0.8 mm. The radiating element is a circular microstrip patch with diameter, $a = 4.5$ mm. The four circular patch antennas are connected to three 3dB $180°$ rat-race couplers via the antenna feed lines, as shown in Figure 8.5. The

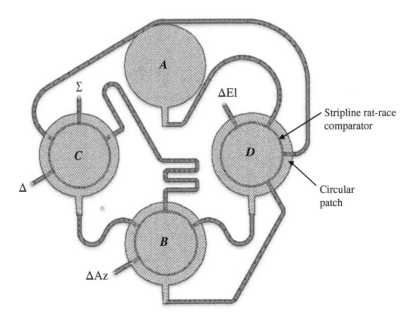

FIGURE 8.5 A microstrip stacked monopulse antenna.

comparator consists of three stripline 3dB 180° rat-race couplers printed on a substrate with relative dielectric constant of 2.2 with thickness of 0.8 mm. The comparator has four output ports—a sum port $\sum$, difference port Δ, elevation difference port ΔEl, and azimuth difference port ΔAz—as shown in Figure 8.5. The antenna bandwidth is 10% for VSWR better than 2 : 1. The antenna beamwidth is around 36°. The measured antenna gain is around 10dBi.

Monopulse comparator specifications

Frequency: 14.5–15.3 GHz
Insertion loss: 0.6dB
VSWR: 1.3 : 1

The comparator losses around 0.7dB. The elevation error in the tracking system is calculated from the difference signal ΔEl and the azimuth error from the difference signal ΔAz. The resulting local minimum in the Δ port at the center of the boresight is very deep, more than -20dB, as shown in Figure 8.6. A high angular accuracy in the tracking process is achieved by comparing the sum and difference signals. A unique tracking algorithm is implemented inside the monopulse processor.

Reflector antenna performance

Frequency: 14.5–15.3 GHz
Gain: 45dBi

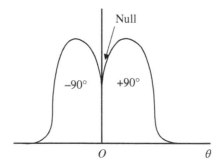

FIGURE 8.6   Monopulse antenna radiation pattern at the Δ port.

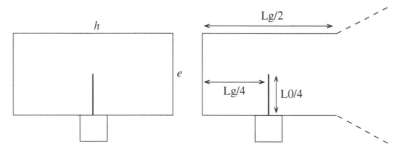

FIGURE 8.7   Ku-band horn antenna feed.

Beamwidth: 0.9°
Dish diameter: 1.6 m

8.1.2.2   Horn Antenna

Horn antenna specifications

Frequency: 14.5–15.3 GHz
Gain: 14 dBi
Dimensions: $160.5 \times 67.6 \times 42.5$ mm

The horn antenna is fed by a coaxial connector as shown in Figure 8.7. The transition from coax to waveguide is presented in Figure 8.7. The Ku-band horn antenna assembly is presented in Figure 8.8. Ku-band horn antenna fabrication drawing is shown in Figure 8.9. The dimensions in inches of a Ku-band waveguide are listed in Table 8.6. The horn antenna dimensions are $160.5 \times 67.6 \times 42.5$ mm. The horn antenna gain is around 14 dBi.

8.1.2.3   Omnidirectional Antenna   The omni antenna is a quarter wavelength monopole antenna. The antenna length is around 5 mm. The monopole ground plane diameter is 120 mm.

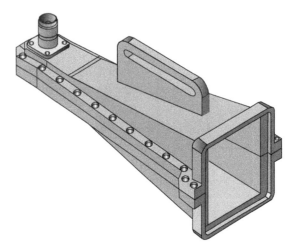

FIGURE 8.8   Ku-band horn antenna assembly.

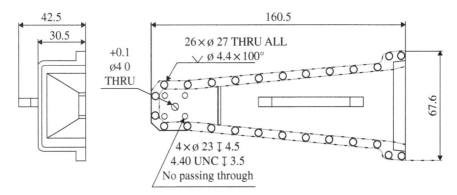

FIGURE 8.9   Ku-band horn antenna fabrication drawing.

TABLE 8.6   Ku-Band Waveguide Dimensions in Inches

WR Size	AMC Model	Frequency (GHz)	Material Type	Inside Dimension	Outside Dimension	Wall Size	Cover Flange
62	207	12.4–18.0	6061 AI	0.622×0.311	0.702×0.391	0.04	UG1665/U

Omnidirectional antenna specifications

Frequency: 14.5–15.3 GHz
Gain: 2dBi
Dimensions: Ø 12 mm × 120 mm

8.1.3   Monopulse Processor

A block diagram of the tracking monopulse system is shown in Figure 8.10.

The tracking monopulse system consists of antennas, low-noise amplifier (LNA), phase comparator array, and a controller. The block diagram of the tracking controller system is shown in Figure 8.11. A detailed block diagram of the monopulse tracking unit is shown in Figure 8.12. The sum, elevation difference, and azimuth difference signals are amplified by an LNA and downconverted by a mixer to 2.4 GHz. The signals are amplified and downconverted to DC and up to 20 MHz. The downconverted signal is fed to a digital processing unit. The digital processing unit supplies the tracking data to the tracking controller shown in Figure 8.11. The monopulse tracking unit (Fig. 8.12) consists of sum and difference channels. Each channel consists of amplifier, filter, IF downconverter mixer, IF amplifier and filter, and downconverter mixer to baseband and filter. The baseband signal from the sum and difference channels is connected to a digital processing unit that provides the tracking commands to the tracking controller.

Pointing system specifications

Elevation range: 0–50°

Azimuth range: continuous in either direction, 360° continuous

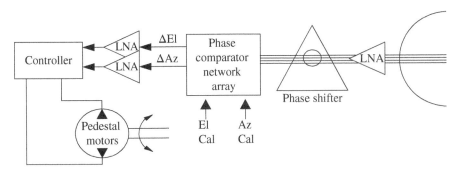

FIGURE 8.10   Monopulse processor.

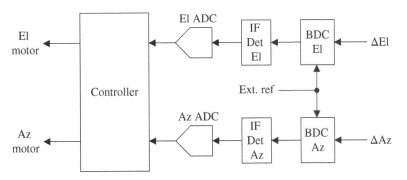

FIGURE 8.11   Tracking system controller.

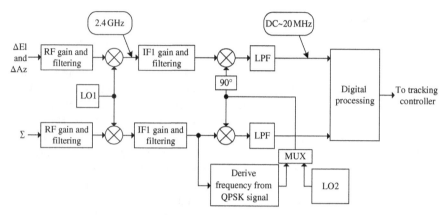

FIGURE 8.12   Monopulse tracking unit.

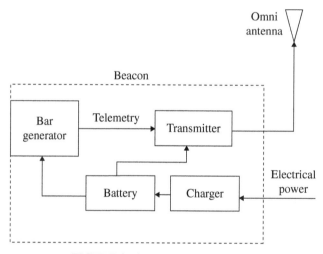

FIGURE 8.13   Omnidirectional link.

Accuracy: ±0.1° azimuth, elevation

Position repeatability: ±0.1°

Position step resolution: continuous

Platform dynamics

Pitch: ±5°/s

Tangential acceleration: ±0.5 g

Turning: 45°/s and 3°/s^2

A block diagram of the omnidirectional link is shown in Figure 8.13. The omnidirectional link consists of a transmitter, an omnidirectional antenna, and an electrical power supply unit.

TABLE 8.7 HPA Specifications

Parameter	Value	Tolerance
TX frequency	14.5–14.8 GHz	
1dB compression point	47dBm	Min. under all conditions
TX gain	80dBm	Min. under all conditions
TX gain adjustment range	+6.0 to −20dB	
TX level flatness	±0.8dB	Over any 36 MHz BW
TX gain stability	±0.8dB	Overall temperature and frequency
TX linearity	−53dBc	2 carriers at 6dB back-off
Input impedance	50 Ω	
Input connector	SMA	
Input VSWR	01 : 01.2	Nominal
Input NF	6dB	Maximum
Output impedance	50 Ω	
Output connector	WR75	W/gasket
Output VSWR	01 : 01.2	Nominal
Output port protection	Open/short internally protected	
Visual indicators	Green LED: power on	
	Red LED: summary alarm	
BIT indications	FWD VSWR	
	REV VSWR	
	High voltage	
	Low voltage	
	AGC SAT (low RF)	
	High temp.	
Power	28VDC/TDB Amp	Not to exceed
Temperature	Operational: −40°C to +55°C	
	Storage: −60°C to +75°C	
ODU	IP65	
Vibration	1 g random	Operational
Shock	10 g	Operational
Weight	5 kg	

8.1.4 High Power Amplifier

The desired output power of 47dBm may be achieved by combining nine 6.5 W packed MMIC power amplifier modules. The HPA specifications are listed in Table 8.7.

8.1.4.1 HPA Design Based on a 6.5 W Ku-Band Power Amplifier

Basic power amplifier module performance

6.5 W Ku-band power amplifier
Frequency range: 13–16 GHz

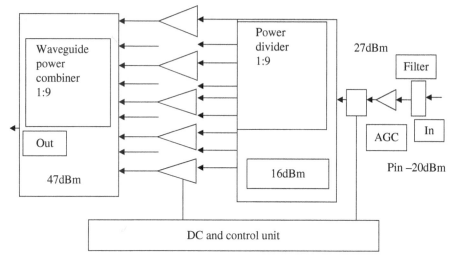

FIGURE 8.14   Power amplifier block diagram with 9 × 6.5 W power modules.

38dBm nominal Psat
24dB nominal gain
14dB nominal return loss
0.25 μm PHEMT MMIC technology
10 lead flange package
Bias conditions: 8 V at 2.6 A Idq.
Package dimension: 0.45 × 0.68 × 0.12 in.

HPA DC power

DC voltage: 8 V
Current: 35 A
Total power: 300 W max.

A block diagram of a power amplifier with nine 6.5 W power modules is presented in Figure 8.14. The −20dBm input power is amplified from around 50dB medium power module to 27dBm. The output of the 27dBm amplifier is connected to a printed nine-way power divider.

The output ports of the power divider are connected to nine 6.5 W modules with 23 dB gain. The output power of each 6.5 W module is around 38dBm.

The nine 6.5 W modules are combined by a waveguide combiner. The output power of the HPA is 47dBm. Transmitter gain and power budget are listed in Table 8.8.

TABLE 8.8 Transmitter Gain and Power Budget

Component	Gain/Loss (dB)	Pout (dBm) Amplifier
Input power min.		−20
Amplifier	19–24	−1
Filter	−1	−2
Amplifier	20	18
Amplifier	9	27
Nine-way power divider	−10.5	16.5
Amplifier	23	38
1 : 9 power combiner	10	47
Total with AGC	70–76	

FIGURE 8.15   Packed MMIC 6.5 W power module.

Packed MMIC 6.5 W power module is shown in Figure 8.15. HPA DC power tree is shown in Figure 8.16. The expected results of the 1 : 9 waveguide power combiner are listed in Table 8.9.

8.1.4.2   HPA Design with 41.5dBm Power Modules   The desired output power of 47dBm may be achieved by combining four 14 W packed MMIC power amplifier modules. The HPA specifications are listed in Table 8.7.

Basic 42dBm power amplifier module performance

f = 13.75–14.5 GHz

Output power at 1dB gain compression point: P1dB 41.5–42.0dBm

Power gain at 1dB gain compression point: G1dB 5.0–6.0dB

Drain current IDS1: 5.5–6.0 A

Power-added efficiency: VDS = 9 V IDSQ = 4.4 A 28%

Third-order intermodulation distortion: IM_3 −25dBc

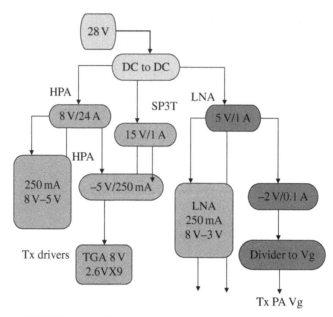

FIGURE 8.16   HPA DC power tree with 9 × 6.5 W modules.

TABLE 8.9   Expected Results of the 1 : 9 Waveguide Power Combiner

Parameter	Measured Results
Frequency range	14.5–15.3 GHz
Insertion loss	0.8dB
Amplitude balance	0.4dB max.
Phase balance	+5° max.
VSWR (input/output)	1.5 : 1

A block diagram of a power amplifier with four 14 W power modules is presented in Figure 8.17. The −20dBm input power is amplified from 50dB medium power module to 27dBm. The output of the 27dBm amplifier is connected to a printed four-way power divider. The output ports of the power divider are connected to four 6.5 W modules with 23dB gain. The output power of each 6.5 W module is around 38dBm.

The four 6.5 W modules are combined to four 14 W power modules. The four 14 W modules are combined by a waveguide combiner. The output power of the HPA is 47dBm. Transmitter gain and power budget are listed in Table 8.10. HPA DC power tree of the 4 × 14 W modules is presented in Figure 8.18.

HPA DC power

DC voltage: 9 V
Current: 35 A
Total power: 315 W max.

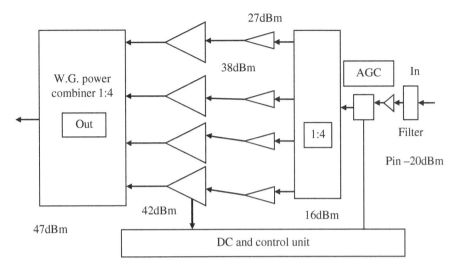

FIGURE 8.17 Power amplifier block diagram with 4×14 W power modules.

TABLE 8.10 Gain and Power budget for HPA with Four-Way Power Combiner

Component	Gain/Loss (dB)	Pout (dBm) Amplifier
Input power min.		−20
Amplifier	19–24	−1
Filter	−1	−2
Amplifier	20	18
Amplifier	9	27
Four-way power divider	−7	20
Amplifier	23	38
Amplifier	5	43
Four-way power combiner	5	48
Total with AGC	70–76	

Transmitter gain and power budget are listed in Table 8.10. The waveguide power divider and combiner of the power amplifier with 9×6.5 W power modules are shown in Figure 8.19.

The waveguide power divider and combiner of the power amplifier with 4×14 W power modules are shown in Figure 8.20.

8.1.5 Tracking System Downconverter/Upconverter

The design of the down- and upconverter unit is presented in this section.

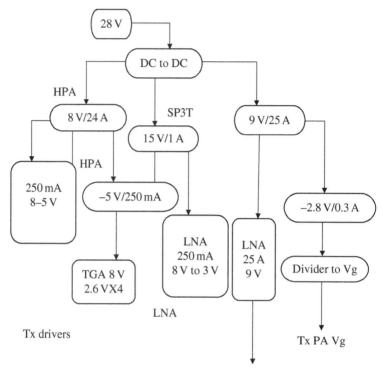

FIGURE 8.18   HPA DC power tree with 4×14 W modules.

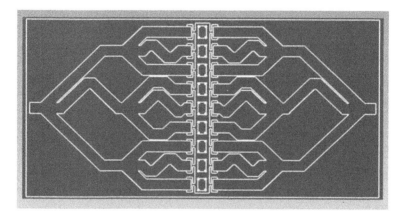

FIGURE 8.19   Power amplifier with 9×6.5 W power modules.

8.1.5.1   Tracking System Upconverter Design   The upconverter consists of an RF amp with moderate NF to provide low overall system NF, high IP_3 double-balanced mixer with 20dBm local oscillator drive to provide linearity, medium power

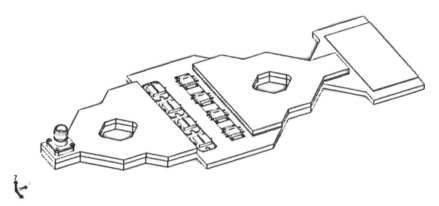

FIGURE 8.20 Power amplifier with 4 × 14 W power modules.

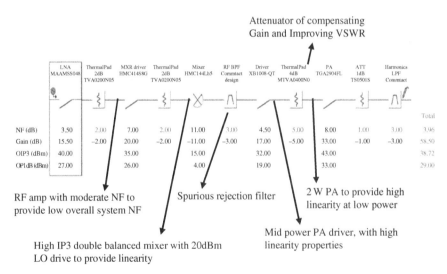

Attenuator of compensating
Gain and Improving VSWR

	LNA MAAMSS048	ThermalPad 2dB TVA0200N05	MXR driver HMC414S8G	ThermalPad 2dB TVA0200N05	Mixer HMC144Lh5	RF BPF Commtact design	Driver XB1008-QT	ThermalPad 4dB MTVA0400N0	PA TGA2904FL	ATT 1dB TS0501S	Harmonics LPF Commtact	Total
NF (dB)	3.50	2.00	7.00	2.00	11.00	3.00	4.50	5.00	8.00	1.00	3.00	3.96
Gain (dB)	15.50	−2.00	20.00	−2.00	−11.00	−3.00	17.00	−5.00	33.00	−1.00	−3.00	58.50
OIP3 (dBm)	40.00		35.00		15.00		32.00		43.00			38.72
OP1dB (dBm)	27.00		26.00		4.00		19.00		33.00			29.00

RF amp with moderate NF to
provide low overall system NF

Spurious rejection filter

2 W PA to provide high
linearity at low power

High IP3 double balanced mixer with 20dBm
LO drive to provide linearity

Mid power PA driver, with high
linearity properties

FIGURE 8.21 Block diagram of the upconverter.

amplifiers, and 2 W power amplifier with high linearity and band-pass filters. A block diagram of the upconverter is shown in Figure 8.21.

Upconverter specifications are listed in Table 8.11.

The block diagram of the upconverter approves that the upconverter gain is around 58dB and the noise figure is around 4dB.

TABLE 8.11 Upconverter Specifications

Parameter	Nominal	Delta
Gain (dB)	58.5	±4.5
NF (dB)	4	±0.8
OP1 (dBm)	29	±1.1
ORR3 at −41 (dBm) input (dBc)	42.5	—

TABLE 8.12 Upconverter Performance

Parameter	Value	Tolerance
IF input frequency	2.38–2.48 GHz	
RF output frequency	14.5–14.8 GHz	
1dB compression point	21dBm	Min. under all conditions
BUC gain	54dBm	Min. under all conditions
BUC level flatness	±0.8dB	Over any 36 MHz BW
BUC gain stability	±0.8dB	Overall temperature and frequency
BUC linearity	−53dBc	2 carriers 300 K apart at 6dB back-off
Input impedance	50 Ω	
Input connector	SMA	
Input VSWR	01 : 01.2	Nominal
Input NF	3dB	Maximum
Output impedance	50 Ω	
Output connector	SMA	W/gasket
Output VSWR	01 : 01.2	Nominal
Visual indicators	Green LED: power on	
	Red LED: summery alarm	
BIT indications	High voltage	
	Low voltage	
	AGC SAT (low RF)	
	High temp.	
Power	28 VDC/TDB Amp	
Temperature	Operational: −40°C to +55°C	
	Storage: −60°C to +75°C	
Vibration	1 g random	Operational
Shock	10 g	Operational
Weight	0.85 kg	

Upconverter performance is listed in Table 8.12. The upconverter gain is around 58.5dB, and the noise figure is around 4dB. Upconverter power consumption is listed in Table 8.13.

8.1.5.2 Tracking System Downconverter Design   The downconverter consists of an RF amplifier with moderate NF to provide low overall system NF, high IP$_3$ double-balanced image reject mixer with 20dBm local oscillator drive to provide linearity,

TABLE 8.13   Upconverter Power Consumption

Component	Voltage (V)	Current (mA)	Power (W)
IF LNA	5	180	0.9
Mixer driver	5	500	2.5
PA driver	5	100	0.5
PA driver	−0.5	10	0.005
PA	7	700	4.9
PA	−0.57	50	0.0285
	Total	1540	8.8335

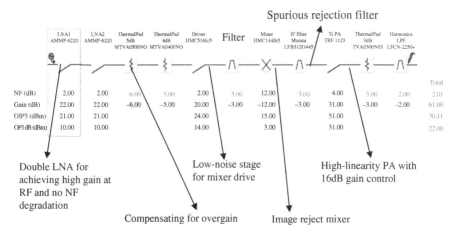

FIGURE 8.22   Block diagram of the downconverter.

and medium power amplifier with high linearity and band-pass filters. A block diagram of the downconverter is shown in Figure 8.22.

Downconverter specifications are listed in Table 8.14.

The block diagram of the downconverter approves that the downconverter gain is around 61 dB and the noise figure is around 2 dB. Downconverter performance is listed in Table 8.15. The upconverter gain is around 61 dB and the noise figure is around 2 dB. Upconverter power consumption is listed in Table 8.16.

8.1.6   Tracking System Interface

A block diagram of tracking system interface is shown in Figure 8.23. The block diagram of the HPA interface is shown in Figure 8.24. Ku-band automatic tracking system assembly is shown in Figure 8.25.

TABLE 8.14   Downconverter Specifications

Parameter	Value	Tolerance
IF output frequency	2.38–2.48 GHz	
RF input frequency	15–15.3 GHz	
1dB compression point	2dBm	Min. under all conditions
BDC gain	58dBm	Min. under all conditions
BDC level flatness	±0.8dB	Over any 36 MHz BW
BDC gain stability	±0.8dB	Overall temperature and frequency
BDC linearity	−53dBc	2 carriers 300 K apart at 6dB back-off
Input impedance	50 Ω	
Input connector	SMA	
Input VSWR	01 : 01.2	Nominal
Input N.F.	2.4dB	Maximum
Output impedance	50 Ω	
Output connector	SMA	W/gasket
Output VSWR	01 : 01.2	Nominal
Visual indicators	Green LED: power on Red LED	
BIT indications	High voltage Low voltage AGC SAT (low RF)	
Power	28 VDC/TDB Amp	Not to exceed
Temperature	Operational: −40°C to +55°C Storage: −60°C to +75°C	
Vibration	1 g random	Operational
Shock	10 g	Operational
Weight	0.45 kg	
LO phase noise	SSB phase noise offset: 10 Hz −35dBc 100 Hz −65dBc 1 kHz −77dBc 10 kHz −85dBc 100 kHz −95dBc 1 MHz −110dBc	
Reference	10 MHz Internal	1.2 ppm TCXO

TABLE 8.15   Downconverter Performance

Parameter	Nominal	Delta
Gain (dB)	61	±3.7
NF (dB)	2	±0.6
OP1 (dBm)	22	±2
ORR3 at −62dBm input (dBc)	62.53	—

TABLE 8.16 Downconverter Power Consumption

Component	Voltage (V)	Current (mA)	Power (W)
LNA1	3	55	0.165
LNA2	3	55	0.165
MXR driver	3	65	0.195
IF PA	7	600	4.2
IF PA	5	25	0.125
IF PA	−5	15	0.075
	Total	815	4.925

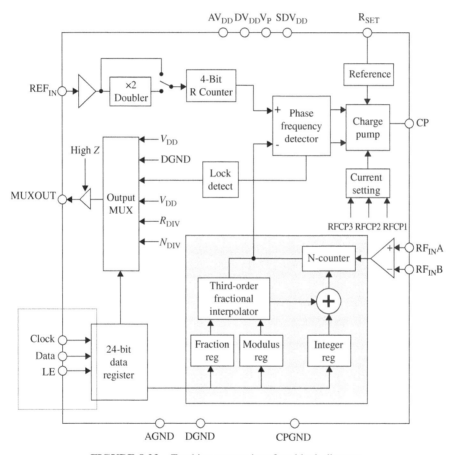

FIGURE 8.23   Tracking system interface block diagram.

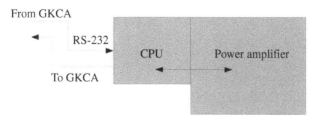

FIGURE 8.24   HPA interface block diagram.

FIGURE 8.25   Ku-band automatic tracking system.

Control

LO frequency set: SPI or RS-232 (both available)
SPI standard protocol
RS-232 standard protocol
Indications (via RS-232):
Overvoltage
Undervoltage
Lock (synthesizer)

Tracking system interface with HPA

Controls (via RS-232):
On/off power
Indications (via RS-232):
High temperature
Overvoltage
Undervoltage
VSWR (FWD, REV)
AGC status

8.2   SUPER COMPACT X-BAND MONOPULSE TRANSCEIVER

The RF head goal is to receive and transmit X-band RF signals via monopulse micro-strip antenna. The RF signal is downconverted to 60 MHz. The RF head may be employed in combined electromagnetic and optical seekers.

8.2.1   RF Head Description

The RF head block diagram is shown in Figure 8.26. The RF head has five input ports: sum, azimuth difference and elevation difference, guard, and BIT. The transmitting channel can transmit 60 W CW signal or 400 W pulse peak power. The sum RF signal, the transmitting channel, is routed through a circulator limiter and a digital attenuator with 0–20dB attenuation level. The output port of the digital attenuator is connected to a very-low-noise amplifier with 26dB gain and 1.2dB noise figure. The sum RF signal is routed to the image-rejected mixer via a 10dB digital attenuator (0 or 10dB) and a band-pass filter. The difference (ΔAz, ΔEl) antenna output ports are connected via a limiter to an SP3T switch. The difference (ΔAz, ΔEl) channels are identical to the sum channel. The same RF components are used in the sum and difference channels. The guard input port is also connected to the SP3T switch via limiter. Downconversion to IF (60 MHz) is performed by 10.64 GHz LO signal. A power divider is used to supply the local oscillator signal to the IRM mixers of the transmitting and receiving chan-nels. An RF built-in-test (BIT) signal is injected to the RF head for BIT and calibration purposes.

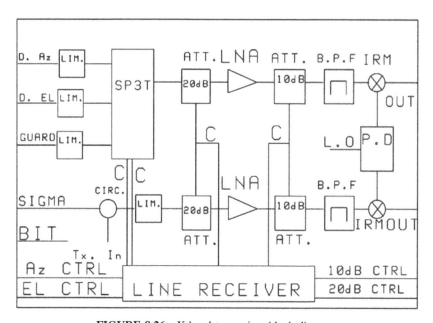

FIGURE 8.26   X-band transceiver block diagram.

8.2.2 RF Head Specifications

RF head specifications are listed in Table 8.17. The difference channel (ΔAz, ΔEl) gain should be around 10dB and the difference channel noise figure should be around 5dB. The sum channel gain should be 11dB, and the sum channel noise figure should be around 5dB.

TABLE 8.17 RF Head Specifications

Parameter	Specification
1. Frequencies	
RF frequency	10,550–10,850 MHz
IF frequency	60 MHz
2. TX input	
Peak power	400 W max.
Average power	60 W max.
PRF	2–220 kHz
Pulse width	0.1–15 μs
Duty factor	15% max.
3. Noise figure NF at +25°C	
Σ channel	NF = 5dB
Δ channel	NF = 5dB
Guard channel	6.0dB max.
4. RF to IF gain	At 25°C at 10,700 MHz
Σ channel	10–12dB
Δ channel	8.5–11.5dB
Gain change versus frequency at 25°C	2dB max.
Gain change versus temp.	3dB max.
RF to IF gain guard	8.5–11.5dB
Amplitude matching	+0.5dB
Phase matching	+8°
5. Isolation	
Sum to Δ	20dB min.
Between Δ	20dB min.
Guard to Σ	30.0dB max.
6. 1 Db C.P.	25°C at 0dB att.
Σ channel	−8.0dBm min.
Δ channel	−9.5dBm min.
25°C at 30dB att.	+12dBm min.
7. IP$_3$	25°C at 0dB att.
Σ channel	−6dBm min.
Δ channel	−7.5dBm min.
At 30dB att.	+12.5dBm min.
8. Recovery time	
Recovery time	200 ns
9. VSWR	
IF port VSWR	1.5 : 1 max.
RF guard VSWR	1.5 : 1 max.

TABLE 8.17   (*Continued*)

Parameter	Specification
Bit VSWR	2.0 : 1 max.
TX VSWR(+750 MHz)	1.6 : 1 max.
LO input	1.6 : 1 max.
Δ and Σ channel	1.8 : 1 max.
10. Circulator isolation	
Circulator isolation	19dB min.
11. Channel power handling	Limiter requirement
10,550–10,850 MHz	
Pulse width	15 μs max.
Duty factor	15% max.
Av. power	5 W max.
Peak power	40 W max.
2,000–18,000 MHz	
CW power	1 W max.
12. switches and attenuators	
Rise/fall time	50 ns
Switching speed	100 ns
13. Attenuator value	
10dB att.	+2.5dB
20dB att.	+2.5dB
14. Image rejection	20dB min.
15. LO input	
Frequency 10,490–10,790 MHz	
Power level	10–15dBm
16. Frequency response, fixed LO	Filter requirement
−3dB points	10.7 + 0.75 GHz
Rejection at +1200 MHz	20dB min.
Rejection at +2400 MHz	30dB min.
17. Switch on time	300 ns max.
18. Power consumption	12 W

8.2.3   RF Head Design

Gain and noise figure budget at the output of the difference channels (ΔAz, ΔEl) is presented in Figure 8.27. The difference channel (ΔAz, ΔEl) gain is around 10.5dB and the difference channel noise figure is around 4.2dB.

Gain and noise figure at the output of the sum channel is presented in Figure 8.28. The sum channel gain is around 11.1dB, and the sum channel noise figure is around 4.5dB.

The compact X-band RF head photo and component arrangement is presented in Figure 8.29. The development and design of the X-band RF head are based on MIC and MMIC technology. The RF head modules are connected by short semirigid coaxial cables to minimize losses in the RF head receiving and transmitting channels. Packed MMIC components are used to minimize system volume and weight. The test

System1

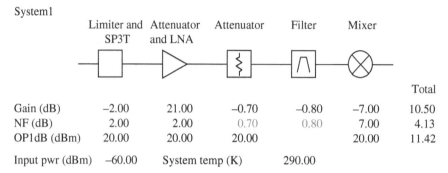

	Limiter and SP3T	Attenuator and LNA	Attenuator	Filter	Mixer	Total
Gain (dB)	−2.00	21.00	−0.70	−0.80	−7.00	10.50
NF (dB)	2.00	2.00	0.70	0.80	7.00	4.13
OP1dB (dBm)	20.00	20.00	20.00		20.00	11.42

Input pwr (dBm)   −60.00   System temp (K)   290.00

FIGURE 8.27   Gain and noise figure budget at the output of the difference channels.

System1

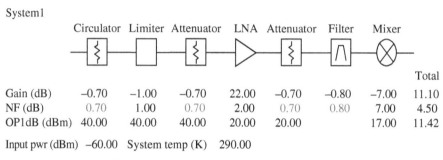

	Circulator	Limiter	Attenuator	LNA	Attenuator	Filter	Mixer	Total
Gain (dB)	−0.70	−1.00	−0.70	22.00	−0.70	−0.80	−7.00	11.10
NF (dB)	0.70	1.00	0.70	2.00	0.70	0.80	7.00	4.50
OP1dB (dBm)	40.00	40.00	40.00	20.00	20.00		17.00	11.42

Input pwr (dBm) −60.00   System temp (K)   290.00

FIGURE 8.28   Gain and noise figure at the output of the sum channel.

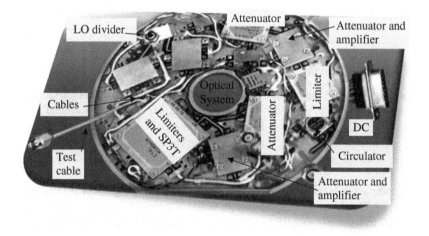

FIGURE 8.29   Compact X-band RF head.

TABLE 8.18   Test Setup

Setup Measurements	Measurements
Network analyzer 851	Phase and amplitude balance VSWR gain isolation
NF meter	NF
Oscilloscope	Rise/fall time
Spectrum analyzer	Image rejection
Power meter	Power handling
Synthesizer	RF + LO source
Power supply	+5 V, ±15 V
Control unit	Control commands

cables are used to test all RF head channels at RF head input and output ports. However, the test cables are removed to integrate the RF head to the monopulse antenna.

8.2.4   RF Head Measurements

Measurement list and test setup are listed in Table 8.18.

Detail measurement list and system requirements are listed as follows:

8.2.4.1   Test List

1. Frequencies		
RF frequency	10550–10850 MHz	
IF frequency	60 MHz	
2. TX input	**Circulator test**	
Peak power	400 W max.	
Average power	60 W max.	
PRF	2–220 kHz	
Pulse width	0.1–15 user	
Duty factor	15 max.	
3. Losses	**Comparator test**	
Bit to guard	3 dB max.	
TX to A, B, C, and D total loss	1.75 dB max.	
4. NF at +25°C	**NF meter**	
A, B, C, D, to Σ	Channel 10.5 dB max.	
	(For A + B + C +D, NF = 5 dB)[a]	
A, B, C, D to Δ	Channel 11.0 dB max.	
	(For A + B + C + D, NF = 5 dB)[a]	
Guard	Channel 6.0 dB max.	

[a]Without comparator.

Network analyzer tests

5. RF to IF gain (A, B, C, D to Σ, El, Az channels)
    25°C at 10,700 MHz   5–7dB

                                         (For A + B + C + D, G = 11/13dB)
    Gain change versus frequency at 25°C   2dB max.
    Gain change versus temperature   3dB max.
    5.1. RF to IF gain (guard)

                                           5–7dB

6. Isolation
    Sum to Δ 20dB min.
    Between Δ 20dB min.
    Guard to Σ 30.0dB max.
    Bit to Σ gain   −30dB max.[a]
    Bit to Δ gain   −30dB max.[a]
    Between A, B, C, and D   20dB min.[a]

[a]Test with antenna.

---

7. 1dB C.P. INPUT (A, B, C, D to Σ, El, Az channels)
    At 25°C at 0dB att.   −9.5dBm min.

                                           (For A + B + C + D, 1dBc =
                                             −6dBm)
    At 25°C at 30dB att   +11.5dBm min.
                                           (For A + B + C + D, 1dBc =
                                             12dBm)

8. IP_3 input (A,B,C,D to Σ, El, Az channels)
    At 25°C at 30dB   −9.5dBm min.
                                           (For A + B + C + D, 1P3 = −6dBm)
    At 25°C at 30dB   +19.5dBm min.
                                           (For A + B + C + D, 1P3 =
                                             12.5dBm)

Spectrum analyzer tests

9. Recovery time
    Recovery time   200 ns

Tests with limiter

10. VSWR
    IF ports VSWR   1.5 : 1 max.
    RF guard VSWR (low-loss state)   1.5 : 1 max.
    Bit VSWR   2.0 : 1 max.
    TX VSWR (±75 MHz)   1.4 : 1 max.
    Lo VSWR   1.3 : 1 max. LO test
    Λ, B, C, and D VSWR   1.5 : 1 max.

Test with antenna

11. Comparator balance
    Phase balance   10° max.
    Amp. balance   1.0dB max.
12. Circulator isolation—circulator test
    Circulator isolation   19dB min.
13. Channel power handling
    13.1. 10,550–10,850 MHz band
      Pulse width   15 μs max.
      Duty factor   15 max.
      Avg. power   6 W max.
      Peak power   40 W max.
    13.2. 000–18,000 MHz band
      CW power   1 W max.
    13.3. BIT power handling   1 W average max.
14. Swatches and attenuators—switch test
    Rise/fall time   50 ns
    Switching speed   100 ns
15. Attenuator value
    10dB att   ±2.5dB
    20dB att   ±2.5dB
16. Image rejection
                                                     20dB min.
17. LO input
    Frequency   10,490–10,790 MHz
    Power level   10–15dBm
    LO to comparator input isolation   40dB
18. Frequency response at fixed LO mixer test
    −3dB points   10.7 ± 0.75 GHz
    Rejection at ±1200 MHz offset   20dB min.
    Rejection at ±2400 MHz offset   30dB min.
19. Switch on time   300 ns max.
20. Power supply current
21. W DC power

8.2.5 X-Band RF Head Test Results

RF head test results are presented in Table 8.19.

The integrated RF head with the microstrip antenna array measured gain of the sum, guard, and difference channels is shown in Figure 8.30. The sum channel gain is around 30dB. The difference channel gain is around 27.5dB. The guard channel gain is around 17dB. RF head measured gain is listed in Table 8.20.

TABLE 8.19 Test Result Summary

Test	Expected Value	Measured	Pass	Fail
1. Frequencies				
RF frequency	10,550–10,850 MHz	10,550–10,850 MHz	+	
IF frequency	60 MHz	60 MHz	+	
2. TX input				
Peak power	400W max.	Circulator	+	
Average power	60 W max.		+	
PRF	2–220 kHz		+	
Pulse width	0.1–15 μs		+	
Duty factor	15% max.		+	
3. Noise figure NF at +25°C				
Σ channel	NF = 5dB	4.9dB	+	
Δ channel	NF = 5dB	5dB	+	
Guard channel	6.0dB max.	5.5dB	+	
4. RF to IF gain at 25°C	At 10,700 MHz			
Σ channel	10–12dB	11dB	+	
Δ channel	8.5–11.5dB	11dB	+	
Gain change versus frequency at 25°C	2dB max.	2dB	+	
Gain change versus temp.	3dB max.			
RF to IF gain guard	8.5–11.5dB	11dB	+	
Amplitude matching	±0.5dB	±0.5dB	+	
Phase matching	±8°			
5. Isolation				
Sum to Δ	20dB min.	25dB	+	
Between Δ	20dB min.	25dB	+	
Guard to Σ	30.0dB max.	30dB	+	
6. 1dB CP at 25°C	At 0dB att.			
Σ channel	−8.0dBm min.	−7.5dBm	+	
Δ channel	−9.5dBm min.	−9.3dBm	+	
25°C at 30dB att.	+12dBm min.	+12dBm	+	
7. IP_3 at 25°C	At 0dB att.			
Σ channel	−6dBm min.	−6dBm min.	+	
Δ channel	−7.5dBm min.	−7.5dBm min.	+	
For 30dB att.	12.5dBm min.	+12.5dBm min.	+	
8. Recovery time	200 ns	200 ns	+	
9. VSWRs				
IF port VSWR	1.5 : 1 max.	1.5 : 1	+	
RF guard VSWR	1.5 : 1 max.	1.5 : 1	+	
Bit VSWR	2.0 : 1 max.	2.0 : 1	+	
LO input	1.6 : 1 max.	1.6 : 1	+	
TX VSWR (+750 MHz)	1.6 : 1 max.	1.6 : 1	+	
Δ and Σ channel	1.8: 1 max.	1.8 : 1	+	
10. Circulator isolation				
Circulator isolation	19dB min.	19dB min.	+	

TABLE 8.19 (*Continued*)

Test	Expected Value	Measured	Pass	Fail
11. Channel power handling				
10,550–10,850 MHz				
Pulse width	15 μs max.	15 μs max.	+	
Duty factor	15% max.	15% max.	+	
Av. power	5 W max.	5 W max.	+	
Peak power	40 W max.	40 W max.	+	
2,000–18,000 MHz				
CW power	1 W max.	1 W max.	+	
12. Switches and attenuators				
Rise/fall time	50 ns	50 ns	+	
Switching speed	100 ns	100 ns	+	
13. Attenuator value				
10dB att.	±2.5dB	10dB att.	+	
20dB att.	±2.5dB	20dB att.	+	
14. Image rejection	20dB min.	20dB min.	+	
15. LO input				
Frequency 10,490–10,790			+	
MHz				
Power level	10–15dBm	10–15dBm	+	
16. Frequency response, fixed LO				
−3dB points	10.7 ± 0.75 GHz		+	
Rejection at +1200 MHz	20dB min.	20dB min.	+	
Rejection at +2400 MHz	30dB min.	30dB min.	+	
17. Switch on time	300 ns max.	300 ns max.	+	
18. Power consumption	12 W	11 W	+	

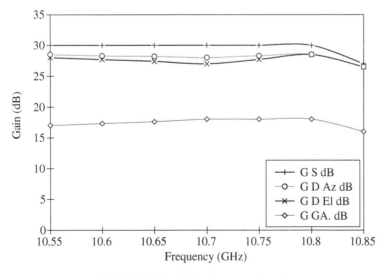

FIGURE 8.30   RF head measured gain.

TABLE 8.20 RF Head Measured Gain

Frequency (GHz)	G Guard (dB)	$G \Delta$El (dB)	$G \Delta$Az (dB)	$G S$ (dB)
10.55	17	28	28.5	30
10.6	17.3	27.7	28.3	30
10.65	17.6	27.4	28.2	30
10.7	18	27	28	30
10.75	18	27.7	28.3	30
10.8	18	28.5	28.5	30
10.85	16	26.5	26.5	27

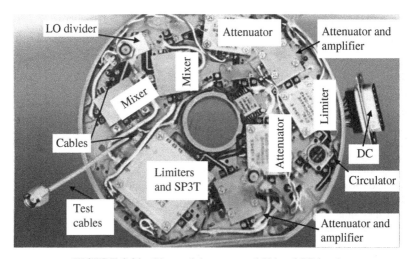

FIGURE 8.31   Photo of the measured X-band RF head.

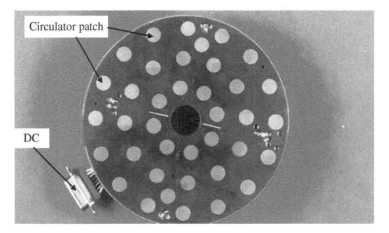

FIGURE 8.32   Photo of the integrated RF head with the microstrip antenna array.

A photo of the measured X-band compact RF head is shown in Figure 8.31. Also a photo of the integrated RF head with the microstrip monopulse antenna array is shown in Figure 8.32. The microstrip monopulse antenna array is printed on RT/duroid 5880 substrate with dielectric constant of 2.2. The antenna consists of two layers. The first layer thickness is around 0.5 mm. The antenna feed network is printed on the first layer. Microstrip circular radiators are printed on the second layer. The antenna array thickness is around 1 mm.

The antenna array has four output ports: sum, azimuth difference, elevation difference, and guard. The antenna output is connected to the RF head input ports.

REFERENCES

[1] Rogers J, Plett C. *Radio frequency Integrated Circuit Design*. Boston: Artech House; 2003.

[2] Mass SA. *Nonlinear Microwave and RF Circuits*. Boston: Artech House; 1997.

[3] Sabban A. *Low Visibility Antennas for Communication Systems*. Boca Raton: Taylor & Francis Group; 2015.

[4] Sabban A. Microstrip Antenna Arrays. In: Nasimuddin N, editor. *Microstrip Antennas*. Rijeka: InTech; 2011. p. 361–384.

[5] Sabban A. Applications of MM wave microstrip antenna arrays. International Symposium on Signals, Systems and Electronics, 2007. ISSSE'07 Conference; July 30 to August 2, 2007; Montreal, QC.

[6] Balanis CA. *Antenna Theory: Analysis and Design*. 2nd ed. Chichester: John Wiley & Sons, Ltd; 1996.

[7] Godara LC, editor. *Handbook of Antennas in Wireless Communications*. Boca Raton: CRC Press LLC; 2002.

[8] Kraus JD, Marhefka RJ. *Antennas for all Applications*. 3rd ed. New York: McGraw Hill; 2002.

[9] James JR, Hall PS, Wood C. *Microstrip Antenna Theory and Design*. London: Peregrinus on Behalf of the Institution of Electrical Engineers; 1981.

9

MIC AND MMIC COMPONENTS AND MODULES DESIGN

9.1 INTRODUCTION

MIC and MMIC technologies was presented in Chapter 4. In design of RF modules, we may gain the advantages of each technology by using both technologies in the development of RF transmitting and receiving modules, see Refs. [1–10]. For example, it is better to design filters by using MIC technology. However, compact and low-cost modules in mass production is achieved by using MMIC design. Design considerations of passive and active devices will be presented in this chapter.

9.2 PASSIVE ELEMENTS

9.2.1 Resistors

Resistors, capacitors, and inductors are basic passive elements that are used in the design of MIC and MMIC modules. The resistor schematic is shown in Figure 9.1. The resistance may be computed by using Equation 9.1. Resistor equivalent circuit is presented in Figure 9.2:

$$R = R_{\text{sh}}\frac{S}{W} + 2R_{\text{c}} \tag{9.1}$$

Wideband RF Technologies and Antennas in Microwave Frequencies, First Edition. Dr. Albert Sabban.
© 2016 John Wiley & Sons, Inc. Published 2016 by John Wiley & Sons, Inc.

R_{sh}—The resistance of the metal film or doped semiconductor film

S—Separation between Ohmic contacts

W—The width of the conductor

R_c—The resistance of the contacts

9.2.2 Capacitor

Capacitor consists of two parallel plate conductors separated by a dielectric layer with dielectric constant ϵ_r and thickness d. The conductor area is A. The capacitor schematic is shown in Figure 9.3. The capacitance may be computed by using Equation 9.2:

$$C = A\frac{\epsilon_0 \epsilon_r}{d} \tag{9.2}$$

where $\epsilon_0 = 8.85 10^{-12} \text{F/m}$

9.2.3 Inductor

A narrow short printed transmission line, as shown in Figure 9.4, behaves as an inductor. The inductor equivalent circuit is presented in Figure 9.5. The inductance L and the capacitance C can be calculated by Equations 9.3 and 9.4.

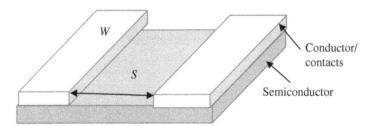

FIGURE 9.1   Schematic of a resistor.

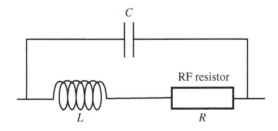

FIGURE 9.2   Resistor equivalent circuit.

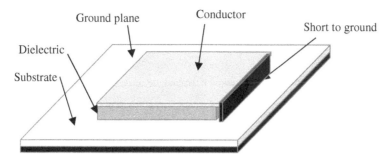

FIGURE 9.3   Capacitor.

FIGURE 9.4   Short printed transmission line.

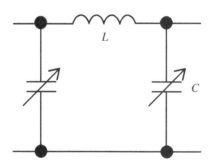

FIGURE 9.5   Inductor equivalent circuit.

Passive elements resistor, capacitor, and inductor in MMIC design are shown in Figure 9.6:

$$L = \frac{Z_0}{2\pi f} \sin\left(\frac{2\pi l}{\lambda}\right) \tag{9.3}$$

$$C = \frac{1}{2\pi f Z_0} \tan\left(\frac{\pi l}{\lambda}\right) \tag{9.4}$$

9.2.4   Couplers

Couplers are used to couple part of the power in the input port to a coupled port. Usually couplers consists of two coupled quarter wavelength transmission lines and have four ports as shown in Figure 9.7. P_1 is the input port, P_2 is the transmitted port, P_3 is

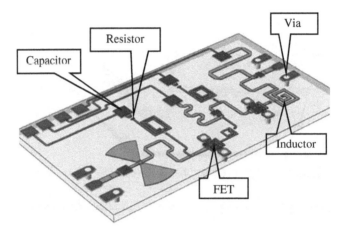

FIGURE 9.6   Passive elements in MMIC design.

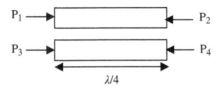

FIGURE 9.7   Coupled lines.

the coupled port, and P_4 is the isolated port. The coupled port may be used to obtain the information about the signal such as frequency and power level without interrupting the main power flow in the device. If the power coupled to the coupled port is half of the input power, 3dB below the input power level, the power on the main transmission line is also 3dB below the input power and equals the coupled power. Such a coupler is called as a 90° hybrid coupler or 3dB coupler. The coupling factor is the ratio between the coupled power to the input power in decibel, when the other ports are terminated, and is given by Equation 9.5. Coupler losses are expressed via the coupler insertion loss given in Equation 9.6. The actual coupler losses are due to coupling loss, dielectric losses, conductor losses, and matching losses:

$$\text{Coupling factor} = \text{CF} = -10\log\frac{P_3}{P_1} \tag{9.5}$$

$$\text{Insertion loss} = \text{IL} = 10\log\left(1 - \frac{P_3}{P_1}\right) \tag{9.6}$$

The isolation factor is the ratio between the power in the isolated port to the input power in decibel, when the other ports are terminated, and is given by Equation 9.7:

$$\text{Isolation} = -10\log\frac{P_4}{P_1} \tag{9.7}$$

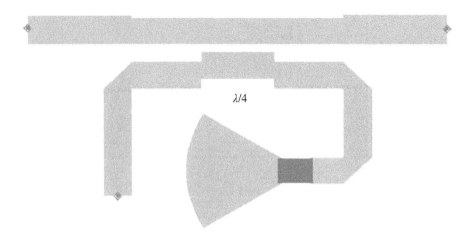

FIGURE 9.8   A wideband mm-wave coupler.

The coupler directivity is the ratio between the power in the isolated port to the coupled power in decibel, when the other ports are terminated, and is given by Equation 9.8:

$$\text{Directivity} = D = -10\log\frac{P_4}{P_3} \qquad (9.8)$$

In several communication systems the amplitude and phase balance between the system ports is important in the signal processing process. Amplitude balance defines the power difference in decibel between two output ports of a 3dB hybrid. In an ideal hybrid circuit, the difference should be 0dB. However, in a practical device, the amplitude balance is frequency dependent. The phase difference between two output ports of a hybrid coupler may be 0, 90, or 180° depending on the coupler type.

9.2.5   A Wideband Millimeter-Wave Coupler

A wideband mm-wave coupler is shown in Figure 9.8 and is printed on alumina with 9.8 dielectric constant and 5.5 mil-inch thickness. The coupler frequency range is 18–40 GHz. The coupler was designed by Momentum ADS software. The coupling value is −13dB and is shown in Figure 9.9. The coupler insertion loss is 0.4dB and is shown in Figure 9.10. The coupler S_{11} parameter is better than −26dB and is shown in Figure 9.11.

9.3   POWER DIVIDERS AND COMBINERS

Power dividers and combiners are used to distribute or combine microwave signals. Power dividers and combiners are passive reciprocal devices. There are several types of power dividers and combiners such as Wilkinson power divider, rat-race power divider, and Gysel power divider.

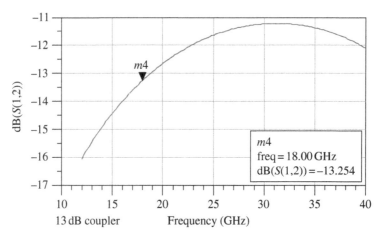

FIGURE 9.9 The coupler coupling value.

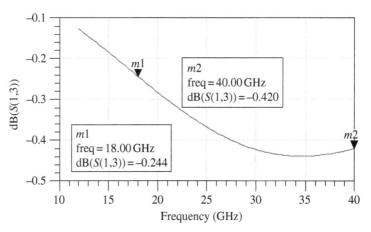

FIGURE 9.10 The coupler insertion loss.

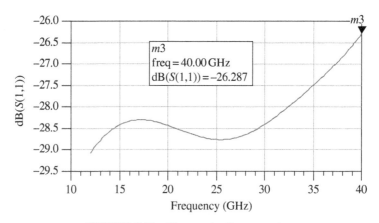

FIGURE 9.11 The coupler S_{11} parameter.

9.3.1 Wilkinson Power Divider

The Wilkinson power divider is shown in Figure 9.12. The impedance of the input port is Z_O. At junction point O the input power is half of the input power. The impedance at junction point O will be $2Z_O$. A quarter wave transformer is used to match $2Z_O$ impedance to Z_O. The bandwidth of the Wilkinson power divider is around 20% bandwidth for VSWR better than 2 : 1. However, by using multiple section quarter wave transformers, we may achieve a bandwidth of 50% for VSWR better than 2 : 1. If a $2Z_O$ resistor is connected between the output ports, the isolation between the output ports will be around 15dB. If a resistor is not connected between the output ports, the isolation between the output ports will be lower than 6dB. The Wilkinson power divider may be used to get unequal power dividers.

For example, two third of the input power at output port A and two third of the input power at output port B, the impedance of the impedance of the quarter wave transformer connecting point O to output port A should be 86.6 Ω. The impedance of the impedance of the quarter wave transformer connecting point O to output port B should be 61.23 Ω.

9.3.2 Rat-Race Coupler

A rat-race coupler is shown in Figure 9.13. The rat-race circumference is 1.5 wavelengths. The distance from A to Δ port is 3λ\4. The distance from A to Σ port is λ\4.

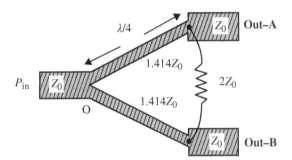

FIGURE 9.12   Wilkinson power divider.

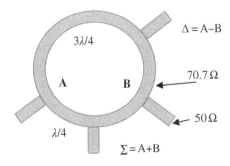

FIGURE 9.13   Rat-race coupler.

For an equal-split rat-race coupler, the impedance of the entire ring is fixed at $1.41 \times Z_O$ or $70.7\,\Omega$ for $Z_O = 50\,\Omega$. For an input signal V, the outputs at ports 2 and 4 are equal in magnitude but $180°$ out of phase.

9.3.3 Gysel Power Divider

The Gysel power divider is shown in Figure 9.14. The Gysel power divider is a wideband power divider, around 60% bandwidth for VSWR better than $2:1$. The Gysel power divider may transmit high power.

9.3.4 Unequal Rat-Race Coupler

The rat-race coupler may be used to get unequal power dividers. The unequal rat-race power divider is shown in Figure 9.15. The power ratio P_A/P_B may be achieved by realizing the impedances Z_{OA} and Z_{OB} as given in Equations 9.9 and 9.10:

$$Z_{OA} = Z_O \left[\frac{1 + (P_A/P_B)}{(P_A/P_B)}\right]^{0.5} \tag{9.9}$$

$$Z_{OB} = Z_O \left[1 + \frac{P_A}{P_B}\right]^{0.5} \tag{9.10}$$

9.3.5 Wideband Three-Way Unequal Power Divider

In Figure 9.16 a 6–18 GHz three-way power divider is presented. The power divider is printed on RT/duroid with dielectric constant of 2.2 and 10 mil-inch thickness. The

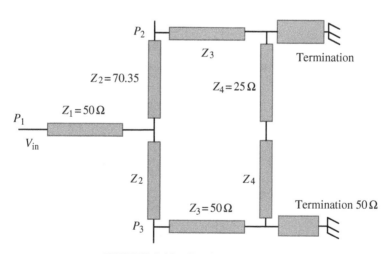

FIGURE 9.14 Gysel power divider.

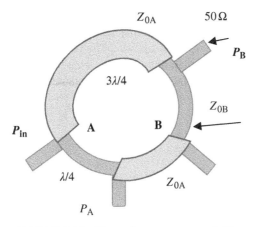

FIGURE 9.15   Unequal rat-race power divider.

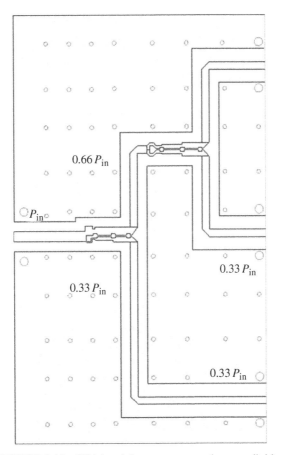

FIGURE 9.16   Wideband three-way unequal power divider.

dimensions of the power divider are around $40 \times 24 \times 0.25$ mm. The power divider consists of three quarter wave transformer sections. The input power is divided to two third and third of the input port power. The two third of the input port power is divided by equal power divider to third of the input port power. The power divider insertion loss is around 1dB. The power divider S_{11} parameter is better than -10dB in the frequency range of 6–18 GHz. The power divider was designed and optimized by Momentum ADS software.

9.3.6 Wideband Six-Way Unequal Power Divider

In Figure 9.17 a 6–18 GHz six-way power divider is presented. The power divider is printed on RT/duroid with dielectric constant of 2.2 and 10 mil-inch thickness. The dimensions of the power divider are around $68.79 \times 23.4 \times 0.25$ mm. The power divider consists of three quarter wave transformer sections. The input power is divided to two third and third of the input port power. The two third of the input port power is divided by equal power divider to third of the input port power. The power divider insertion loss is around 1.5dB. The power divider S_{11} parameter is better than -10dB in the frequency range of 6–18 GHz. The power divider was designed and optimized by Momentum ADS software.

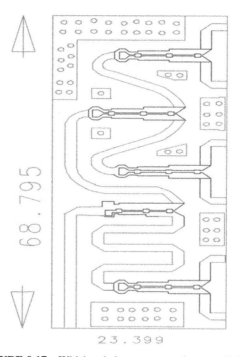

FIGURE 9.17 Wideband six-way unequal power divider.

9.3.7 Wideband Five-Way Unequal Power Divider

In Figure 9.18 a 6–18 GHz five-way power divider is presented. The power divider is printed on RT/duroid with dielectric constant of 2.2 and 10 mil-inch thickness. The dimensions of the power divider are around $25.5 \times 22.1 \times 0.25$ mm. The input power is divided to two fifth and to three fifth of the input port power. The two fifth of the input port power is divided by equal power divider to fifth of the input port power. The three fifth of the input port power is divided by unequal power divider to fifth of the input port power and to two fifth of the power. The two fifth of the input port power is divided by equal power divider to fifth of the input port power. The power at each output port is fifth of the input power. The power divider insertion loss is around 1.5dB. The power divider S_{11} parameter is better than −10dB in the frequency range of 6–18 GHz.

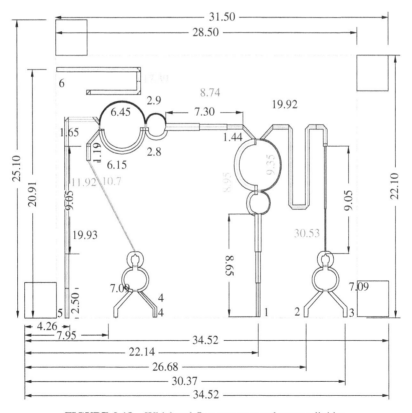

FIGURE 9.18 Wideband five-way unequal power divider.

9.4   RF AMPLIFIERS

Amplifiers are specified by their gain, frequency, efficiency, linearity, power, and size. Power amplifiers are designed to transmit high-power values. However, low-noise amplifiers (LNA) are designed to get low-noise figure values. Gain amplifiers are designed to get higher gain values. A typical transceiver block diagram is shown in Figure 9.19. The transceiver consists of power amplifier in the transmitting channel, LNA in the receiving channel, and IF amplifiers.

There are several definitions to amplifier gain. Power gain is the ratio of power dissipated in the load, Z_L, to the power delivered to the input of the amplifier. Available gain is the ratio of the amplifier output power to the available power from the generator source; it depends on Z_G but it does not depends on Z_L. Exchangeable gain is the ratio of the output exchangeable power to the input exchangeable power as written in Equation 9.11. Insertion gain is the ratio of the output power that would be dissipated in the load if the amplifier was not connected. There is a problem in applying this definition to mixers of parametric upconverters where the input and output frequencies are different:

$$P = \frac{|V|^2}{4\Re\{Z_G\}} \quad \Re\{Z_G\} \neq 0 \tag{9.11}$$

Transducer gain is the ratio of the power delivered to the load to the available power from the source. A two-port equivalent network is presented in Figure 9.20. This definition depends on both Z_G and Z_L. It gives positive gain for negative resistance amplifiers as well since the characteristics of real amplifiers impedance change

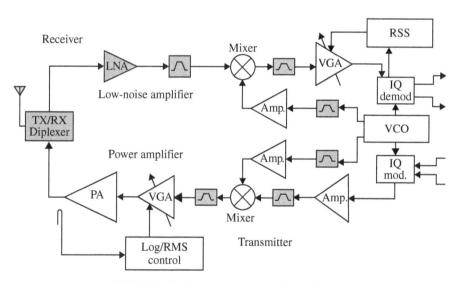

FIGURE 9.19   Typical transceiver block diagram.

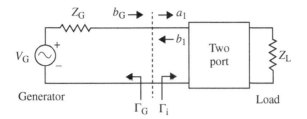

FIGURE 9.20   Two-port equivalent network.

when either the load or generator impedance is changed. Thus the transducer power
gain definition is found to be the most useful and is given in Equation 9.12:

$$\text{Load impedance is} \rightarrow V_2 = -I_2 Z_L$$

$$\text{Input impedance is} \rightarrow Z_{in} = \frac{V_1}{I_1} = z_{11} - \frac{z_{12}z_{21}}{z_{22} + Z_L}$$

$$\text{Power delivered to the load is} \rightarrow P_2 = \frac{1}{2}|I_2|^2 \Re\{Z_L\}$$

$$\text{Power available from the source is} \rightarrow P_{1a} = \frac{|V_G|^2}{8\Re\{Z_G\}}$$

$$\underbrace{G_T}_{\substack{\text{Transducer} \\ \text{power} \\ \text{gain}}} = \frac{P_2}{P_{1a}} = \frac{4\Re\{Z_L\}\Re\{Z_G\}|z_{21}|^2}{|(Z_G + z_{11})(Z_L + z_{22}) - z_{21}z_{21}|^2} \tag{9.12}$$

The available power, P_a, is written in Equation 9.13:

$$P_a = \frac{\frac{1}{2}|b_G|^2}{1 - |\Gamma_G|^2} \tag{9.13}$$

The power delivered to the load is written in Equation 9.14:

$$P_L = \frac{1}{2}|b_2|\left(1 - |\Gamma_L|^2\right) \tag{9.14}$$

The transducer gain is written in Equation 9.15. In Equation 9.16 the ratio b_2/b_g is
written. The transducer gain G_T is written in Equation 9.17:

$$G_T = \frac{|b_2|^2}{|b_G|^2}\left(1 - |\Gamma_L|^2\right)\left(1 - |\Gamma_G|^2\right) \tag{9.15}$$

$$b_2 = \frac{\dfrac{b_G}{1 - S_{11}\Gamma_G}}{1 - \dfrac{S_{21}S_{12}\Gamma_G\Gamma_L}{(1 - S_{11}\Gamma_G)(1 - S_{22}\Gamma_L)}} \frac{S_{21}}{(1 - S_{22}\Gamma_L)} \tag{9.16}$$

$$G_T = \frac{|S_{21}|^2 \left(1 - |\Gamma_G|^2\right)\left(1 - |\Gamma_L|^2\right)}{|(1 - S_{11}\Gamma_G)(1 - S_{22}\Gamma_L) - \Gamma_G\Gamma_L S_{21}S_{12}|^2} \tag{9.17}$$

9.4.1 Amplifiers Stability

A stable amplifier is an amplifier that does not oscillate. Oscillations may occur when there is some feedback path from the output back to the input or under certain environmental conditions. The criteria for unconditional stability require that for any passive terminating loads. It is helpful to find the borderline between the stable and the unstable regions. The Smith chart may be helpful in defining stability circles for the amplifier input and output impedances. The center and the radius of the stability circles for the amplifier input and output ports are given in Equations 9.18 and 9.19:

$$C_L = \frac{\Delta S_{11} - S_{22}}{|\Delta|^2 - |S_{22}|^2} \text{(load center)} \quad C_G = \frac{\Delta S_{22} - S_{11}}{|\Delta|^2 - |S_{22}|^2} \text{(generator center)} \tag{9.18}$$

$$r_L = \left|\frac{S_{21}S_{12}}{|\Delta|^2 - |S_{22}|^2}\right| \text{(stability radius)} \quad r_G = \left|\frac{S_{21}S_{12}}{|\Delta|^2 - |S_{11}|^2}\right| \tag{9.19}$$

where $|\Delta| = |S_{11}S_{22} - S_{12}S_{21}|$

These two stability circles represent the borderline between stability and instability, and they are drawn on a Smith chart. Unconditional stability requires that both $|\Gamma_i| < 1$ and $|\Gamma_O| < 1$ for any passive load impedance and generator impedance connected to the ports.

The Rollett criteria determines if a given transistor is unconditionally stable for any passive load impedance and generator impedance connected to the ports.

There are two conditions written in Equations 9.20 and 9.21:

$$k = \frac{1 - |S_{11}|^2 - |S_{22}|^2 + |\Delta|^2}{2|S_{12}S_{21}|} \geq 1 \tag{9.20}$$

$$|\Delta| \leq 1 \tag{9.21}$$

where

$$\{|\Delta| = |S_{11}S_{22} - S_{12}S_{21}| < |S_{11}S_{22}| + |S_{12}S_{21}|\}$$

9.4.2 Stabilizing a Transistor Amplifier

There are a variety of approaches to stabilize an amplifier. We can stabilize the amplifier by choosing the amplifier terminating impedances inside the stable regions at all frequencies. We can also load the amplifier with an additional shunt or series resistor on either the generator or load ports. However it is better to load the output port to optimize the amplifier NF. Another approach is to introduce an external feedback path to minimize the internal feedback of the transistor.

9.4.3 Class A Amplifiers

Class A amplifiers amplifies the incoming signal in a linear fashion. A linear class A amplifier will introduce low harmonic frequency components and low intermodulation distortion (IMD). The amplifier output voltage is given by Equation 9.22:

$$\frac{V}{2}\cos(\omega_c + \omega_m)t + \frac{V}{2}\cos(\omega_c - \omega_m)t \tag{9.22}$$

where ω_c is the high-frequency carrier frequency. Where ω_m is the low-frequency modulation, frequency IMD would result in frequencies at $\omega_c + n\omega_m$. Harmonic distortion would cause frequency generation at $k\omega_c + n\omega_m$.

Reduction of IMD depends on efficient power combining methods and careful design of the amplifier. Example of common emitter class A amplifier is presented in Figure 9.21. The amplifier currents and output voltage are given in Equations 9.23 and 9.24:

$$i_c = i_{dc} - i_0 \sin \omega t \tag{9.23}$$

where $i_{dc} = i_Q = i_{outmax}$
i_Q-quiescent current

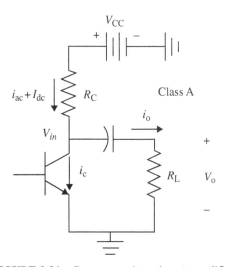

FIGURE 9.21   Common emitter class A amplifier.

$$V_{ce} = V_{cc} + V_0 \sin \omega t \qquad (9.24)$$

where $V_{outmax} = i_{outmax} \times R_L = V_{cc}$

The maximum efficiency of the class A amplifier shown in Figure 9.21 is around
25%. By replacing the collector resistor by an inductor, the efficiency may be
improved up to 50%.

9.4.4 Class B Amplifier

The class B amplifier is biased so that the transistor is on only during half of the
incoming cycle. The other half of the cycle is amplified by another transistor so that
at the output the full wave is reconstituted. While each transistor is clearly operating in
a nonlinear mode, the total input wave is directly replicated at the output. The class
B amplifier is therefore classed as a linear amplifier. In this case the bias current,
$IC = 0$. Since only one of the transistors is cut off when the total voltage is less than
0, only the positive half of the wave is amplified. The conduction angle is 180°.

Figure 9.22 shows a complementary type of class B amplifier. In this case transistor
Q1 is biased so that it is in the active mode when the input voltage $V_{in} > 0.7$ and cut off
when the input signal $V_{in} < 0.7$. The other half of the signal is amplified by transistor
Q2 when $V_{in} < 0.7$. When no input signal is present, no power is drawn from the bias

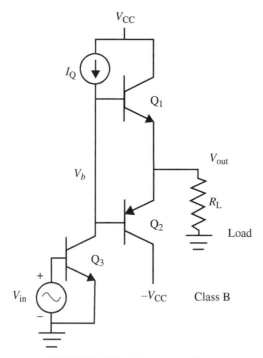

FIGURE 9.22   Class B amplifier.

supply through the collectors of Q1 or Q2, so the class B operation is attractive when low standby power consumption is an important consideration. There is a small region of the input signal for which neither Q1 nor Q2 is on. The amplifier output will suffer some distortion. One of the primary problems in using this type of class B amplifier is the requirement for obtaining two equivalent complementary transistors. The problem fundamentally arises because of the greater mobility of electrons by over a 3 : 1 factor over that of holes in silicon. The symmetry of the gain in this circuit depends on the two output transistors having the same short circuit base to collector current gain, $\beta = i_c/i_b$. When it is not possible to obtain a high βpnp transistor, it is sometimes possible to use a composite transistor connection. A high-power npn transistor, Q1, is connected to a low-power low βpnp transistor, Q2.

The maximum efficiency of a class B amplifier is found by calculating the ratio of the output power delivered to the load to the required DC power from the bias voltage supply. By calculating efficiency in this way, power losses caused by nonzero base currents and crossover distortion compensation circuits are neglected. Furthermore the power efficiency rather than the power-added efficiency is calculated so as to form a basis for comparison for alternative circuits. It is sufficient to do the calculation during the part of the cycle when Q1 is on and Q2 is off. The load resistance is R_L.

The maximum output power occurs when the output voltage is $V_{cc} - V_{ce}$ (sat) and is given in Equation 9.25:

$$P_{L(max)} = 0.5 \frac{\left(V_{cc} - V_{ce(sat)}\right)^2}{R_L} \tag{9.25}$$

$$\eta_{max} = \frac{\pi}{4} \frac{V_{cc} - V_{ce(sat)}}{V_{cc}} \approx 78.5\% \tag{9.26}$$

This efficiency for the class B amplifier is around 78.5%, given in Equation 9.26, which should be compared with the maximum efficiency of a class A amplifier, where η_{max} is 25% when the bias to the collector is supplied through a resistor and η_{max} is 50% when the bias to the collector is supplied through an RF choke.

9.4.5 Class C Amplifier

The class C amplifier is useful for providing a high-power continuous wave (CW) or frequency modulation (FM) output. When it is used in amplitude modulation schemes, the output variation is done by varying the bias supply. There are several characteristics that distinguish the class C amplifier from the class A or B amplifiers. Class C amplifier is biased so that the transistor conduction angle is <180°. Consequently the class C amplifier is clearly nonlinear in that it does not directly replicate the input signal like the class A and B amplifiers do. The class A amplifier requires one transistor, the class B amplifier requires two transistors, and the class C amplifier uses one transistor. Topologically it looks similar to the class A except for the DC bias levels. It was noted that in the class B amplifier, an output filter is used optionally to help clean up the output signal. In the class C amplifier, such a tuned output is necessary in order to recover the sine wave. Finally, class C operation is capable of higher efficiency than class A and class B. Class C amplifier, for the appropriate signal types, is very attractive as power amplifiers.

9.4.6 Class D Amplifier

In class D operation the transistors act as near ideal switches that are on half of the time and off half of the time. The input ideally is excited by a square wave. If the transistor switching time is near zero, then the maximum drain current flows while the drain–source voltage, $V_{ds} = 0$. As a result 100% efficiency is theoretically possible. In practice the switching speed of a bipolar transistor is not sufficiently fast for square waves above a few megahertz, and the switching speed for field-effect transistors is not adequate for frequencies above a few tens of megahertz.

9.4.7 Class F Amplifier

A class F amplifier is characterized by a load network that has resonances at one or more harmonic frequencies as well as at the carrier frequency. The class F amplifier was one of the early methods used to increase amplifier efficiency and has attracted some renewed interest recently. When the transistor is excited by a sinusoidal source, it is on for approximately half the period time and off for half of the period time.

9.4.8 Feedforward Amplifiers

The feedback approach is an attempt to correct an error after it has occurred. A 180° phase difference in the forward and reverse paths in a feedback system can cause unwanted oscillations. In contrast, the feedforward design is based on cancellation of amplifier errors in the same time frame in which they occur. Signals are handled by wideband analog circuits, so multiple carriers in a signal can be controlled simultaneously. Feedforward amplifiers are inherently stable, but this comes at the price of a somewhat more complicated circuit. Consequently feedforward circuitry is sensitive to changes in ambient temperature, input power level, and supply voltage variation. Nevertheless, feedforward offers many advantages that have brought it increased interest. The major source of distortion, such as harmonics, IMD, and noise, in a transmitter is the power amplifier. This distortion can be greatly reduced using feedforward design.

9.5   LINEARITY OF RF AMPLIFIERS AND ACTIVE DEVICES

Any continuous analytical function may described as given in Equation 9.27:

$$V_O = \sum_{i=1}^{\infty} K_i V_{in}^i \qquad (9.27)$$

In the linear zone of the function, V_O may be approximated by the first three elements of the Taylor series as written in Equation 9.28:

$$V_O \approx K_1 V_{in} + K_2 V_{in}^2 + K_3 V_{in}^3 \qquad (9.28)$$

For a sinusoidal signal, the second-order element in Equation 9.2 influences the DC and the second harmonic at the output. In this analysis we assume that the magnitude of the input signal does not change the component parameters. K_1 is the first-order parameter and usually represent gain or attenuation. K_2 is the second-order parameter.

K_3 is the third-order parameter. For a narrowband frequency range, V_O may be written as in 9.29:

$$V_{O\text{-narrow band}} \approx K_1 V_{in} + K_3 V_{in}^3 \qquad (9.29)$$

In active RF components when the input power is increased, we get a deviation from linearity of the output signal. If the input power is increased by X(dB), the output power is increased by Y(dB). Where $Y < X$. The third-order component causes the deviation from linearity and compress the output signal (Fig. 9.23). Compression is obtained if

$$K_1 K_3 < 0$$

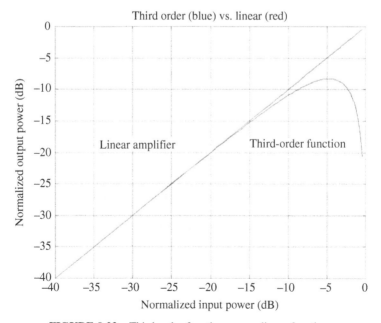

FIGURE 9.23 Third-order function versus linear function.

For a narrowband frequency range, V_O may be written as in Equation 9.30:

$$|V_O| \approx |K_1|\, V_{in} - |K_3|\, V_{in}^3 \qquad (9.30)$$

After normalization of Equation 9.4 by K_1, V_O may be written as in Equation 9.31:

$$|\hat{V}_O| \approx V_{in} - K V_{in}^3$$

$$\text{where } K = \left|\frac{K_3}{K_1}\right| \qquad (9.31)$$

The output power of a linear device with gain of $G_{[dB]}$ is given in Equation 9.32:

$$P_{O-\text{Lin}[dBm]} = P_{in[dBm]} + G_{[dB]}$$

$$G_{[dB]} = 20\log(K_1) \qquad (9.32)$$

However in an active device, the output power of the device with gain of $G_{[dB]}$ is given in Equation 9.33:

$$P_{out[dBm]} = P_{O-\text{Lin}[dBm]} + 20\log\left(1 - K p_{in[W]}\right)$$

$$\text{where } K = \left|\frac{K_3}{K_1}\right| \qquad (9.33)$$

The compression may be written as in Equation 9.34:

$$\text{Comp}_{[dB]} = 20\log\left(1 - K p_{in[W]}\right) \qquad (9.34)$$

Example
The amplifier small-signal gain (SSG) is 12dB. The output 1dBc power is 16dBm. What is the input power at the 1dBc point?

Solution
The amplifier gain at compression point is 11dB:

$$P_{in} = P_{out} - G = +5\text{dBm}$$

$$\text{At 1dB Comp}\quad 0.89125 \approx 1 - K p_{in\text{-Comp}[W]}$$

$$\text{Comp}_{[dB]} \approx -20\log(1 - 0.109\, d)$$

$$\text{where } d = 10^{\Delta/10}$$

$$\text{and } \Delta = P_{in[dBm]} - P_{in\text{-Comp}[dBm]}$$

where $P_{in\text{-Comp}}$ is the input power at 1dBc power. Where $\Delta_{[dB]}$ is called as the input backoff. The maximum point of the function given at Equation 9.30 is written in Equation 9.35. The 1dBc point is written in Equation 9.36:

$$V_{in\text{-MAX}} = \sqrt{\frac{1}{3K}}$$

$$V_{O\text{-MAX}} = \frac{2}{3}\sqrt{\frac{1}{3K}}$$

(9.35)

$$V_{in\text{-CP}} \approx \frac{0.33}{\sqrt{K}}$$

$$V_{O\text{-CP}} \approx \frac{0.29}{\sqrt{K}}$$

(9.36)

From Equations 9.35 and 9.36, we may conclude that for the maximum saturated output power the amplifier is 3.52dB compressed, relative to the linear line. The maximum saturated output power of the amplifier is 2.34dB from 1dBc output power. We have to increase the input power by 4.87dB from the 1dBc input power to reach to the maximum saturated output power as shown in Figure 9.24.

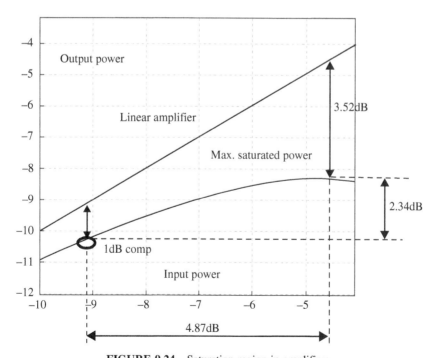

FIGURE 9.24 Saturation region in amplifiers.

9.5.1 Third-Order Model for a Single Tone

Third-order model for a single tone was presented in Equation 9.2. The input signal may be presented by Equation 9.37. The output signal consists of a V_{dc} component and the fundamental signal V_{fund} and two harmonics, h_2 and h_3, as written in Equation 9.38:

$$V_{in} = a\ \cos(\omega_0 t) \tag{9.37}$$

$$V_O = V_{dc} + V_{fund} + h_2 + h_3 \tag{9.38}$$

where the output signals are written in Equations 9.38–9.40:

$$V_{dc} = \frac{1}{2}K_2 a^2 \tag{9.39}$$

$$V_{fund} = \left(K_1 + \frac{3}{4}K_3 a^2\right)V_{in} \tag{9.40}$$

$$h_2 = \frac{1}{2}K_2 a^2\ \cos(2\omega_0 t) \tag{9.41}$$

$$h_3 = \frac{1}{4}K_3 a^3\ \cos(3\omega_0 t) \tag{9.42}$$

Figure 9.25 presents the output spectrum for a single tone.

The voltage gain is given in Equation 9.43. The gain $G_{[dB]}$ is written in Equation 9.44:

$$\sqrt{g} = \left|\frac{V_{fund}}{V_{in}}\right| = |K_1|\left(1 - \frac{3}{4}Ka^2\right)$$
$$\text{where }\ K = \left|\frac{K_3}{K_1}\right| \tag{9.43}$$

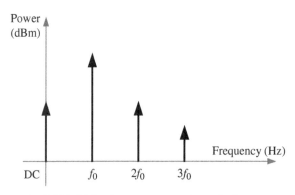

FIGURE 9.25 Output spectrum for a single tone.

$$G_{[dB]} = 10\log(g) \tag{9.44}$$

The SSG is written in Equation 9.45. The gain $G_{[dB]}$ is written in Equation 9.46:

$$SSG_{[dB]} = 20\log|K_1| \tag{9.45}$$

$$G_{[dB]} = SSG - Comp$$
$$Comp = 20\log\left(1 - \frac{3}{4}K\,a^2\right) \tag{9.46}$$

where $p_{in[W]} = \dfrac{a^2}{2}$

The gain $G_{[dB]}$ may be written as in Equation 9.47:

$$G_{[dB]} = SSG_{[dB]} - Comp_{[dB]}$$
$$Comp_{[dB]} = 20\log\left(1 - \frac{3}{2}K\,p_{in[W]}\right) \tag{9.47}$$

9.5.2   Third-Order Model for Two Tones

For two tones the input signal is written as $V_{in} = A + B$. For two tones the output signal is given by Equation 9.48:

$$V_O = V_{out-a} + V_{out-b} + V_{IM_2} + V_{IM_3}$$
$$V_{out-a} = K_1\,A + K_2\,A^2 + K_3\,A^3$$
$$V_{out-b} = K_1\,B + K_2\,B^2 + K_3\,B^3 \tag{9.48}$$
$$V_{IM_2} = 2K_2\,A\,B$$
$$V_{IM_3} = 3K_3\,A^2\,B + 3K_3\,A\,B^2$$

where IM_2—second-order intermodulation and IM_3—third-order intermodulation.

If the magnitude level of the two input tones is equal, the ratio between the voltage level of the second-order intermodulation and the voltage of the second harmonic is 2. The power level of the second-order intermodulation is greater by 6dB from the power level of the second harmonic. If the magnitude level of the two input tones is equal, the ratio between the voltage level of the third-order intermodulation and the voltage of the third harmonic is 3. The power level of the third-order intermodulation is greater by 9.54dB from the power level of the third harmonic as given in Equation 9.49:

$$IM_{2[dBm]} = h_{2[dBm]} + 6dB$$
$$IM_{3[dBm]} = h_{3[dBm]} + 9.54dB \tag{9.49}$$

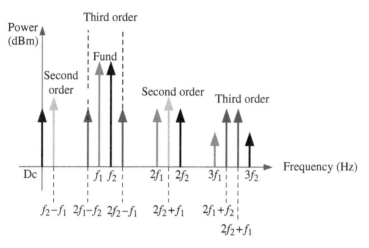

FIGURE 9.26   Two tones output spectrum.

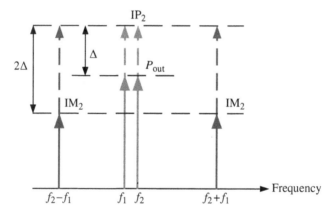

FIGURE 9.27   Second-order intermodulation.

For the input signals given in Equation 9.50, the output frequency spectrum is shown in Figure 9.26:

$$V_{in} = A + B$$
$$A = E \quad \cos(\omega_1 t) \qquad (9.50)$$
$$B = E \quad \cos(\omega_2 t)$$

For a decrease of $\Delta_{[dB]}$ in the signals A and B, the level of the second intermodulation signal will decrease by $2\Delta_{[dB]}$ as given in Equation 9.51. The intermodulation signal IM_2 is written in Equation 9.52. Figure of the second-order intermodulation calculation is shown in Figure 9.27:

$$\text{IP}_{2[\text{dBm}]} - P_{\text{out}[\text{dBm}]} = \Delta_{[\text{dB}]}$$
$$\text{IP}_{2[\text{dBm}]} - \text{IM}_{2[\text{dBm}]} = 2\Delta_{[\text{dB}]}$$

(9.51)

$$\text{IM}_{2[\text{dBm}]} = 2P_{\text{out}[\text{dBm}]} - \text{IP}_{2[\text{dBm}]}$$

(9.52)

9.5.3 Third-Order Intercept Point

For a decrease of $\Delta_{[\text{dB}]}$ in the signals A and B, the level of the third intermodulation signals will decrease by $3\,\Delta_{[\text{dB}]}$ as given in Equation 9.53. The intermodulation signal IM_3 is written in Equation 9.54. Figure of the third-order intermodulation calculation is shown in Figure 9.28:

$$\text{IP}_{3[\text{dBm}]} - P_{\text{out}[\text{dBm}]} = \Delta_{[\text{dB}]}$$
$$\text{IP}_{3[\text{dBm}]} - \text{IM}_{3[\text{dBm}]} = 3\Delta_{[\text{dB}]}$$

(9.53)

$$\text{IM}_{3[\text{dBm}]} = 3P_{\text{out}[\text{dBm}]} - 2\text{IP}_{3[\text{dBm}]}$$

(9.54)

9.5.4 IP$_2$ and IP$_3$ Measurements

The setup for IP$_2$ and IP$_3$ measurements is shown in Figure 9.29. Second-order intermodulation results are shown in Figure IP$_2$ and may be computed by Equation 9.55:

$$\text{IP}_{2[\text{dBm}]} = P_{\text{out}[\text{dBm}]} + \Delta_{[\text{dB}]}$$

(9.55)

Second-order intermodulation results are shown in Figure 9.30.

Third-order intermodulation results are shown in Figure 9.31. IP$_3$ may be computed by Equation 9.56:

$$\text{IP}_{3[\text{dBm}]} = P_{\text{out}[\text{dBm}]} + \frac{\Delta_{[\text{dB}]}}{2}$$

(9.56)

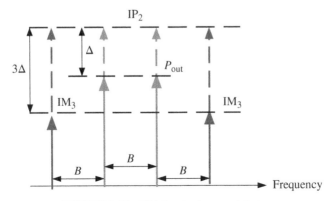

FIGURE 9.28   Third-order intermodulation.

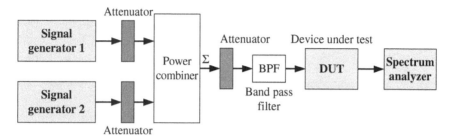

FIGURE 9.29 Setup for IP$_2$ and IP$_3$ measurements.

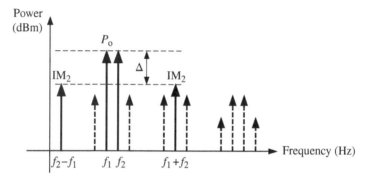

FIGURE 9.30 Second-order intermodulation results.

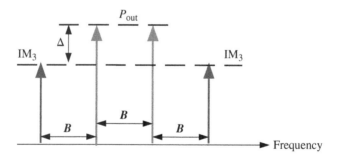

FIGURE 9.31 Third-order intermodulation results.

9.6 WIDEBAND PHASED ARRAY DIRECTION FINDING SYSTEM

9.6.1 Introduction

The wideband direction finding system consists of an active phased array and a wideband receiving direction finding system. A block diagram of the system is shown in Figure 9.32. The system consists of 11 receiving and transmitting modules for vertical scanning. The received signals in the six receiving antennas are

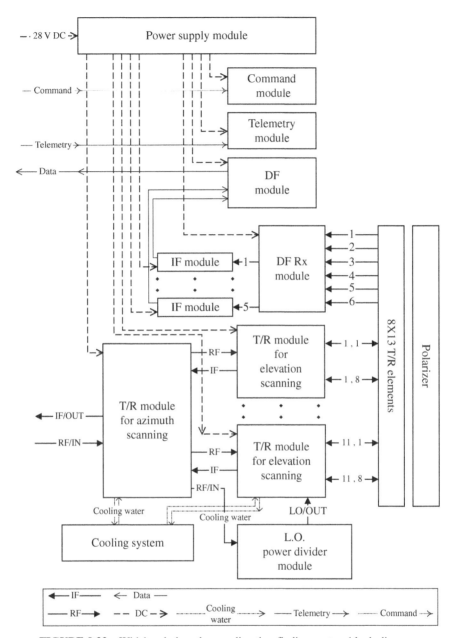

FIGURE 9.32 Wideband phased array direction finding system block diagram.

amplified by LNA and down converted to 70 MHz IF frequency. Input port number six provides the local oscillator (LO) signal to the module. The LO signal is amplified by LNA and is routed to the single side band (SSB) unit. A 70 MHz voltage-controlled oscillator supplies a RF signal to the SSB unit. The LO signal is amplified to 10dBm. A wideband six-way power divider supplies the LO signals to the mixers. The antennas are wideband notch antennas, 6–18 GHz, with VSWR better than 3 : 1.

9.6.2 Wideband Receiving Direction Finding System

The system consists of 11 receiving identical modules for vertical scanning. The receiving channel module consists of LNA and downconverter mixers to 70 MHz IF frequency. The received signals in the six receiving ports are amplified by LNA and down converted to 70 MHz IF frequency. Input port number six provides the LO signal to the module. The LO signal is amplified by LNA and is routed to the SSB unit. A 70 MHz voltage-controlled oscillator supplies a RF signal to the SSB unit. The LO signal is amplified to 10dBm. A wideband six-way power divider supplies the LO signals to the mixers. The module dimensions are $279 \times 130 \times 15.5$ mm. The module hot spots are cooled by a cold water cooling system that is located in the module backside.

9.6.2.1 Module Specifications   The wideband receiving direction finding system specifications are listed in Table 9.1.

**TABLE 9.1   Wideband Receiving Direction Finding
System Specifications**

Parameter	Specifications
RF frequency	6–18 GHz
IF frequency	70 MHz
Gain	4dB
Gain flatness	±2dB
Phase flatness	±6°
Noise figure	7.5dB
Output power	0dBm
Max. input RF power	−7dBm
Max. input LO power	−7dBm
60 MHz input RF power	10dBm
IF ports	5
VSWR	3 : 1
Impedance	50 Ω
Dimensions	$279 \times 130 \times 15.5$ mm
Temperature	$71°C \div -54°C$
Weight	700 g
DC power	20 W

9.6.3 Receiving Channel Design

The wideband receiving direction finding system block diagram is shown in Figure 9.33.

The RF components are located on the upper side of the module.

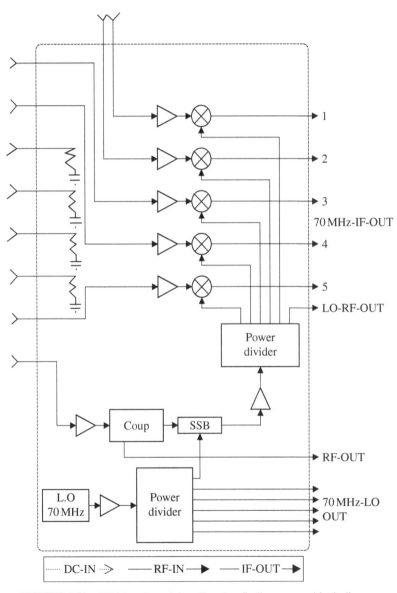

FIGURE 9.33 Wideband receiving direction finding system block diagram.

System1

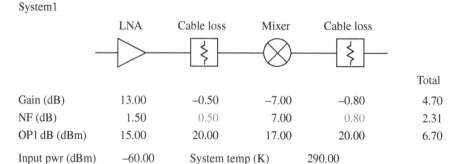

	LNA	Cable loss	Mixer	Cable loss	Total
Gain (dB)	13.00	−0.50	−7.00	−0.80	4.70
NF (dB)	1.50	0.50	7.00	0.80	2.31
OP1 dB (dBm)	15.00	20.00	17.00	20.00	6.70

Input pwr (dBm)	−60.00	System temp (K)	290.00

FIGURE 9.34 Wideband receiving direction finding system analysis design.

The received signals in the six receiving ports are amplified by LNA and down converted to 70 MHz IF frequency. The LNA amplifiers are packed MMICS with 1dB noise figure. The mixers are compact double balance mixers with 7dB insertion loss. Input port number six provides the LO signal to the module. The LO signal is amplified by LNA and is routed to the SSB unit. A 70 MHz voltage-controlled oscillator supplies a RF signal to the SSB unit. The LO signal is amplified to 10dBm. A wideband six-way power divider supplies the LO signals to the mixers. The IF and DC components are located on the back side of the module. Figure 9.34 presents the receiving direction finding system analysis results. The analysis results prove that the module specifications may be satisfied.

9.6.3.1 Cooling System   The cooling system is located on the amplifiers back side. The cooling systems consist of a heat sink pipe lines. The cooling liquid flows via the heat sink pipe lines and return to the system through a cooling return connector.

9.6.3.2 DC Unit   A 24-pin DC connector supplies the DC voltages to the receiving system. The connector consists of 17 DC pins and 7 RF connectors. A list of the DC connector pins and DC power consumption is given in Table 9.2.

9.6.4 Measured Results of the Receiving Channel

Measured results versus requirements of the receiving module are listed in Table 9.3.

A photo of the wideband receiving direction finding system is shown in Figure 9.35.

Photo of the wideband receiving module with the module cover is shown in Figure 9.36. Absorbing material is glued on the cover to improve isolation between

TABLE 9.2 DC Connector

Pin	Function	Voltage (V)	Current (A)
1	VCO	15	0.9
2	LNA	6	0.1
3	LNA	6	0.1
4	LNA	6	0.1
5	IF amp.	15	0.1
6	Limiter	−15 V	
7		N.C	
8		N.C	
9		N.C	
10		N.C	
11	Limiter	15	
12		N.C	
13		GND	
14		N.C	
15		N.C	
16		N.C	
17		N.C	

TABLE 9.3 Measured Results versus Requirements of the Receiving Module

Parameter	Specifications	Measured Results
RF frequency	6–18 GHz	6–18 GHz
IF frequency	70 MHz	70 MHz
Gain	4dB	4dB
Gain flatness	±2dB	±2dB
Phase flatness	±6°	±6°
Noise figure	7.5dB	7.5dB
Output power	0dBm	0dBm
Max. input RF power	−7dBm	−7dBm
Max. input LO power	−7dBm	−7dBm
60 MHz input RF power	10dBm	10dBm
1dBc compression point	−10dBm	−9dBm
IP3	4dBm	4dBm
Spurious level	−40dB	−45dB
VSWR	3 : 1	3 : 1
Impedance	50 Ω	50 Ω
Dimensions	$279 \times 130 \times 15.5$ mm	$279 \times 130 \times 15.5$ mm
Temperature	71°C ÷ −54°C	71°C ÷ −54°C
Weight	700 g	700 g
DC power	20 W	20 W

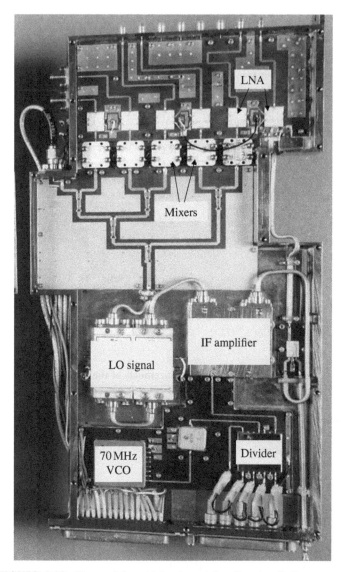

FIGURE 9.35   Photo of the wideband receiving direction finding system.

the module components. Photo of the wideband receiving direction finding system interface and connectors is shown in Figure 9.37.

The RF module dimension is $279 \times 130 \times 15.5$ mm. The compact dimensions were achieved by using compact-packed LNA and mixers. However, a much more compact module may be designed by combining the LNAs, mixers, and the power divider on the same MMIC module. The electrical performance may degrade but a MMIC

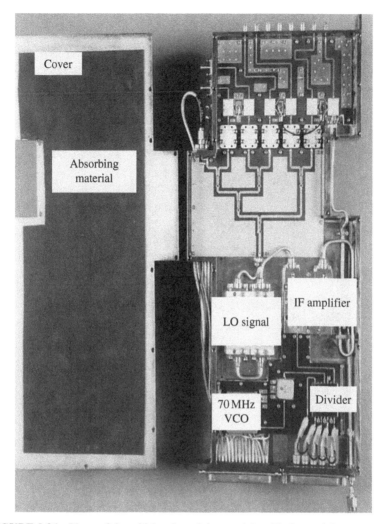

FIGURE 9.36 Photo of the wideband receiving module with the module cover ports.

module can be designed to meet the electrical specifications. The cost of the MMIC module in mass production will be cheaper than the presented module in Figure 9.35.

9.7 CONCLUSIONS

MIC and MMIC components and modules design was presented in this chapter.

In design of RF modules, we may gain the advantages of each technology by using both technologies in the development of RF transmitting and receiving modules. For

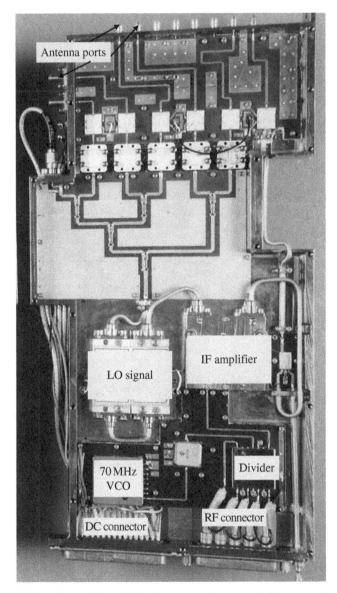

FIGURE 9.37 Photo of the wideband receiving direction finding system interface.

example, it is better to design filters by using MIC technology. However, compact and low-cost modules in mass production is achieved by using MMIC design. Design considerations of passive and active devices were presented in this chapter.

The cost of MMIC modules in mass production will be much cheaper than MIC modules. However, there are technology limitations in designing LNAs and high-power amplifiers.

REFERENCES

[1] Rogers J, Plett C. *Radio Frequency Integrated Circuit Design*. Boston: Artech House; 2003.

[2] Maluf N, Williams K. *An Introduction to Microelectromechanical System Engineering*. Boston: Artech House; 2004.

[3] Sabban A. Microstrip antenna arrays. In: Nasimuddin N, editor. *Microstrip Antennas*. Rijeka: InTech; 2011. p 361–384.

[4] Sabban A. Applications of MM wave microstrip antenna arrays. ISSSE 2007 Conference; Montreal, Canada; August 2007.

[5] Gauthier GP, Raskin GP, Rebiez GM, Kathei PB. A 94 GHz micro-machined aperture-coupled microstrip antenna. *IEEE Trans Antenna Propag* 1999;47 (12):1761–1766.

[6] Milkov MM. Millimeter-wave imaging system based on antenna-coupled bolometer [MSc thesis], University of California Los Angeles; 2000.

[7] de Lange G, Konistis K, Hu Q. A 3*3 mm-wave micro machined imaging array with sis mixers. *Appl Phys Lett* 1999;75 (6):868–870.

[8] Rahman A, de Lange G, Hu Q. Micro-machined room temperature micro bolometers for MM-wave detection. *Appl Phys Lett* 1996;68 (14):2020–2022.

[9] Mass SA. *Nonlinear Microwave and RF Circuits*. Boston: Artech House; 1997.

[10] Sabban A. *Low Visibility Antennas for Communication Systems*. Boca Raton: Taylor & Francis Group; 2015.

10

MICROELECTROMECHANICAL SYSTEMS (MEMS) TECHNOLOGY

10.1 INTRODUCTION

Communication and radar industry in microwave and mm-wave frequencies is currently in continuous growth. Radio-frequency (RF) modules such as front end, filters, power amplifiers, antennas, passive components, and limiters are important modules in radar and communication links (see Refs. [1–6]). The electrical performance of the modules determines if the system will meet the required specifications. Moreover, in several cases the modules' performance limits the system performance. Minimization of the losses, size, and weight of the RF modules is achieved by employing microelectromechanical systems (MEMS) technology. However, integration of MEMS components and modules raise several technical challenges. Design parameters that may be neglected at low frequencies cannot be ignored in the design of wideband integrated RF modules. Powerful RF design software, such as ADS and HFSS, are required to achieve accurate design of RF modules in mm-wave frequencies. Accurate design of mm-wave RF modules is crucial. It is an impossible mission to tune mm-wave RF modules in the fabrication process.

10.2 MEMS TECHNOLOGY

MEMS is the integration of mechanical elements, sensors, actuators, and electronics on a common silicon substrate through microfabrication technology. These devices replace bulky actuators and sensors with micron-scale equivalent that can be produced

Wideband RF Technologies and Antennas in Microwave Frequencies, First Edition. Dr. Albert Sabban.
© 2016 John Wiley & Sons, Inc. Published 2016 by John Wiley & Sons, Inc.

in large quantities by fabrication process used in integrated circuits (IC) in photolithography. They reduce cost, bulk, weight, and power consumption while increasing performance, production volume, and functionality by orders of magnitude.

The electronics are fabricated using IC process sequences (e.g., CMOS, bipolar, or BiCMOS processes), and the micromechanical components are fabricated using compatible "micromachining" processes that selectively etch away parts of the silicon wafer or add new structural layers to form the mechanical and electromechanical devices.

10.2.1 MEMS Technology Advantages

- Low insertion loss less than 0.1dB
- High isolation greater than 50dB
- Low distortion
- High linearity
- Very high Q
- Size reduction, system-on-a-chip
- High power handling approximately 40dBm
- Low power consumption (~milliwatt and no LNA)
- Low-cost high-volume fabrication

10.2.2 MEMS Technology Process

Bulk micromachining fabricates mechanical structures in the substrate by using orientation-dependent etching. Bulk-micromachined substrate is presented in Figure 10.1.

Surface micromachining fabricates mechanical structures above the substrate surface by using sacrificial layer.

Surface-micromachined substrate is presented in Figure 10.2.

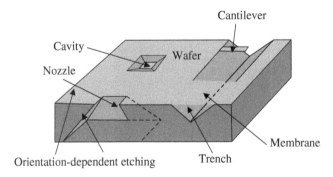

FIGURE 10.1 Bulk micromachining.

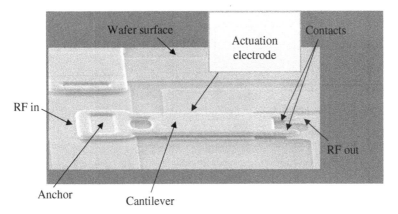

FIGURE 10.2 Surface micromachining.

In bulk micromachining process silicon is machined using various etching processes. Surface micromachining uses layers deposited on the surface of a substrate as the structural materials rather than using the substrate itself. Surface micromachining technique is relatively independent of the substrate used and therefore can be easily mixed with other fabrication techniques that modify the substrate first. An example is the fabrication of MEMS on a substrate with embedded control circuitry, in which MEMS technology is integrated with IC technology. This is being used to produce a wide variety of MEMS devices for many different applications. On the other hand, bulk micromachining is a subtractive fabrication technique, which converts the substrate, typically a single-crystal silicon, into the mechanical parts of the MEMS device. MEMS device is first designed with a computer-aided design (CAD) tool. The design outcome is a layout and masks that are used to fabricate the MEMS device. In Figure 10.3 MEMS fabrication process is presented. A summary of MEMS fabrication technology is presented in Table 10.1. In Figure 10.4 the block diagram of a MEMS bolometer-coupled antenna array is presented.

10.2.2.1 X-Ray Resist Packaging of the device tends to be more difficult, but structures with increased heights are easier to fabricate when compared to surface micromachining.

This is because the substrates can be thicker, resulting in relatively thick unsupported devices. Applications of RF MEMS technology are as follows:

- Tunable RF MEMS inductor
- Low-loss switching matrix
- Tunable filters
- Bolometer-coupled antenna array
- Low-cost W-band detection array

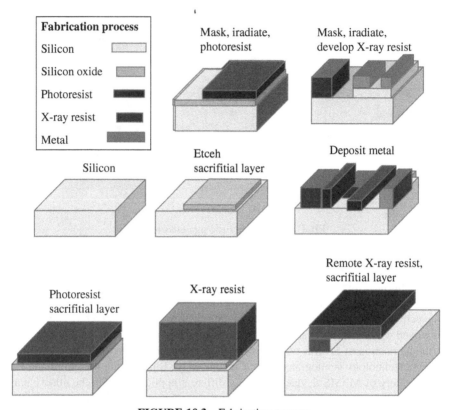

FIGURE 10.3   Fabrication process.

TABLE 10.1   Fabrication Technology

Fabrication Technology	Process
Surface micromachining	Release and drying systems to realize free-standing microstructures
Bulk micromachining	Dry etching systems to produce deep 2D free-form geometries with vertical sidewalls in substrates. Anisotropic wet etching systems with protection for wafer front sides during etching. Bonding and aligning systems to join wafers and perform photolithography on the stacked substrates

10.2.3   MEMS Components

MEMS components are categorized in one of several applications:

1. **Sensors** are a class of MEMS that are designed to sense changes and interact with their environments. These classes of MEMS include chemical, motion, inertia, thermal, RF sensors, and optical sensors. Microsensors are useful because of their small physical size, which allows them to be less invasive.

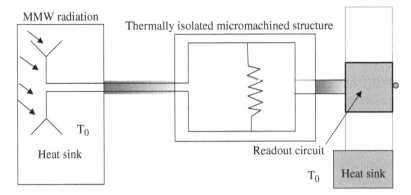

MMW radiation

Thermally isolated micromachined structure

T_0

Heat sink

Readout circuit

T_0 Heat sink

FIGURE 10.4   Bolometer-coupled antenna array.

2. **Actuators** are a group of devices designed to provide power or stimulus to other components or MEMS devices. MEMS actuators are either electrostatically or thermally driven.

3. **RF MEMS** are a class of devices used to switch or transmit high-frequency RF signals. Typical devices include metal contact switches, shunt switches, tunable capacitors, antennas, and so on.

4. **Optical MEMS** are devices designed to direct, reflect, filter, and/or amplify light. These components include optical switches and reflectors.

5. **Microfluidic MEMS** are devices designed to interact with fluid-based environments. Devices such as pumps and valves have been designed to move, eject, and mix small volumes of fluid.

6. **Bio-MEMS** are devices that, much like microfluidic MEMS, are designed to interact specifically with biological samples. Devices such as these are designed to interact with proteins, biological cells, medical reagents, and so on and can be used for drug delivery or other *in situ* medical analysis.

10.3   W-BAND MEMS DETECTION ARRAY

In this section we present the development of millimeter-wave radiation detection array. The detection array may employ around 256–1024 patch antennas. These patches are coupled to a resistor. Optimization of the antenna structure, feed network dimensions, and resistor structure allow us to maximize the power rate dissipated on the resistor. Design considerations of the detection antenna array are given in this section. Several imaging approaches are presented [7–12]. The common approach is based on an array of radiators (antennas) that receives radiation from a specific direction by using a combination of electronic and mechanical scanning. Another approach is based on a steering array of radiation sensors at the focal plane of a lens of reflector. The sensor can be an antenna coupled to a resistor.

10.3.1   Detection Array Concept

Losses in the microstrip feed network are very high in the W-band frequency range. In W-band frequencies we may design a detection array. The array concept is based on an antenna coupled to a resistor. A direct antenna-coupling surface to a micromachined microbridge resistor is used for heating and sensing. The feed network determines the antenna efficiency. The insertion loss of a gold microstrip line with width of 1 and 188 μm length is 4.4dB at 95 GHz. The insertion loss of a gold microstrip line with width of 10 and 188 μm length is 3.6dB at 95 GHz. The insertion loss of a gold microstrip line with width of 20 and 188 μm length is 3.2dB at 95 GHz. To minimize losses the feed line dimension was selected as $60 \times 10 \times 1$ μm. Analog CMOS readout circuit may be employed as a sensing channel per pixel. Figure 10.5 presents a pixel block diagram.

10.3.2   The Array Principle of Operation

The antenna receives effective mm-wave radiation. The radiation power is transmitted to a thermally isolated resistor coupled to a Ti resistor. The electrical power raises the structure temperature with a short response time. The same resistor changes its temperature and therefore its electrical resistance. Figure 10.6 shows a single-array pixel. The pixel consist a patch antenna, a matching network, printed resistor, and DC pads.

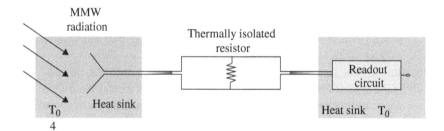

FIGURE 10.5   Antenna coupled to a resistor.

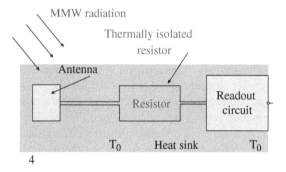

FIGURE 10.6   A single-array pixel.

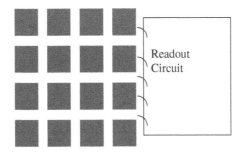

FIGURE 10.7   Array concept.

The printed resistor consist titanium lines and a titanium resistor coupled to an isolated resistor.

The operating frequency range of 92–100 GHz is the best choice. In the frequency range of 30–150 GHz there is a proven contrast between land, sky, and high transmittance of clothes. Size and resolution considerations promote higher frequencies above 100 GHz. Typical penetration of clothing at 100 GHz is 1dB and 5–10dB at 1 THz. Characterization and measurement considerations promote lower frequencies. The frequency range of 100 GHz allows sufficient bandwidth when working with illumination. The frequency range of 100 GHz is the best compromise. Figure 10.7 presents the array concept. Several types of printed antennas may be employed as the array element such as bow-tie dipole, patch antenna, and ring resonant slot.

10.3.3   W-Band Antenna Design

The bow-tie dipole and a patch antenna have been considered as the array element. Computed results shows that the directivity of the bow-tie dipole is around 5.3dBi and the directivity of a patch antenna is around 4.8dBi. However the length of the bow-tie dipole is around 1.5 mm and the size of the patch antenna is around 700 × 700 μm. We used a quartz substrate with thickness of 250 μm. The bandwidth of the bow-tie dipole is wider than that of a patch antenna. However, the patch antenna bandwidth meets the detection array electrical specifications. We chose the patch antenna as the array element since the patch size is significantly smaller than that of the bow-tie dipole. This feature allows us to design an array with a higher number of radiating elements. The resolution of detection array with a higher number of radiating elements is improved. We also realized that the matching network between the antenna and the resistor has smaller size for a patch antenna than that for a bow-tie dipole. The matching network between the antenna and the resistor consists of microstrip open stubs. Figure 10.8 shows the 3D radiation pattern of the bow-tie dipole. Figure 10.9 presents the S_{11} parameter of the patch antenna. The electrical performance of the bow-tie dipole and the patch antenna was compared. The VSWR of the patch antenna is better than 2 : 1 for 10% bandwidth. Figure 10.10 presents the 3D radiation pattern of the patch antenna at 95 GHz.

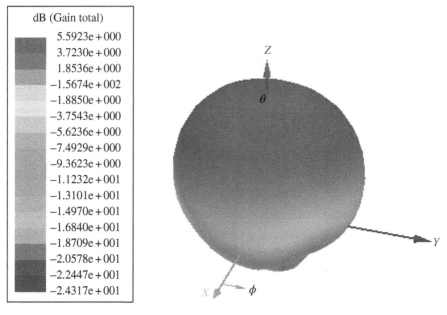

FIGURE 10.8 Dipole 3D radiation pattern.

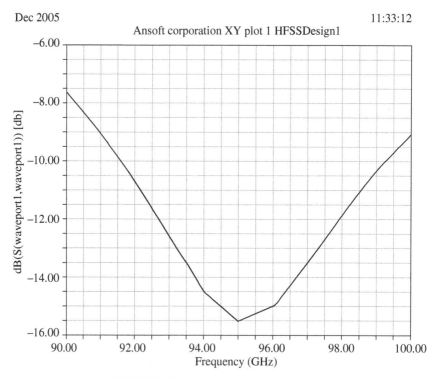

FIGURE 10.9 Patch S_{11} computed results.

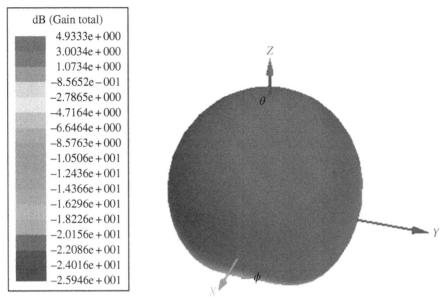

FIGURE 10.10   Patch 3D radiation pattern.

TABLE 10.2   Material Properties

Property	Units	siNi	Ti
Conductivity (K)	W/m/K	1.6	7
Capacity (C)	J/Kg/K	770	520
Density (ρ)	Gr/cm^3	2.85	4.5
Resistance	Ω/ (= square)	>1e8	90
Thickness	μm	0.1	0.1

10.3.4   Resistor Design

As described by Milkov [7], the resistor is thermally isolated from the patch antenna by using a sacrificial layer. Optimizations of the resistor structure maximize the power rate dissipated on the resistor. Ansoft HFSS software is employed to optimize the height of the sacrificial layer and the transmission line width and length. Dissipated power on a titanium resistor is higher than the dissipated power on a platinum resistor. The rate of the dissipated power on the titanium resistor is around 25%. The rate of the dissipated power on the platinum resistor is around 4%. Material properties are given in Table 10.2.

The sacrificial layer thickness may be 2–3 μm. Figure 10.11 shows the resistor configuration.

Figure 10.12 presents the MEMS bolometer layout.

The patch coupled to a bolometer is shown in Figure 10.13.

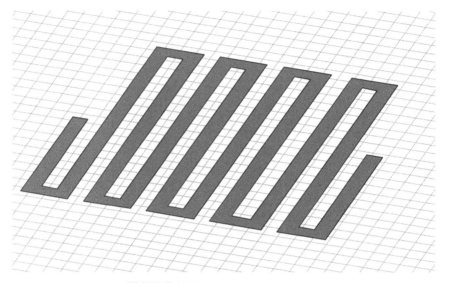

FIGURE 10.11   Resistor configuration.

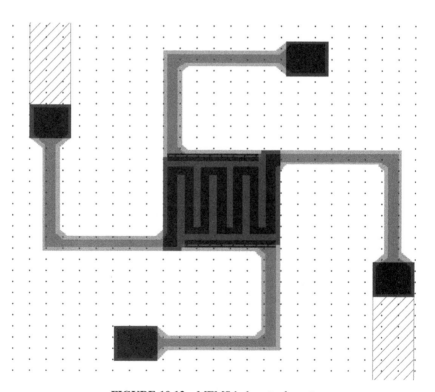

FIGURE 10.12   MEMS bolometer layout.

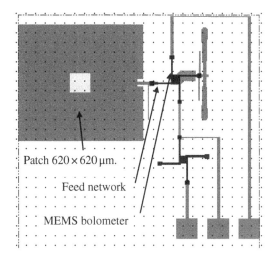

FIGURE 10.13   Patch antenna coupled to bolometer.

TABLE 10.3   Mask Process

Masks	Layer	Process	Layer Thickness (μm)
1	L 1 lift or etch	Gold reflector Au	1
2	L 2 etch	Streets open S.L.	3
3	L 3 etch	S.L. contacts	
4	L 4 etch	SiN + contacts	0.1
5	L 5 etch	Ti_1	0.1
		SiN	0.15
6	L 6 etch	VOx	0.1
7	L 7 etch	Contacts for Ti_2	
		Ti_2	0.1
	L 3 lift	Metal cap	0.1–0.5
8	L 8 etch	Ti_2	
		SiN	0.1
9	L 9 etch	Membrane definition	

10.4   ARRAY FABRICATION AND MEASUREMENT

Nine masks are used to fabricate the detection array. Mask process and layer thickness are listed in Table 10.3. Layer thickness has been determined as the best compromise between technology limits and design consideration.

Dimensions of detection array elements have been measured in several array pixels as part of visual test of the array after fabrication of the array, and some of the measured results are listed in Table 10.4. From results listed in Table 10.4 we may conclude

TABLE 10.4   Comparison of Design and Fabricated Array Dimensions

Element	Design (μm)	Pixel 1 (μm)	Pixel 2 (μm)
Patch width	600	599.5	600.5
Patch length	600	600.3	600.5
Hole width	100	99.8	100
Hole length	100	100	99.8
Feed line	10	10	10
Feed line	10	9.8	10
Stub width	2	2	1.8
Tapered line	15	15.2	14.8
Stub width	2	1.8	2
Tapered line	25	25.3	25.2

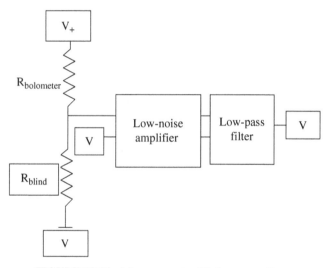

FIGURE 10.14   Measurements of bolometer voltage.

that the fabrication process is very accurate. There is a good agreement between computed and measured results of the array pixel electrical and mechanical parameters.

Figure 10.14 presents how we measure the bolometer output voltage. V_{ref} is the bolometer output voltage when no radiated power is received by the detection array. The voltage difference between bolometer voltage and V_{ref} is amplified by a low-noise differential amplifier. The rate of the dissipated power on the titanium resistor is around 25–30%.

Figure 10.15 presents the operational concept of the detection array.

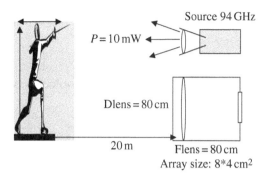

FIGURE 10.15 Detection array.

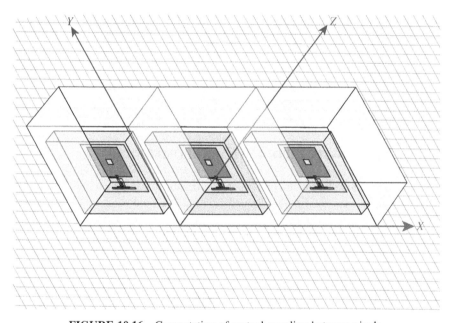

FIGURE 10.16 Computation of mutual coupling between pixels.

10.5 MUTUAL COUPLING EFFECTS BETWEEN PIXELS

HFSS software has been used to compute mutual coupling effects between pixels in the detection array as shown in Figure 10.16. Computation results indicate that the power dissipated on the centered pixels in the array is higher by 1–2% than the pixels located at the corners of the array.

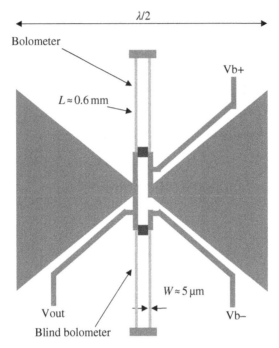

FIGURE 10.17   MEMS bow-tie dipole with bolometer.

10.6   MEMS BOW-TIE DIPOLE WITH BOLOMETER

A bow-tie dipole with bolometer printed on quartz substrate is shown in Figure 10.17.
The length of the bow-tie dipole is around 1.5 mm. The bolometer length is 0.6 mm.
The bolometer line width is 5 μm. Figure 10.18 presents the S_{11} parameter of the bow-
tie dipole. Figure 10.19 shows the 3D radiation pattern of the bow-tie dipole.
   Computed results shows that the directivity of the bow-tie dipole is around 5.3dBi.

10.7   220 GHz MICROSTRIP PATCH ANTENNA

Quartz substrate with a thickness of 50–100 μm has been used to fabricate microstrip
antennas at frequencies higher than 200 GHz. The size of the patch antenna is around
300×300 μm. Figure 10.20 presents the S_{11} parameter of the patch antenna.
Figure 10.21 shows the 3D radiation pattern of the patch antenna.

10.8   CONCLUSIONS

Losses in the microstrip feed network form a significant limit on the possible applications of microstrip antenna arrays in mm-wave frequencies. Array of patch antenna
coupled to a bolometer decreases the effect of losses in the array feed network.

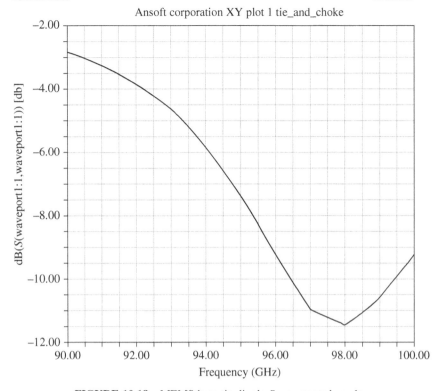

FIGURE 10.18   MEMS bow-tie dipole S_{11} computed results.

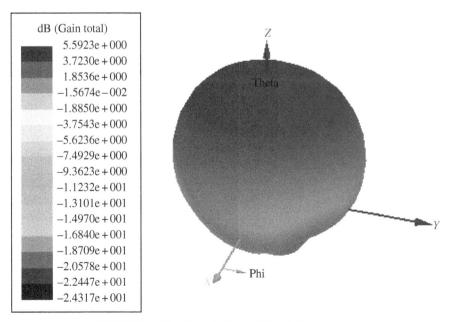

FIGURE 10.19   Bow-tie dipole 3D radiation pattern.

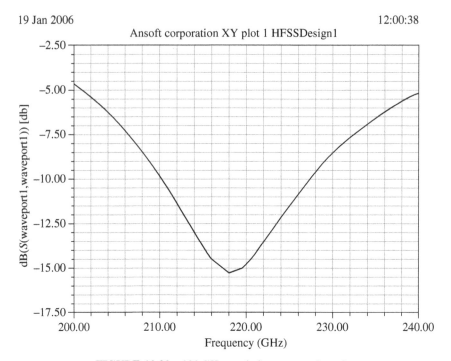

FIGURE 10.20   220 GHz patch S_{11} computed results.

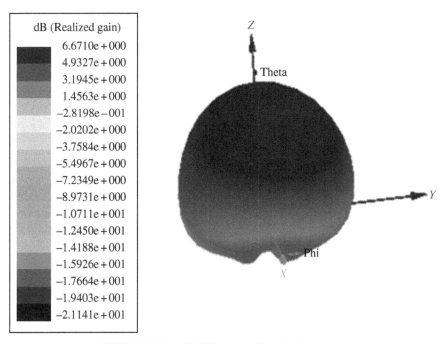

FIGURE 10.21   220 GHz patch 3D radiation pattern.

The array may be constructed from 256 to 1024 elements at 94 or at 220 GHz. Design considerations of the antenna and the feed network are given in this chapter. Optimization of the antenna structure and feed network allows us to design and fabricate microstrip antenna arrays with high efficiency. Millimeter-wave detection arrays may be applied to detect explosive materials and weapons underneath terrorist clothes. MEMS technology has been applied to produce low-cost millimeter-wave detection arrays. Dimension of the detection array elements has been measured in several array pixels as part of visual test of the array after fabrication of the array. The fabrication process was very accurate. There is a good agreement between computed and measured results of the array pixel electrical and mechanical parameters.

REFERENCES

[1] Rogers J, Plett C. *Radio frequency Integrated Circuit Design*. Boston: Artech House; 2003.

[2] Maluf N, Williams K. *An Introduction to Microelectromechanical System Engineering*. Boston: Artech House; 2004.

[3] Sabban A. Microstrip antenna arrays. In: Nasimuddin N, editor. *Microstrip Antennas*. Rijeka: InTech; 2011. p 361–384.

[4] Sabban A. Applications of MM wave microstrip antenna arrays. ISSSE 2007 Conference; August 2007; Montreal, Canada.

[5] Gauthier GP, Raskin GP, Rebiez GM, Kathei PB. A 94 GHz micro-machined aperture-coupled microstrip antenna. IEEE Trans Antenna Propag 1999;47 (12):1761–1766.

[6] Mass SA. *Nonlinear Microwave and RF Circuits*. Boston: Artech House; 1997.

[7] Milkov MM. Millimeter-wave imaging system based on antenna-coupled bolometer [MSc thesis], University of California Los Angeles; 2000.

[8] de Lange G, Konistis K, Hu Q. A 3∗3 mm-wave micro machined imaging array with sis mixers. Appl Phys Lett 1999;75 (6):868–870.

[9] Rahman A, de Lange G, Hu Q. Micro-machined room temperature micro bolometers for MM-wave detection. Appl Phys Lett 1996;68 (14):2020–2022.

[10] Luukanen HA, Espoo EH, Veikkola VV. Imaging system functioning on submillimeter waves. US patent 6,242,740; June 2001.

[11] Sinclair GN, Appleby R, Coward P, Price S. Passive millimeter wave imaging in security scanning. Proc SPIE 2000;4032:40–45.

[12] Sabban A. *Low Visibility Antennas for Communication Systems*. Boca Raton: Taylor & Francis Group; 2015.

11

LOW-TEMPERATURE COFIRED CERAMIC (LTCC) TECHNOLOGY

11.1 INTRODUCTION

Communication and radar industry in microwave and mm-wave frequencies is currently in continuous growth. Radio-frequency (RF) modules such as front end, filters, power amplifiers, antennas, passive components, and limiters are important modules in radar and communication links (see Refs. [1–7]). The electrical performance of the modules determines if the system will meet the required specifications. Moreover, in several cases the modules' performance limits the system performance. Minimization of the losses, size, and weight of the RF modules is achieved by employing low-temperature cofired ceramic (LTCC) technology (see Refs. [3, 4]). However, integration of LTCC components and modules raise several technical challenges. Design parameters that may be neglected at low frequencies and in microwave integrated circuit (MIC) components cannot be ignored in the design of compact integrated LTCC RF modules. Powerful RF design software, such as ADS, HFSS, and CST, are required to achieve accurate design of RF modules in mm-wave frequencies. Accurate design of mm-wave RF modules is crucial. It is an impossible mission to tune integrated mm-wave RF modules in the fabrication process.

Wideband RF Technologies and Antennas in Microwave Frequencies, First Edition. Dr. Albert Sabban.
© 2016 John Wiley & Sons, Inc. Published 2016 by John Wiley & Sons, Inc.

11.2　LTCC AND HTCC TECHNOLOGY FEATURES

Cofired ceramic devices are monolithic, ceramic microelectronic devices where the entire ceramic support structure and any conductive, resistive, and dielectric materials are fired in a kiln at the same time. Typical devices include capacitors, inductors, resistors, transformers, and hybrid circuits. The technology is also used for a multilayer packaging for the electronics industry, such as military electronics. Cofired ceramic devices are made by processing a number of layers independently and assembling them into a device as a final step. Cofiring can be divided into low-temperature (LTCC) and high-temperature (high-temperature cofired ceramic (HTCC)) applications: low temperature means that the sintering temperature is below 1000°C (1830°F), while high temperature is around 1600°C (2910°F). There are two types of raw ceramics to manufacture multilayer ceramic (MLC) substrate:

- Ceramics fired at high temperature ($T \geq 1500°C$): HTCC
- Ceramics fired at low temperature ($T \leq 1000°C$): LTCC

The base material of HTCC is usually Al_2O_3. HTCC substrates are raw ceramic sheets. Because of the high firing temperature of Al_2O_3, the material of the embedded layers can only be high-melting-temperature metals: wolfram, molybdenum, or manganese. The substrate is unsuitable to bury passive elements, although it is possible to produce thick-film networks and circuits on the surface of HTCC.

The breakthrough for LTCC fabrication was when the firing temperature of ceramic–glass substrate was reduced to 850°C. The equipment for conventional thick-film process could be used to fabricate LTCC devices. LTCC technology evolved from HTCC technology combined the advantageous features of thick-film technology. Because of the low firing temperature (850°C), the same materials are used for producing buried and surface wiring and resistive layers as thick-film hybrid IC (i.e., Au, Ag, Cu wiring RuO_2-based resistive layers). It can be fired in

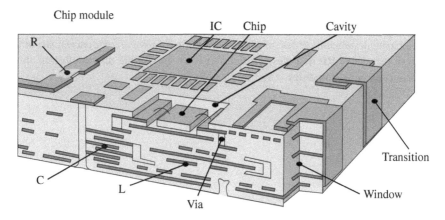

FIGURE 11.1　LTCC module.

TABLE 11.1 Dielectric Properties at 9 GHz of LTCC Substrates

Material	ε_r	Tan $\delta \times 10^{-3}$
99.5% AL	9.98	0.1
LTCC1	7.33	3.0
LTCC2	6.27	0.4
LTCC3	7.2	0.6
LTCC4	7.44	1.2
LTCC5	6.84	1.3
LTCC6	8.89	1.4

an oxygen-rich environment unlike HTCC boards, where reduced atmosphere is used. During cofiring the glass melts, and the conductive and ceramic particles are sintered. On the surface of LTCC substrates, hybrid integrated circuits can be realized, shown in Figure 11.1. Passive elements can be buried into the substrate, and we can place semiconductor chips in a cavity. Dielectric properties at 9 GHz of LTCC substrates are listed in Table 11.1.

11.3 LTCC AND HTCC TECHNOLOGY PROCESS

• Low Temperature	LTCC 875°C
• **High T**emperature	HTCC 1400–1600°C
• Cofired	Cofiring of (di)electric pastes
	LTCC: precious metals (Au, Ag, Pd, Cu)
	HTCC: refractory metals (W, Mo, MoMn)
• Ceramic	Mix of alumina Al_2O_3
	Glasses SiO_2–B_2O_3–CaO–MgO
	Organic binders
	HTTC: essentially Al_2O_3

In Table 11.2 LTCC process steps are listed. LTCC raw material comes as sheets or rolls. Material manufacturers are DuPont, ESL, Ferro, and Heraeus.

In Figure 11.2 LTCC process block diagram is presented. In Table 11.3 several electrical, thermal, and mechanical characteristics of several LTCC materials are listed.

In Table 11.4 LTCC line losses at 2 GHz are listed for several LTCC materials. For LTTC1 material losses are 0.004dB/mm.

11.4 DESIGN OF HIGH-PASS LTCC FILTERS

The trend in the wireless industry toward miniaturization, cost reduction, and improved performance drives microwave designer to develop microwave components in LTCC technology. A significant reduction in the size and the cost of microwave components may be achieved by using LTCC technology. In LTCC technology

TABLE 11.2   LTCC Process List

LTCC Process
Tape casting
Sheet cutting
Laser punching
Printing
Cavity punching
Stacking
Bottom side printing
Pressing
Side hole formation
Side hole printing
Snap line formation
Pallet firing
Plating Ni/Au

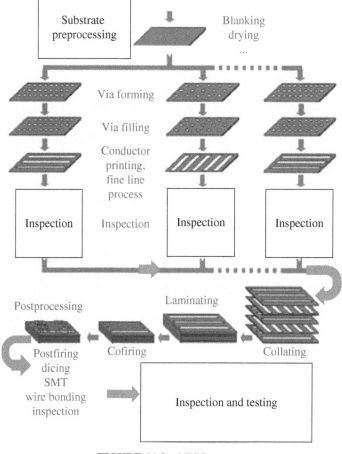

FIGURE 11.2   LTCC process.

TABLE 11.3 LTCC Material Characteristics

Material	LTCC DP951	Al$_2$O$_3$ 96%	BeO	AIN 98%
Electrical characteristics at 10 MHz				
Dielectric constant, ε_r	7.8	9.6	6.5	8.6
Dissipation factor, tan δ	0.00015	0.0003	0.0002	0.0005
Thermal characteristics				
Thermal expansion $10^{-6}/°C$	5.8	7.1	7.5	4.6
Thermal conductivity W/mk 25–300°C	3	20.9	251	180
Mechanical characteristics				
Density	3.1	3.8	2.8	3.3
Flexural strength (MPa)	320	274	241	340
Young's modulus (GPa)	120	314	343	340

TABLE 11.4 LTCC Line Loss

Material	Dissipation Factor, Tan $\delta \times 10^{-3}$	Line Loss dB/mm at 2 GHz
LTTC1	3.8	0.004
LTTC2	2.0	0.0035
LTTC6-CT2000	1.7	0.0033
Alumina 99.5%	0.65	0.003

discrete surface-mounted components such as capacitors and inductors are replaced by integrated printed components. LTCC technology allows the designer to use multilayer design if needed to reduce the size and cost of the circuit. However, multilayer design results in more losses due to connections and due to parasitic coupling between different parts of the circuit. To improve the filter performance, all the filter parameters have been optimized. Package effects were taken into account in the design.

11.4.1 High-Pass Filter Specification

Frequency	1.5–2.5 GHz
Insertion loss 1.1 Fo	1dB
Rejection 0.9 Fo	3dB
Rejection 0.75 Fo	20dB
Rejection 0.5 Fo	40dB
VSWR	2 : 1
Case dimensions	$700 \times 300 \times 25.5$ mil inch

The filters are realized by using lumped elements. The filter inductor and capacitor parameters were optimized by using HP ADS software. The filter consists of five

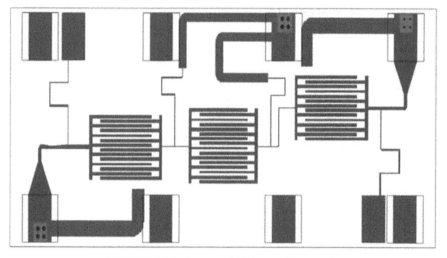

FIGURE 11.3 Layout of high-pass filter no. 1.

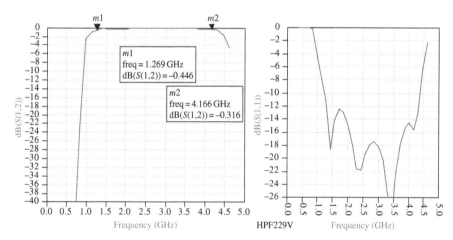

FIGURE 11.4 S_{12} and S_{11} results of high-pass filter no. 1.

layers of 5.1 mil substrate with $\varepsilon_r = 7.8$. Package effects were taken into account in the design. Changes in the design were made to compensate and minimize package effects. In Figure 11.3 the filter layout is presented. S_{11} and S_{12} momentum simulation results are shown in Figure 11.4. In Figure 11.5 the filter 2 layout is presented. S_{11} and S_{12} momentum simulation results are shown in Figure 11.6. Simulation results of tolerance check are shown in Figure 11.7. The parameters that were tested in the tolerance check are inductor and capacitor line width and length and spacing between capacitor fingers.

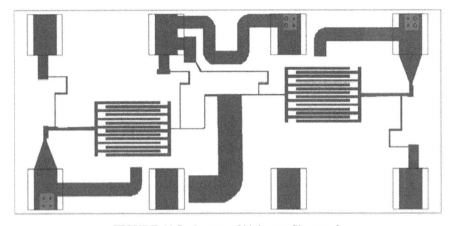

FIGURE 11.5   Layout of high-pass filter no. 2.

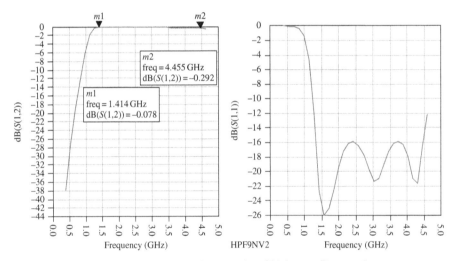

FIGURE 11.6 S_{12} and S_{11} results of high-pass filter no. 2.

11.5   COMPARISON OF SINGLE-LAYER AND MULTILAYER MICROSTRIP CIRCUITS

In a single-layer microstrip circuit all conductors are in a single layer. Coupling between conductors is achieved through edge or end proximity (across narrow gaps). Single-layer microstrip circuits are cheap in production. In Figure 11.8 single-layer microstrip edge coupled filter is shown.

Figure 11.9 presents the layout of single-layer microstrip directional coupler. Figure 11.10 presents the structure of a multilayer microstrip coupler.

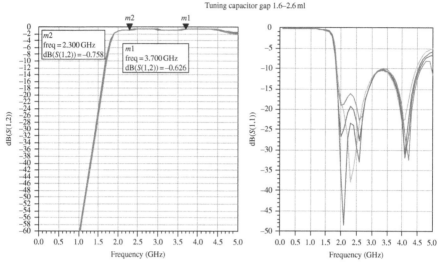

FIGURE 11.7   Tolerance simulation for spacing between capacitor fingers.

FIGURE 11.8   Edge-coupled filter.

FIGURE 11.9   Single-layer microstrip directional coupler.

In multilayer microwave circuits conductors are separated by dielectric layers and stacked on different layers. This structure allows for (strong) broadside coupling. Registration between layers is not difficult to achieve as narrow gaps between strips in single-layer circuits. Multilayer structure technique is well suited to thick-film print technology and also suitable for LTCC technology. Thick-film print technology provides low-cost multilayer microwave circuits with good feasibility for mass

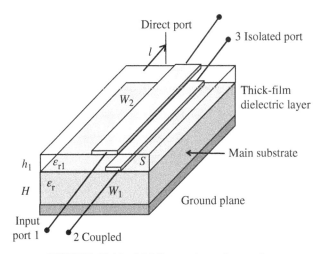

FIGURE 11.10   Multilayer microstrip coupler.

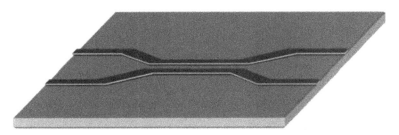

FIGURE 11.11   Single-layer planar side-coupled directional coupler.

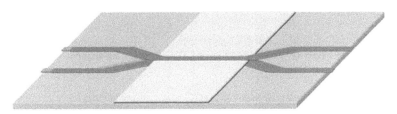

FIGURE 11.12   Multilayer printed directional coupler.

production. However, fabrication of narrow printed lines and gaps has limited quality. Thick-film technology provides an important fabrication process for multilayer microwave and mm-wave circuits. Figure 11.11 presents the layout of single-layer microstrip side-coupled directional coupler. Figure 11.12 presents the structure of a multilayer microstrip directional coupler.

Figure 11.13 presents the layout of single-layer microstrip DC block. Figure 11.14 presents the structure of a multilayer DC block.

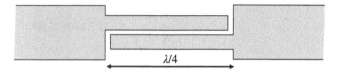

FIGURE 11.13   Single-layer microstrip DC block.

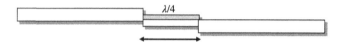

FIGURE 11.14   Multilayer microstrip DC block.

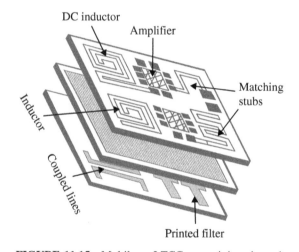

FIGURE 11.15   Multilayer LTCC transmitting channel.

Figure 11.15 presents a multilayer LTCC transmitting channel.

The transmitting channel consist of power amplifier, filters, coupler, and matching network.

11.6   LTCC MULTILAYER TECHNOLOGY DESIGN CONSIDERATIONS

MIC consist of discrete component, interconnection board, and a package. MIC structure complicates the assembly of the circuit and increases losses, weight, and volume. Multilayer LTCC technology reduces the circuit losses, weight, and volume. HTCC

TABLE 11.5   Physical Characteristics of Conductors

Conductor	Resistivity ($\mu\Omega$/cm)	Melting Point (°C)
Tungsten	5.5	3370
Gold	2.44	1063
Copper	1.72	1083
Silver	1.63	961

technology may be used to fabricate hermetic packages. However, the high temperature of the HTCC process limits the use of metals with low resistivity and low melting point. Tungsten with high resistivity (5.5 $\mu\Omega$/cm) increases the losses of printed transmission lines. Physical characteristics of conductors are listed in Table 11.5.

Gold and silver have low resistivity and a melting point lower than 1400°C as required in the HTCC process. Gold and silver cannot be used in HTCC fabrication process. However, they can be used in LTCC process. In LTCC fabrication process each layer is processed separately. This is a significant advantage over other multilayer technologies. In LTCC fabrication process each layer can be as small as 3 mil inch in diameter. This feature provides high interconnect density, good RF ground, and shielding. LTCC may provide around 40 layers with a single firing process. In the LTCC process we may construct various geometries by layer cutouts. This features allow us to form high-frequency structures and transitions between different transmission lines. In the LTCC process microstrip line, stripline, and coplanar transmission lines may be integrated in the same multilayer structure. The three-dimensional LTCC process provides impedance control and shielding of RF signals and DC control signals via a low-loss medium within the same multilayer structure with high isolation between signals.

11.6.1   Buried Components

11.6.1.1   Resistors   Resistors are characterized by the conductor resistivity, ρ, thickness, t, and the resistor dimensions as given in Equation 11.1 (resistor length, l, and width, W):

$$R = \rho \frac{l}{tW} \tag{11.1}$$

11.6.1.2   Capacitors   Capacitors are characterized by the dielectric constant, ε, the capacitor area, A, and the distance, d, between the conductors. The capacitance value may be calculated by using Equation 11.2:

$$C = \varepsilon \frac{A}{d} \tag{11.2}$$

Resistor and capacitor expected tolerances are listed in Table 11.6.

TABLE 11.6   Resistor and Capacitor Expected Tolerances

Component Tolerance (±%)	Resistor Tolerance (±%)	Capacitor Tolerance (±%)
Resistivity ρ	10	
Thickness, t	10	
Length, l	2	
Width, W	2	
Dielectric constant, ε	2	2
Capacitor area, A	1	1
Distance, d	3	3
Total tolerance (±%)	27	7

TABLE 11.7   Toroidal Inductor Expected Tolerances

Component Tolerance (±%)	Inductor Tolerance (±%)
Permeability, μ	2
Thickness, h	3
$2\pi \ln\left(b/a\right)$	1
Total tolerance (±%)	5

11.6.1.3   Toroidal Inductors   Inductors are characterized by the permeability, μ, number of turns N, conductor thickness, h, and conductor inner diameter, a, and outer diameter, b:

$$L = \mu h N^2 / 2\pi \ln\left(\frac{b}{a}\right) \tag{11.3}$$

Toroidal inductor expected tolerances are listed in Table 11.7.

11.7   CAPACITOR AND INDUCTOR QUALITY (Q) FACTOR

The Q factor is defined as the total energy stored in the circuit compared to the total energy dissipated in the circuit in one cycle:

$$Q = 2\pi \frac{E_{\max}}{E_{\text{dis}}} \tag{11.4}$$

The maximum power stored in the RLC circuit, shown in Figure 11.16, is given in Equation 11.5:

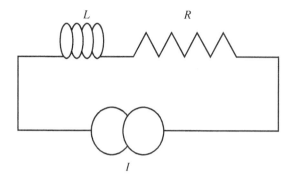

FIGURE 11.16   RLC resonant circuit.

$$E_{max} = \frac{L I^2_{MAX}}{2} \tag{11.5}$$

where $I = I_{max} \cos \omega t$.

The dissipated power stored in the RLC circuit is given in Equation 11.6:

$$E_{max} = \tau \frac{I^2_{MAX}}{2} \tag{11.6}$$

where $\tau = 1/f$ is the time period and $\tau = 2\pi/\omega$.

The quality factor, Q, is given in Equation 11.7:

$$Q = \frac{\omega L}{R} \tag{11.7}$$

At the circuit resonant circuit the quality factor, Q_0, is given in Equation 11.8:

$$Q_0 = \frac{\omega_0 L}{R} \text{ where } \omega_0 = \frac{1}{\sqrt{LC}} \tag{11.8}$$

The quality factor, Q_0, as function of circuit bandwidth is given in Equation 11.9:

$$Q_0 = \frac{\omega_0}{B} \text{ where } B = \omega_2 - \omega_1 \tag{11.9}$$

The quality factor, Q_0, of 10 pf embedded capacitor in HTCC circuit, with tan δ of 0.002, with tungsten conductors is around 340. The quality factor, Q_0, of 10 pf embedded capacitor in LTCC circuit, with tan δ of 0.002, with silver conductors is around 400. The quality factor, Q_0, of 10 pf embedded capacitor in HTCC circuit, with tan δ of 0.001, with tungsten conductors is around 500. The quality factor, Q_0, of 10 pf embedded capacitor in LTCC circuit, with tan δ of 0.002, with silver conductors is around 600.

The quality factor, Q_0, of 10 nH embedded capacitor in HTCC circuit, with tungsten conductors is around 50. The quality factor, Q_0, of 10 pf embedded capacitor in LTCC circuit, with silver conductors is around 200.

11.8   SUMMARY OF LTCC PROCESS ADVANTAGES AND LIMITATIONS

Several advantages and limitation of LTCC technology are listed in this section.

11.8.1   LTCC Process Advantages

- Low permittivity tolerance
- Good thermal conductivity
- Low TCE (adapted to silicon and GaAs)
- Excellently suited for multilayer modules
- Integration of cavities and passive elements such as R, L, and C components
- Very robust against mechanical and thermal stress (hermetically sealed)
- Composable with fluidic, chemical, thermal, and mechanical functionalities
- Low material costs for silver conductor paths
- Low production costs for medium and large quantities

Advantages for high-frequency applications are:

- Parallel processing (high yield, fast turnaround, lower cost)
- Precisely defined parameters
- High-performance conductors
- Potential for multilayer structures
- High interconnect density

11.8.2   LTCC Limitations

The main limitation of LTCC limitation is associated with the limit of forming precision geometries of components with the fabrication processes. This limitation imposes limit on the shape, size, and density of components and interconnections. These facts narrow the range of impedances that can be realized within the LTCC process.

11.9   CONCLUSIONS

Dimension and losses of microwave systems are minimized by using MIC, MMIC, MEMS, and LTCC technology. Dimension and losses of microwave systems are minimized by using multilayer structure technique. Multilayer structure technique is well suited to thick-film print technology and for LTCC technology. LTCC technology allows the integration of cavities and passive elements such as R, L, and C components as part of the LTCC circuits. Sensors, actuators, and RF switches may be manufactured by using MEMS technology. Losses of MEMS components are considerably lower than MIC and MMIC RF components. MMICs are circuits in which active and passive elements are formed on the same dielectric substrate. MMICs are dimensionally small (from around 1 to 10 mm^2) and can be mass-produced. MMIC components cannot be tuned. Accurate design is crucial in the design of MMICs. MMIC, MEMS, and LTCC designers' goal is to comply with customer specifications in one design iteration.

REFERENCES

[1] Rogers J, Plett C. *Radio Frequency Integrated Circuit Design*. Boston: Artech House; 2003.

[2] Maluf N, Williams K. *An Introduction to Microelectromechanical System Engineering*. Boston: Artech House; 2004.

[3] Sabban A. *Low Visibility Antennas for Communication Systems*. Boca Raton: Taylor & Francis Group; 2015.

[4] Brown L, Ehlert ME. A flexible manufacturing approach for low-temperature co-fired ceramic MCM. Proceedings of International Symposium on Microelectronics, International Society for Hybrid Microelectronics; San Francisco, CA; 1992. p 70–75.

[5] Mass SA. *Nonlinear Microwave and RF Circuits*. Boston: Artech House; 1997.

[6] Sabban A. Microstrip antenna arrays. In: Nasimuddin N, editor. *Microstrip Antennas*. Rijeka: InTech; 2011. p 361–384.

[7] Sabban A. Applications of MM wave microstrip antenna arrays. ISSSE 2007 Conference; August 2007; Montreal, Canada.

12

ADVANCED ANTENNA TECHNOLOGIES FOR COMMUNICATION SYSTEM

Communication and biomedical industry is in rapid growth in the last decade. Low-profile small antennas are crucial in the development of commercial compact systems. Small printed antennas suffer from low efficiency.

12.1 NEW WIDEBAND WEARABLE METAMATERIAL ANTENNAS FOR COMMUNICATION APPLICATIONS

Metamaterial technology is used to design small wideband wearable antennas with high efficiency. Design considerations and computed and measured results of printed metamaterial antennas with high efficiency are presented in this chapter. The proposed antenna may be used in communication and Medicare systems. The antennas S_{11} results for different positions on the human body are presented in this chapter. The gain and directivity of the patch antenna with split ring resonator (SRR) are higher by 2.5dB than the patch antenna without SRR. The resonant frequency of the antenna with SRR on human body is shifted by 3%.

12.1.1 Introduction

Microstrip antennas are widely used in communication systems. Microstrip antennas have several advantages such as low profile, flexible, light weight, small volume, and

Wideband RF Technologies and Antennas in Microwave Frequencies, First Edition. Dr. Albert Sabban.
© 2016 John Wiley & Sons, Inc. Published 2016 by John Wiley & Sons, Inc.

low production cost. Compact printed antennas are presented in journals and books, as referred in Refs. [1–4]. However, small printed antennas suffer from low efficiency. Metamaterial technology is used to design small printed antennas with high efficiency. Printed wearable antennas were presented in Ref. [5]. Artificial media with negative dielectric permittivity were presented in Ref. [6]. Periodic SRR and metallic post structures may be used to design materials with dielectric constant and permeability less than 1 as presented in Refs. [6–14]. In this chapter metamaterial technology is used to develop small antennas with high efficiency. RF transmission properties of human tissues have been investigated in several papers such as Refs. [15, 16]. Several wearable antennas have been presented in papers in the last years as referred in Refs. [17–24]. New wearable printed metamaterial antennas with high efficiency are presented in this chapter. The bandwidth of the metamaterial antenna with SRR and metallic strips is around 50% for VSWR better than 2.3 : 1. Computed and measured results of meta-materials antennas on the human body are discussed in this chapter.

12.1.2 New Antennas with Split Ring Resonators

A microstrip dipole antenna with SRR is shown in Figure 12.1. The microstrip loaded dipole antenna with SRR provides horizontal polarization. The slot antenna provides vertical polarization. The resonant frequency of the antenna with SRR is 400 MHz. The resonant frequency of the antenna without SRR is 10% higher. The antennas shown in Figure 12.1 consist of two layers. The dipole feed network is printed on the first layer. The radiating dipole with SRR is printed on the second layer. The thickness of each layer is 0.8 mm. The dipole and the slot antenna create dual-polarized antenna. The computed S_{11} parameters are presented in Figure 12.2.

The length of the dual-polarized antenna with SRR shown in Figure 12.1 is 19.8 cm. The length of the dual-polarized antenna without SRR shown in Figure 12.3 is 21 cm. The ring width is 1.4 mm and the spacing between the rings is 1.4 mm. The antennas have been analyzed by using Agilent ADS software. The matching stub locations and dimensions have been optimized to get the best VSWR results. The length of the stub L in Figures 12.1 and 12.3 is 10 mm. The locations and number of the coupling stubs may vary the antenna axial ratio from 0 to 30dB. The number of coupling stubs

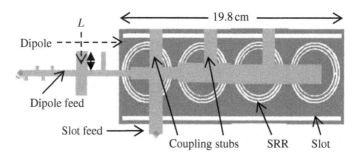

FIGURE 12.1 Printed antenna with split ring resonators.

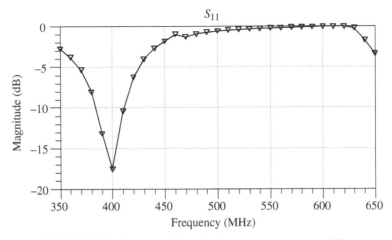

FIGURE 12.2   Antenna with split ring resonators, computed S_{11}.

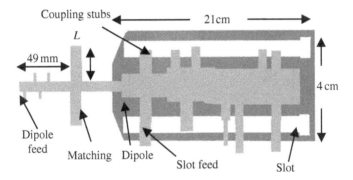

FIGURE 12.3   Dual-polarized microstrip antenna.

may be minimized. The number of coupling stubs in Figure 12.1 is three. The antenna axial ratio value may be adjusted also by varying the slot feed location. The dimensions of the antenna shown in Figure 12.3 are presented in Ref. [5]. The bandwidth of the antenna shown in Figure 12.3 is around 10% for VSWR better than 2 : 1. The antenna beamwidth is 100°. The antenna gain is around 2 dBi. The computed S_{11} parameters are presented in Figure 12.4. Figure 12.5 presents the antenna measured S_{11} parameters. There is a good agreement between measured and computed results. The antenna presented in Figure 12.1 has been modified as shown in Figure 12.6. The location and the dimension of the coupling stubs have been modified to get two resonant frequencies. The first resonant frequency is 370 MHz and is lower by 20% than the resonant frequency of the antenna without the SRR (Figure 12.7).

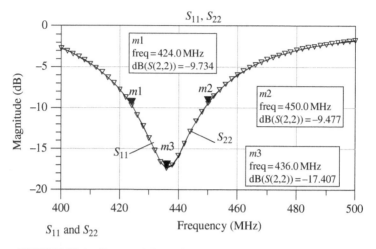

FIGURE 12.4   Computed S_{11} and S_{22} results of antenna without SRR.

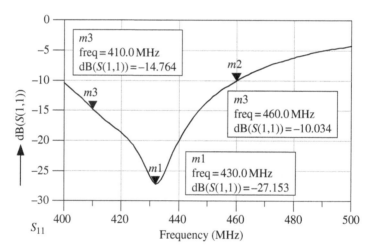

FIGURE 12.5   Measured S_{11} of the antenna without SRR.

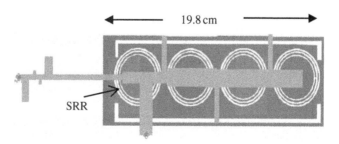

FIGURE 12.6   Antenna with SRR with two resonant frequencies.

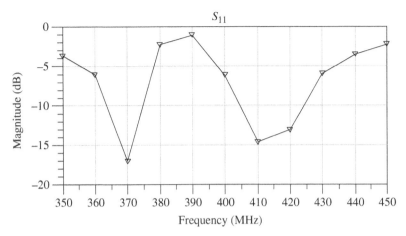

FIGURE 12.7 S_{11} for antenna with two resonant frequencies.

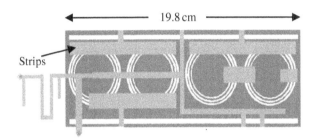

FIGURE 12.8   Antenna with SRR and metallic strips.

Metallic strips have been added to the antenna with SRR as presented in Figure 12.8. The computed S_{11} parameter of the antenna with metallic strips is presented in Figure 12.9. The antenna bandwidth is around 50% for VSWR better than $3:1$. The computed radiation pattern is shown in Figure 12.10. The 3D computed radiation pattern is shown in Figure 12.11. The directivity and gain of the antenna with SRR are around 5 dBi (see Figure 12.12). The directivity of the antenna without SRR is around 2 dBi. The length of the antennas with SRR is smaller by 5% than the antennas without SRR. Moreover, the resonant frequency of the antennas with SRR is lower by 5–10%.

The feed network of the antenna presented in Figure 12.8 has been optimized to yield VSWR better than $2:1$ in frequency range of 250–440 MHz. Optimization of the number of the coupling stubs and the distance between the coupling stubs may be used to tune the antenna resonant frequency. An optimized antenna with two coupling stubs has two resonant frequencies. The first resonant frequency is 370 MHz and the second resonant frequency is 420 MHz. Antenna with SRR with two coupling stubs is presented in Figure 12.13.

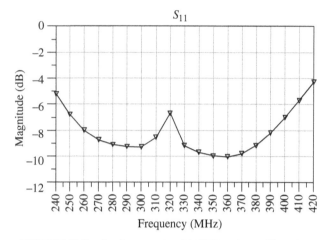

FIGURE 12.9 S_{11} for antenna with SRR and metallic strips.

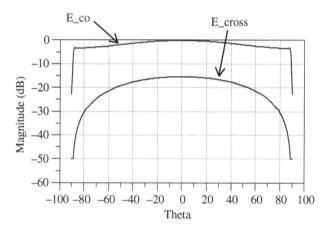

FIGURE 12.10   Radiation pattern for antenna with SRR.

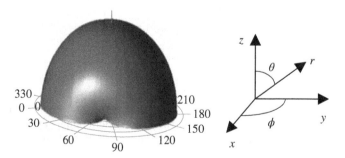

FIGURE 12.11   3D radiation pattern for antenna with SRR.

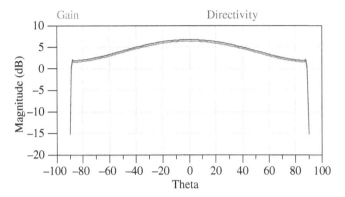

FIGURE 12.12   Directivity of the antenna with SRR.

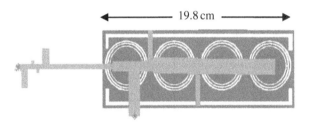

FIGURE 12.13   Antenna with SRR with two coupling stubs.

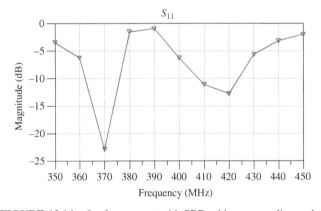

FIGURE 12.14 S_{11} for antenna with SRR with two coupling stubs.

The computed S_{11} parameter of the antenna with two coupling stubs is presented in Figure 12.14. The 3D radiation pattern for antenna with SRR and two coupling stubs is shown in Figure 12.15.

The antenna with metallic strips has been optimized to yield wider bandwidth as shown in Figure 12.16. The computed S_{11} parameter of the modified antenna with metallic strips is presented in Figure 12.17. The antenna bandwidth is around 50% for VSWR better than 2.3 : 1.

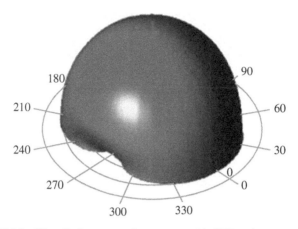

FIGURE 12.15   3D radiation pattern for antenna with SRR and two coupling stubs.

FIGURE 12.16   Wideband antenna with SRR and metallic strips.

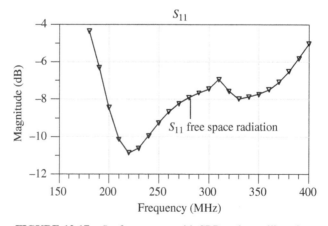

FIGURE 12.17 S_{11} for antenna with SRR and metallic strips.

12.1.3 Folded Dipole Metamaterial Antenna with SRR

The length of the antenna shown in Figure 12.3 may be reduced from 21 to 7 cm by folding the printed dipole as shown in Figure 12.18. Tuning bars are located along the feed line to tune the antenna to the desired frequency. The antenna bandwidth is around 10% for VSWR better than $2:1$ as shown in Figure 12.19. The antenna beamwidth is around 100°. The antenna gain is around 2dBi. The size of the antenna with SRR shown in Figure 12.6 may be reduced by folding the printed dipole as shown in Figure 12.20. The dimensions of the folded dual-polarized antenna with SRR presented in Figure 12.20 are $11 \times 11 \times 0.16$ cm. Figure 12.21 presents the antenna computed S_{11} parameters. The antenna bandwidth is 10% for VSWR better than $2:1$. The computed radiation pattern of the folded antenna with SRR is shown in Figure 12.22.

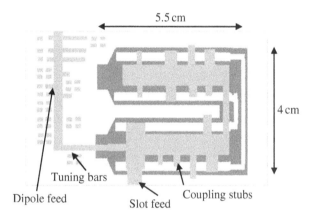

FIGURE 12.18   Folded dipole antenna, $7 \times 5 \times 0.16$ cm.

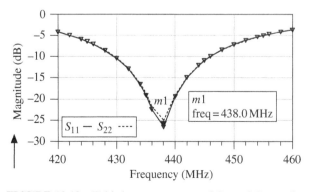

FIGURE 12.19   Folded antenna computed S_{11} and S_{22} results.

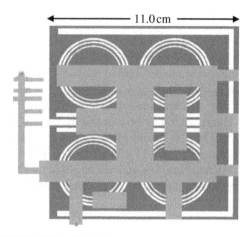

FIGURE 12.20   Folded dual-polarized antenna with SRR.

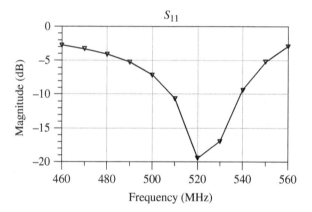

FIGURE 12.21   Folded antenna with SRR, computed S_{11}.

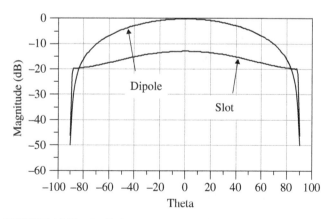

FIGURE 12.22   Radiation pattern of the folded antenna with SRR.

12.2   STACKED PATCH ANTENNA LOADED WITH SRR

At first a microstrip stacked patch antenna [1–3] has been designed. The second step was to design the same antenna with SRR. The antenna consists of two layers. The first layer consists of FR4 dielectric substrate with dielectric constant of 4 and 1.6 mm thick. The second layer consists of RT/duroid 5880 dielectric substrate with dielectric constant of 2.2 and 1.6 mm thick. The dimensions of the microstrip stacked patch antenna shown in Figure 12.23 are $33 \times 20 \times 3.2$ mm. The antenna has been analyzed by using Agilent ADS software. The antenna bandwidth is around 5% for VSWR better than $2.5 : 1$. The antenna beamwidth is around $72°$. The antenna gain is around 7 dBi. The computed S_{11} parameters are presented in Figure 12.24. Radiation pattern of the microstrip stacked patch is shown in Figure 12.25. The antenna with SRR is shown in Figure 12.26. This antenna has the same structure as the antenna shown in

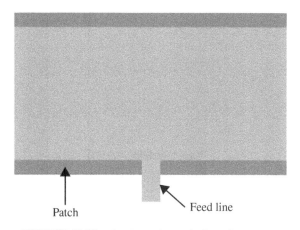

FIGURE 12.23   A microstrip stacked patch antenna.

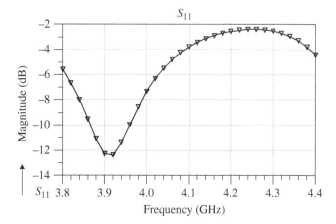

FIGURE 12.24   Computed S_{11} of the microstrip stacked patch.

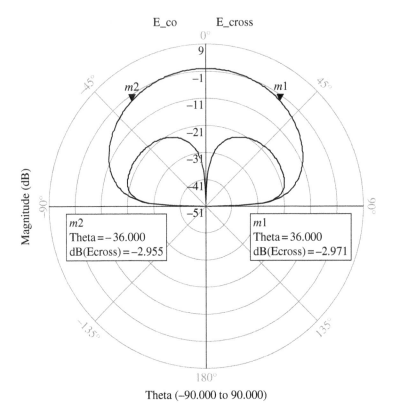

FIGURE 12.25   Radiation pattern of the microstrip stacked patch.

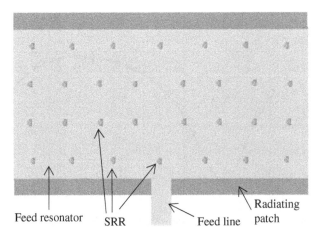

FIGURE 12.26   Printed antenna with split ring resonators.

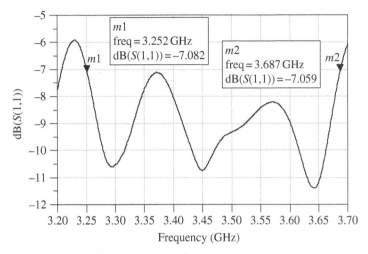

FIGURE 12.27 Patch with split ring resonators, measured S_{11}.

Figure 12.23. The ring width is 0.2 mm and the spacing between the rings is 0.25 mm. Twenty-eight SRR are placed on the radiating element. There is a good agreement between measured and computed results. The measured S_{11} parameters of the antenna with SRR are presented in Figure 12.27. The antenna bandwidth is around 12% for VSWR better than 2.5 : 1. By adding an air space of 4 mm between the antenna layers, the VSWR was improved to 2 : 1. The antenna gain is around 9–10dBi. The antenna efficiency is around 95%. The antenna computed radiation pattern is shown in Figure 12.28. The patch antenna with SRR performs as a loaded patch antenna. The effective area of a patch antenna with SRR is higher than the effective area of a patch antenna without SRR. The resonant frequency of a patch antenna with SRR is lower by 10% than the resonant frequency of a patch antenna without SRR.

The antenna beamwidth is around 70°. The gain and directivity of the stacked patch antenna with SRR are higher by 2–3dB than patch the antenna without SRR.

12.3 PATCH ANTENNA LOADED WITH SPLIT RING RESONATORS

A patch antenna with SRR has been designed. The antenna is printed on RT/duroid 5880 dielectric substrate with dielectric constant of 2.2 and 1.6 mm thick. The dimensions of the microstrip patch antenna shown in Figure 12.29 are $36 \times 20 \times 1.6$ mm. The antenna bandwidth is around 5% for S_{11} lower than −9.5dB. However, the antenna bandwidth is around 10% for VSWR better than 3 : 1. The antenna beamwidth is around 72°. The antenna gain is around 7.8dBi. The directivity of the antenna is 8. The antenna gain is 6.03. The antenna efficiency is 77.25%. The computed S_{11}

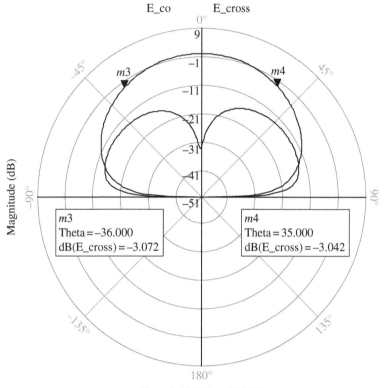

FIGURE 12.28   Radiation pattern for patch with SRR.

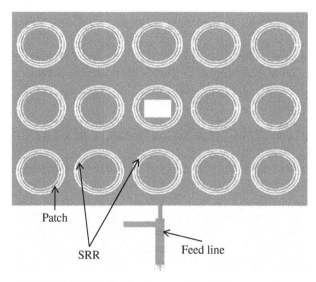

FIGURE 12.29   Patch antenna with split ring resonators.

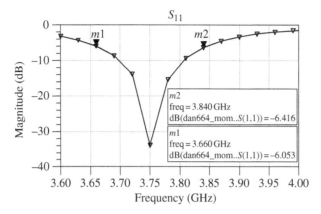

FIGURE 12.30 Patch with split ring resonators, computed S_{11}.

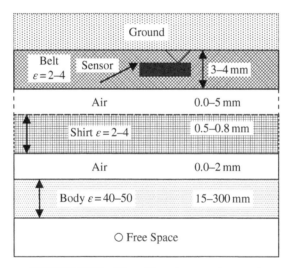

FIGURE 12.31 Wearable antenna environment.

parameters are presented in Figure 12.30. The gain and directivity of the patch antenna with SRR is higher by 2.5dB than the patch antenna without SRR.

12.4 METAMATERIAL ANTENNA CHARACTERISTICS IN VICINITY TO THE HUMAN BODY

The antennas' input impedance variation as function of distance from the human body had been computed by using the structure presented in Figure 12.31. Electrical properties of human body tissues are listed in Table 12.1 (see Ref. [15]). The antenna

TABLE 12.1   Electrical Properties of Human Body Tissues

Tissue	Property	434 MHz	600 MHz
Prostate	σ	0.75	0.90
	ε	50.53	47.4
Skin	σ	0.57	0.6
	ε	41.6	40.43
Stomach	σ	0.67	0.73
	ε	42.9	41.41
Colon, muscle	σ	0.98	1.06
	ε	63.6	61.9
Lung	σ	0.27	0.27
	ε	38.4	38.4

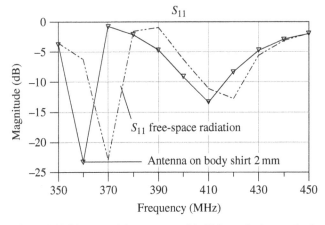

FIGURE 12.32 S_{11} of the antenna with SRR on the human body.

location on human body may be taken into account by calculating S_{11} for different dielectric constant of the body. The variation of the dielectric constant of the body from 43 at the stomach to 63 at the colon zone shifts the antenna resonant frequency by 2%. The antenna was placed inside a belt with thickness between 1 and 4 mm with dielectric constant from 2 to 4.

The air spacing between the belt and the patient shirt is varied from 0 to 8 mm. The dielectric constant of the patient shirt was varied from 2 to 4.

Figure 12.32 presents the S_{11} results of the antenna with SRR shown in Figure 12.13 on the human body. The antenna resonant frequency is shifted by 3%. Figure 12.33 presents the S_{11} results of the antenna with SRR and metallic strips, shown in Figure 12.16, on the human body. The antenna resonant frequency is shifted by 1%.

Figure 12.34 presents the S_{11} results (of the antenna shown in Fig. 12.3) for different air spacing between the antennas and human body and belt thickness and shirt thickness. Results presented in Figure 12.33 indicate that the antenna has VSWR

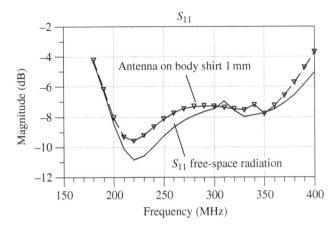

FIGURE 12.33   Antenna with SRR, S_{11} results on the human body.

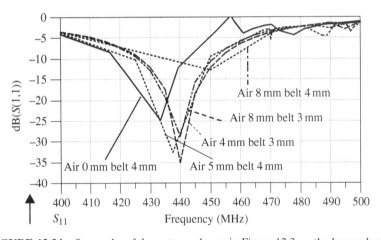

FIGURE 12.34 S_{11} results of the antenna shown in Figure 12.3 on the human body.

better than 2.5 : 1 for air spacing up to 8 mm between the antennas and the body. Figure 12.35 presents the S_{11} results for different position relative to the human body of the folded antenna shown in Figure 12.6. Explanation of Figure 12.35 is given in Table 12.2. If the air spacing between the antennas and the human body is increased from 0 to 5 mm, the antenna resonant frequency is shifted by 5%. Tunable wearable antenna may be used to control the antenna resonant frequency at different positions on the human body (see Ref. [25]).

Figure 12.36 presents the S_{11} results of the folded antenna with SRR, shown in Figure 12.8, on the human body. The antenna resonant frequency is shifted by 2%. The radiation pattern of the folded antenna with SRR on human body is presented in Figure 12.37.

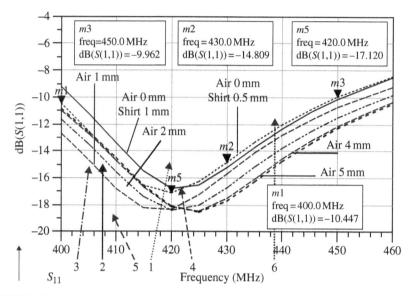

FIGURE 12.35 S_{11} results for different locations relative to the human body for the antenna shown in Figure 12.6.

TABLE 12.2 Explanation of Figure 12.35

Picture	Line Type	Sensor Position
1	Dot··············	Shirt thickness 0.5 mm
2	Line———	Shirt thickness 1 mm
3	Dash dot—·—·—	Air spacing 2 mm
4	Dash—————·	Air spacing 4 mm
5	Long dash— — —	Air spacing 1 mm
6	Big dots··············	Air spacing 5 mm

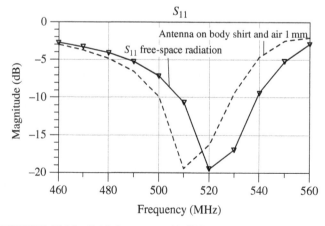

FIGURE 12.36 Folded antenna with SRR, S_{11} results on the body.

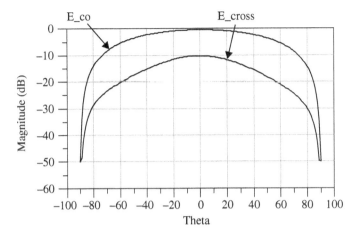

FIGURE 12.37   Radiation pattern of the folded antenna with SRR on human body.

12.5   METAMATERIAL WEARABLE ANTENNAS

The proposed wearable metamaterial antennas may be placed inside a belt as shown in Figure 12.38. Three to four antennas may be placed in a belt and attached to the patient stomach. More antennas may be attached to the patient back to improve the level of the received signal from different locations in the human body. The cable from each antenna is connected to a recorder. The received signal is transferred via a SP8T switch to the receiver. The antennas receive a signal that is transmitted from various positions in the human body. The medical system selects the signal with the highest power.

In several systems the distance separating the transmitting and receiving antennas is in the near-field zone. In these cases the electric field intensity decays rapidly with distance. The near fields only transfer energy to close distances from the antenna and do not radiate energy to far distances. The radiated power is trapped in the region near to the antenna. In the near-field zone the receiving and transmitting antennas are magnetically coupled. The inductive coupling value between two antennas is measured by their mutual inductance. In these systems we have to consider only the near-field electromagnetic coupling.

In Figures 12.39, 12.40, 12.41, and 12.42, several photos of printed antennas for medical applications are shown. The dimensions of the folded dipole antenna are $7 \times 6 \times 0.16$ cm. The dimensions of the compact folded dipole presented in Ref. [5] and shown in Figure 12.39 are $5 \times 5 \times 0.5$ cm. The antenna electrical characteristics on human body have been measured by using a phantom. The phantom has been designed to represent the human body electrical properties as presented in Ref. [5]. The tested antenna was attached to the phantom during the measurements of the antenna electrical parameters.

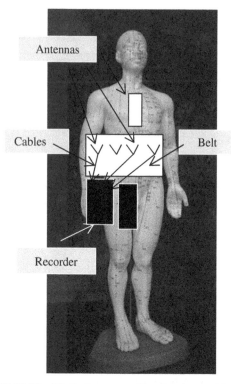

FIGURE 12.38   Medical system with printed wearable antenna.

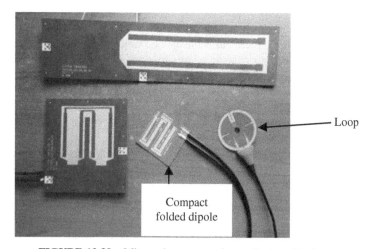

FIGURE 12.39   Microstrip antennas for medical applications.

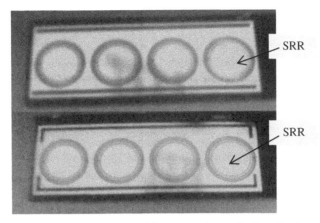

FIGURE 12.40 Metamaterial antennas for medical applications.

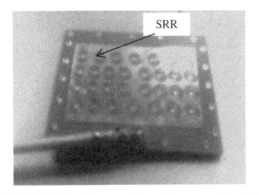

FIGURE 12.41 Metamaterial patch antenna with SRR.

FIGURE 12.42 Metamaterial stacked patch antenna with SRR.

12.6   WIDEBAND STACKED PATCH WITH SRR

A wideband microstrip stacked patch antenna with air spacing [1–3] has been designed. The antenna has been designed with SRR. The antenna consists of two layers. The first layer consists of FR4 dielectric substrate with dielectric constant of 4 and 1.6 mm thick. The second layer consists of RT/duroid 5880 dielectric substrate with dielectric constant of 2.2 and 1.6 mm thick. The layers are separated by air spacing. The dimensions of the microstrip stacked patch antenna shown in Figure 12.43 are $33 \times 20 \times 3.2$ mm. The antenna has been analyzed by using Agilent ADS software. The antenna bandwidth is around 10% for VSWR better than $2.0 : 1$. The antenna beamwidth is around 72°. The antenna gain is around 90–10dBi. The antenna efficiency is around 95%. The computed S_{11} parameters are presented in Figure 12.44. Radiation pattern of the stacked patch is shown in Figure 12.45. There is a good agreement between measured and computed results.

Figure 12.32 presents the S_{11} results of the antenna with SRR shown in Figure 12.13 on the human body. The antenna resonant frequency is shifted by 3%. Figure 12.33 presents the S_{11} results of the antenna with SRR and metallic strips, shown in Figure 12.16, on the human body. The antenna resonant frequency is shifted by 1%.

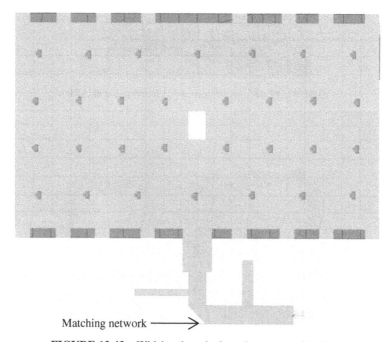

Matching network $\longrightarrow$

FIGURE 12.43   Wideband stacked patch antenna with SRR.

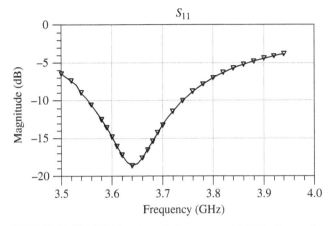

FIGURE 12.44 Wideband stacked antenna with SRR, S_{11} results.

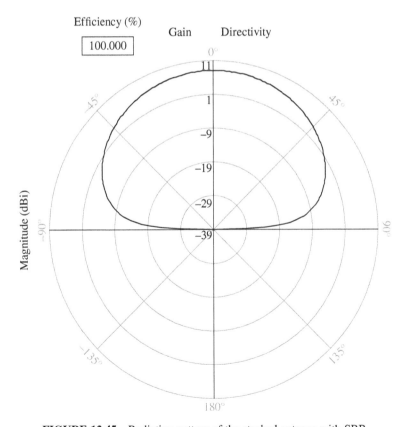

FIGURE 12.45 Radiation pattern of the stacked antenna with SRR.

12.7 FRACTAL PRINTED ANTENNAS

12.7.1 Introduction

A **fractal antenna** is an antenna that uses antenna design with similar fractal segments to maximize the antenna effective area. Fractal antennas are also referred as multilevel structure with space-filling curves (SFC). The key aspect lies in a repetition of a motif over two or more scale sizes or "iterations." Fractal antennas are very compact, multiband, or wideband and have useful applications in cellular telephone and microwave communications. Several fractal antennas was presented in books, papers, and patents (see Refs. [26–39]).

12.7.2 Fractal Structures

A curve, with endpoints, is represented by a continuous function whose domain is the unit interval [0, 1]. The curve may line in a plane or in a 3D space. A fractal curve is a densely self-intersecting curve that passes through every point of the unit square. A fractal curve is a continuous mapping from the unit interval to the unit square.

In mathematics, an SFC is a curve whose range contains the entire two-dimensional unit square. Most SFC are constructed iteratively as a limit of a sequence of piecewise linear continuous curves, each one closely approximating the space-filling limit. For SFC, where two subcurves intersect (in the technical sense), there is self-contact without self-crossing. An SFC can be (everywhere) self-crossing if its approximation curves are self-crossing. An SFC's approximations can be self-avoiding, as presented in Figure 12.46. In three dimensions, self-avoiding approximation curves can even contain joined ends. SFC are special cases of fractal constructions. No differentiable SFC can exist.

The term "fractal curve" was introduced by B. Mandelbrot [26, 27] to describe a family of geometrical objects that are not defined in standard Euclidean geometry. Fractals are geometric shapes that repeat itself over a variety of scale sizes. One of the key properties of a fractal curve is the self-similarity. A self-similar object is unchanged after increasing or shrinking its size. An example of a repetitive geometry is the Koch curve, presented in 1904 by Helge von Koch, and is shown in Figure 12.46b. Koch generated the geometry by using a segment of straight line

(a)   (b)   (c)   (d)

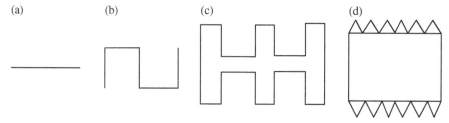

FIGURE 12.46 (a) Line, (b) motif of bended line, (c) bended line fractal structures, and (d) fractal structure.

and raises an equilateral triangle over its middle third. Repeating once more the process of erecting equilateral triangles over the middle thirds of straight line results what is presented in Figure 12.47a. Iterating the process infinitely many times results in a "curve" of infinite length. This geometry is continuous everywhere but is nowhere differentiable. Applying the Koch process to an equilateral triangle, after many iterations, converges to the Koch snowflake shown in Figure 12.47. This process can be applied to several geometries as shown in Figures 12.48 and 12.49. Many variations of these geometries are presented in several papers.

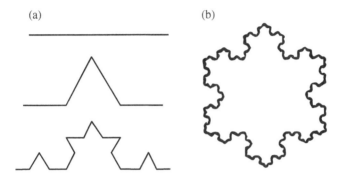

FIGURE 12.47   (a) Koch fractal structures and (b) Koch snowflakes.

FIGURE 12.48   Folded fractal structures.

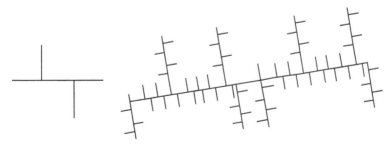

FIGURE 12.49   Variations of Koch fractal structures.

12.7.3   Fractal Antennas

Fractal geometries may be applied to design antennas and antenna arrays. The advantages of printed circuit technology and printed antennas enhance the design of fractal printed antennas and microwave components. The effective area of a fractal antenna is significantly higher than the effective area of a regular printed antenna. Fractal antenna may operate with good performance at several different frequencies simultaneously. Fractal antennas are compact multiband antennas. Directivity of fractal antennas is usually higher than directivity of a regular printed antenna. The number of element in a fractal antenna array may be reduced by around a quarter of the number of elements in a regular array. A fractal antenna could be considered as a nonuniform distribution of radiating elements. Each of the elements contributes to the total radiated power density at a given point with a given amplitude and phase. By spatially superposing these line radiators, we can study the properties of a fractal antenna array.

Small antenna features are:

- A large input reactance (either capacitive or inductive) that usually has to be compensated with an external matching network
- A small radiating resistance
- Small bandwidth and low efficiency

This means that it is highly challenging to design a resonant antenna in a small space in terms of the wavelength at resonance.

The use of microstrip antennas is well known in mobile telephony handsets [30]. The planar inverted-F antenna (PIFA) configuration is popular in mobile communication systems. The advantages of PIFA antennas are their low profile, low fabrication costs, and easy integration within the system structure. One of the miniaturization techniques used in this antenna system is based on SFC. In some particular case of antenna configuration system, the antenna shape may be described as a multilevel structure. Multilevel technique has been already proposed to reduce the physical dimensions of microstrip antennas. The present integrated multiservice antenna system for communication system comprises the following parts and features.

The antenna includes a conducting strip or wire shaped by an SFC, composed of at least 200 connected segments forming a substantially right angle with each adjacent segment smaller than a hundredth of the free-space operating wavelength. The important reduction size of such antennas system is obtained by using space-filling geometries. An SFC can be described as a curve that is large in terms of physical length but small in terms of the area in which the curve can be included. An SFC can be fitted over a flat or curved surface, and due to the angles between segments, the physical length of the curve is always larger than that of any straight line that can be fitted in the same area (surface).

Additionally, to properly shape the structure of a miniature antenna, the segments of the SFC must be shorter than a 10th of the free-space operating wavelength. The antenna is fed with a two-conductor structure such as a coaxial cable, with one of the conductors connected to the lower tip of the multilevel structure and the other

(a)   (b)

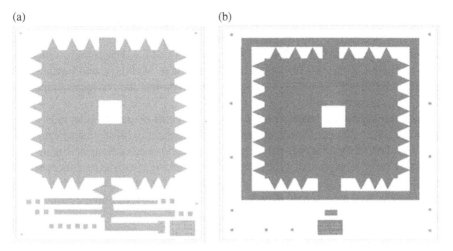

FIGURE 12.50 (a) Patch with space-filling perimeter of the conducting sheet and (b) microstrip patch with space-filling perimeter of the conducting sheet.

conductor connected to the metallic structure of the system that acts as a ground plane. This antenna type features a size reduction below a 20% than the typical size of a conventional external quarter-wave whip antenna. This feature together with the small profile of the antenna, which can be printed in a low-cost dielectric substrate, allows a simple and compact integration of the antenna structure.

Reducing the size of the radiating elements can be achieved by using a PIFA configuration, consisting of connecting two parallel conducting sheets, separated either by air or a dielectric, magnetic, or magnetodielectric material. The sheets are connected through a conducting strip near one of the sheet corners and orthogonally mounted to both sheets.

The antenna is fed through a coaxial cable, having its outer conductor connected to the first sheet, with the second sheet being coupled either by direct contact or capacitive to inner conductor of the coaxial cable.

In Figure 12.50a and b two examples of space-filling perimeter of the conducting sheet to achieve an optimized miniaturization of the antenna are presented.

12.8 ANTIRADAR FRACTALS AND/OR MULTILEVEL CHAFF DISPERSERS

Chaff was one of the forms of countermeasure employed against radar. It usually consists of a large number of electromagnetic dispersers and reflectors, normally arranged in form of strips of metal foil packed in a bundle. Chaff is usually employed to foil or to confuse surveillance and tracking radar.

12.8.1   Geometry of Dispersers

Here a new geometry of the dispersers or reflectors that improves the properties of radar chaff is presented [31]. Some of the geometries of the dispersers or reflectors presented here are related with some forms for antennas. Multilevel and fractal structure antennas are distinguished in being of reduced size and having a multiband behavior, as presented in Refs. [32–39].

The main electrical characteristic of a radar chaff disperser is its radar cross section (RCS), which is related with the reflective capability of the disperser.

A fractal curve for a chaff disperser is defined as a curve comprising at least 10 segments that are connected so that each element forms an angle with its neighbors and no pair of these segments defines a longer straight segment, these segments being smaller than a 10th part of the resonant wavelength in free space of the entire structure of the dispenser.

In many of the configuration presented, the size of the entire disperser is smaller than a quarter of the lowest operating wavelength.

The SFC (or fractal curves) can be characterized by:

1. They are long in terms of physical length but small in terms of area in which the curve can be included. The dispersers with a fractal form are long electrically but can be included in a very small surface area. This means it is possible to obtain a smaller packaging and a denser chaff cloud using this technique.
2. Frequency response: Their complex geometry provides a spectrally richer signature when compared with rectilinear dispersers known in the stat of the art.

The fractal structure properties of disperser not only introduce an advantage in terms of reflected radar signal response but also in terms of aerodynamic profile of dispersers. It is known that a surface offers greater resistance to air than a line or a one-dimensional form.

Therefore, giving a fractal form to the dispersers with a dimension greater than unity ($D > 1$) increases resistance to the air and improves the time of suspension.

12.9   DEFINITION OF MULTILEVEL FRACTAL STRUCTURE

Multilevel structures are a geometry related with fractal structures. In the case of radar chaff, a multilevel structure is defined as structure that includes a set of polygons, which are characterized in having the same number of sides, wherein these polygons are electromagnetically coupled either by means of capacitive coupling or by means of an ohmic contact. The region of contact between the directly connected polygons is smaller than 50% of the perimeter of the polygons mentioned in at least 75% of the polygons that constitute the defined multilevel structure.

A multilevel structure:

- Provides both a reduction in the size of dispensers and an enhancement of their frequency response
- Can resonate in a nonharmonic way and can even cover simultaneously and with the same relative bandwidth at least a portion of numerous bands

The fractal structures (SFC) are preferred when a reduction in size is required, while multilevel structures are preferred when it is required that the most important considerations be given to the spectral response of radar chaff.

The main advantages for configuring the form of the chaff dispersers are the following:

1. The dispersers are small; consequently more dispersers can be encapsulated in a same cartridge, rocket, or launch vehicle.
2. The dispersers are also lighter; therefore they can remain more time floating in the air than the conventional chaff.
3. Due to the smaller size of the chaff dispersers, the launching devices (cartridges, rockets, etc.) can be smaller with regard to chaff systems in the state of the art providing the same RCS.
4. Due to lighter weight of the chaff dispersers, the launching devices can shot the packages of chaff farther from the launching devices and locations.
5. Chaff dispersers constituted by multilevel and fractal structures provide larger RCS at longer wavelengths than conventional chaff dispersers of the same size.
6. The dispersers with long wavelengths can be configured and printed on light dielectric supports having a nonaerodynamic form and opposing a greater resistance to the air and thereby having a longer time of suspension.
7. The dispersers provide a better frequency response with regard to dispersers of the state of the art. In the following images such size compression structures based on fractal curves are presented (Figure 12.51).

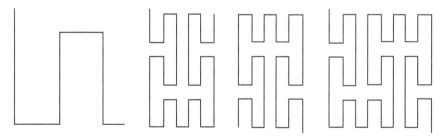

FIGURE 12.51   Fractal curves that can be used to configure a chaff disperser.

FIGURE 12.52   Hilbert fractal curves.

Figure 12.52 shows several examples of Hilbert fractal curves (with increasing iteration order) that can be used to configure the chaff disperser.

12.10 ADVANCED ANTENNA SYSTEM

The main advantage of advanced antenna system lies in the multiband and multiservice performance of the antenna. This enables convenient and easy connection of a simple antenna for most communication systems and applications. The main advantages addressed by advanced antennas featured similar parameters (input impedance, radiation pattern) at several bands maintaining their performance, compared with conventional antennas. Fractal shapes permit to obtain compact antenna of reduced dimensions compared to other conventional antennas. Multilevel antennas introduced a higher flexibility to design multiservice antennas for real applications, extending the theoretical capabilities of ideal fractal antennas to practical, commercial antennas.

12.10.1 Comparison between Euclidean Antenna and Fractal Antenna

Most conventional antennas are Euclidean design/geometry, where the closed antenna area is directly proportional to the antenna perimeter. Thus, for example, when the length of a Euclidean square is increased by a factor of 3, the enclosed area of the antenna is increased by a factor of 9. Gain, directivity, impedance, and efficiency of Euclidean antennas are a function of the antenna's size to wavelength ratio.

Euclidean antennas are typically desired to operate within a narrow range (e.g., 10–40%) around a central frequency f_c, which in turn dictates the size of the antenna (e.g., half or quarter wavelength). When the size of a Euclidean antenna is made much smaller than the operating wavelength (λ), it becomes very inefficient because the antenna's radiation resistance decreases and becomes less than its ohmic resistance (i.e., it does not couple electromagnetic excitations efficiently to free space). Instead, it stores energy reactively within its vicinity (reactive impedance X_c). These aspects of Euclidean antennas work together to make it difficult for small Euclidean antennas to couple or match to feeding or excitation circuitry and cause them to have a high Q factor (lower bandwidth). Q (quality) factor may be defined as approximately the ratio of input reactance X_{in} to radiation resistance R_r: $Q = X_{in}/R_r$.

The Q factor may also be defined as the ratio of average stored electric energies (or magnetic energies stored) to the average radiated power. Q can be shown to be inversely proportional to bandwidth.

Thus, small Euclidean antennas have very small bandwidth, which is of course undesirable (matching network may be needed). Many known Euclidean antennas are based upon closed-loop shapes.

Unfortunately, when small in size, such loop-shaped antennas are undesirable because, as discussed previously, the radiation resistance decreases significantly when the antenna size is decreased. This is because the physical area (A) contained within the loop-shaped antenna's contour is related to the loop perimeter.

Radiation resistance R_r of a circular loop-shaped Euclidean antenna is defined by R_r, as given in Equation 12.1, k being a constant:

$$R_r = \eta \pi \left(\frac{2}{3}\right) \left(\frac{KA}{\lambda}\right)^2 = 20\pi^2 \left(\frac{C}{\lambda}\right) \tag{12.1}$$

Since the resistance R_C is only proportional to perimeter (C), then for $C < 1$, the resistance R_C is greater than the radiation resistance R_r and the antenna is highly inefficient. This is generally true for any small circular Euclidean antenna. A small-sized antenna will exhibit a relatively large ohmic resistance and a relatively small radiation resistance R_r. This low efficiency limits the use of the small antennas.

Fractal geometry is a non-Euclidean geometry, which can be used to overcome the problems with small Euclidean antennas. Radiation resistance R_r of a fractal antenna decreases as a small power of the perimeter (C) compression, which a fractal loop or island always having a substantially higher radiation resistance than a small Euclidean loop antenna of equal size. Fractal geometry may be grouped into:

- Random fractals, which may be called chaotic or Brownian fractals.
- Deterministic or exact fractals. In deterministic fractal geometry, a self-similar structure results from the repetition of a design or motif (generator) with self-similarity and structure at all scales. In deterministic or exact self-similarity, fractal antennas may be constructed through recursive or iterative means. In other words, fractals are often composed of many copies of themselves at different scales, thereby allowing them to defy the classical antenna performance constraint, which is the size to wavelength ratio.

12.10.2 Multilevel and Space-Filling Ground Planes for Miniature and Multiband Antennas

A new family of antenna ground planes of reduced size and enhanced performance based on an innovative set of geometries is presented in Figure 12.55.

These new geometries are known as multilevel and space-filling structures, which had been previously used in the design of multiband and miniature antennas.

One of the key issues of the present antenna system is considering the ground plane of an antenna as an integral part of the antenna that mainly contributes to its radiation and impedance performance (impedance level, resonant frequency, and bandwidth).

The multilevel and space-filling structures are used in the ground plane of the antenna obtaining this way a better return loss or VSWR, a better bandwidth, a multiband behavior, or a combination of all these effects. The technique can be seen as well a means of reducing the size of the ground plane and therefore the size of the overall antenna. The key point of the present antenna system is shaping the ground plane of an antenna in such a way that the combined effect of the ground plane and the radiating element enhances the performance and characteristics of the whole antenna device, either in terms of bandwidth, VSWR, multiband, efficiency, size, or gain.

12.10.2.1 Multilevel Geometry

The resulting geometry is no longer a solid, conventional ground plane, but a ground plane with a multilevel or space-filling geometry, at least in a portion of ground plane.

A multilevel geometry for a ground plane consists of a conducting structure including a set of polygons, featuring the same number of sides, electromagnetically coupled either by means of a capacitive coupling or ohmic contact. The contact region between directly connected polygons is narrower than 50% of the perimeter of polygons in at least 75% of polygons defining the conducting ground plane. In this definition of multilevel geometry, circles and ellipses are included as well, since they can be understood as polygons with infinite number of sides.

12.10.2.2 Space-Filling Curve

A space-filling curve (hereafter SFC) is a curve that is large in terms of physical length but small in terms of the area in which the curve can be included.

This is also a curve composed of at least 10 segments that are connected in such a way that each segment forms an angle with their neighbors, that is, no pair of adjacent segments defines a larger straight segment, and wherein the curve can be optionally periodic along a fixed straight direction of space if and only if the period is defined by a nonperiodic curve composed of at least 10 connected segments and no pair of adjacent and connected segments defines a straight longer segment.

An SFC can be fitted over a flat or curved surface, and due to the angles between segments, the physical length of the curve is always larger than that of any straight line that can be fitted in the same area (surface).

Additionally, to properly shape the ground plane, the segments of the SFC curves included in the ground plane must be shorter than a 10th of the free-space operating wavelength.

Figure 12.53 shows several examples of fractal geometries that can be used as SFC.

Figure 12.54 shows several examples of Hilbert fractal curves that can be used as SFC.

The curves shown in the Figure 12.53 are some examples of such SFC. Due to the special geometry of multilevel and space-filling structure, the current distributes over the ground plane in such a way that it enhances the antenna performance and features in terms of:

- Reduced size compared to antennas with a solid ground plane
- Enhanced bandwidth compared to antennas with a solid ground plane

- Multifrequency performance
- Better VSWR feature at the operating band or bands
- Better radiation efficiency
- Enhanced gain

Figure 12.55a shows a patch antenna above a particular example of a new ground-plane structure formed by both multilevel and space-filling geometries. Figure 12.55b shows a monopole antenna above a ground-plane structure formed by both multilevel and space-filling geometries.

Figure 12.56 shows several examples of different contour shapes for multilevel ground planes, such as rectangular 56a and 56b and circular 56c.

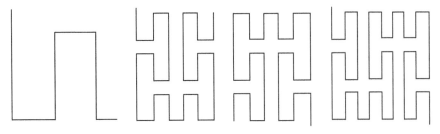

FIGURE 12.53   Fractal curves that can be used as space-filling curves.

FIGURE 12.54   Hilbert fractal curves that can be used as space-filling curves.

(a)   (b)

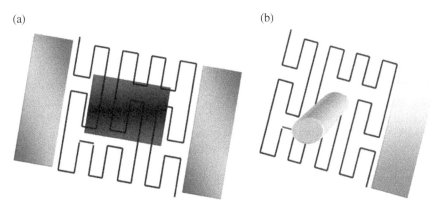

FIGURE 12.55   (a) Patch antenna above a new ground-plane structure and (b) monopole antenna above a ground-plane structure formed by both multilevel and space-filling geometries.

(a)   (b)   (c)

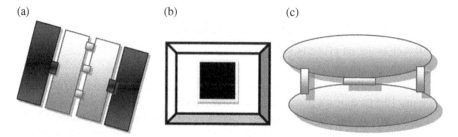

FIGURE 12.56   Examples of different contour shaped multilevel ground planes. (a) Rectangular ground planes, (b) multilevel rectangular ground planes, and (c) circular ground planes.

12.11   APPLICATIONS OF FRACTAL PRINTED ANTENNAS

In this chapter several designs of fractal printed antennas are presented for communication applications. These fractal antennas are compact and efficient. The antenna gain is around 8dBi with 90% efficiency.

12.11.1   New 2.5 GHz Fractal Antenna with Space-Filling Perimeter on the Radiator

A new fractal microstrip antenna was designed as presented in Figure 12.57. The antenna was printed on duroid substrate 0.8 mm thick with 2.2 dielectric constant. The antenna dimensions are $5.2 \times 48.8 \times 0.08$ cm. The antenna was designed by using ADS software.

The antenna bandwidth is around 2% around 2.5 GHz for VSWR better than 3 : 1. The antenna bandwidth may be improved to 5%, for VSWR better than 2 : 1, by adding a second layer above the resonator. A patch radiator is printed on the second layer as presented in Figure 12.58. The radiator was printed on FR4 substrate 0.8 mm thick with 4.5 dielectric constant. The electromagnetic fields radiated by the resonator are electromagnetic coupled to the patch radiator. The patch radiator dimensions are $45.2 \times 48.8 \times 0.08$ cm. The stacked fractal antenna structure is shown in Figure 12.59. The spacing between the two layers may be varied to get wider bandwidth. The stacked fractal antenna S_{11} parameters with 8 mm air spacing between the layers are presented in Figure 12.60. The S_{11} parameter of the fractal stacked patch antenna with 10 mm air spacing is given in Figure 12.61. The antenna bandwidth is improved to 5% for VSWR better than 2 : 1. The fractal stacked patch antenna radiation pattern is shown in Figure 12.62. The antenna beamwidth is around 76°, with 8dBi gain and 91% efficiency.

A photo of the stacked fractal patch antenna is shown in Figure 12.63. The antenna resonator is shown in Figure 12.63a. The antenna radiator is shown in Figure 12.63b. A modified version of the antenna is shown in Figure 12.64. The S_{11} parameter of the modified fractal stacked patch antenna with 8 mm air spacing is given in Figure 12.65.

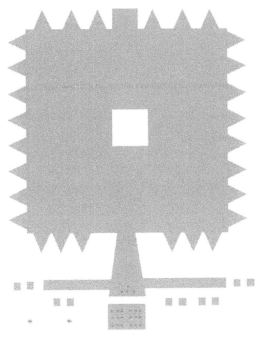

FIGURE 12.57 Fractal antenna resonators.

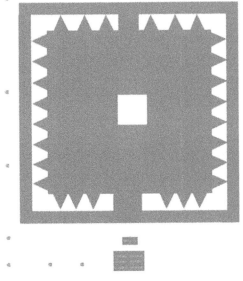

FIGURE 12.58 Fractal antenna patch radiator.

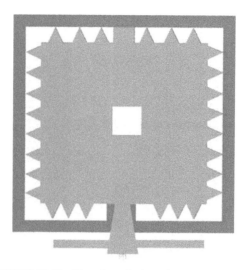

FIGURE 12.59 Fractal stacked patch antenna structure.

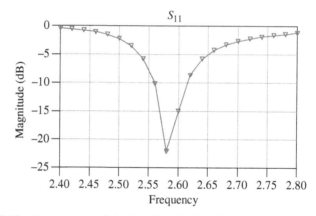

FIGURE 12.60 S_{11} parameter of the fractal stacked patch antenna with 8 mm air spacing.

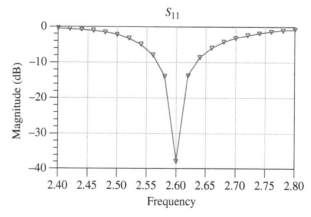

FIGURE 12.61 S_{11} parameter of the fractal stacked patch antenna with 10 mm air spacing.

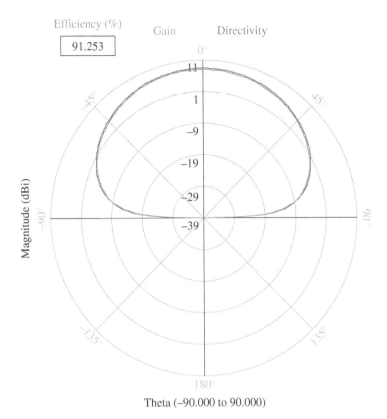

Theta (−90.000 to 90.000)

FIGURE 12.62 Fractal stacked patch antenna radiation pattern with 10 mm air spacing.

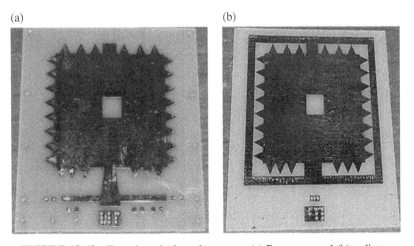

FIGURE 12.63 Fractal stacked patch antenna. (a) Resonator and (b) radiator.

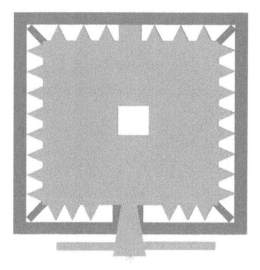

FIGURE 12.64   A modified fractal stacked patch antenna structure.

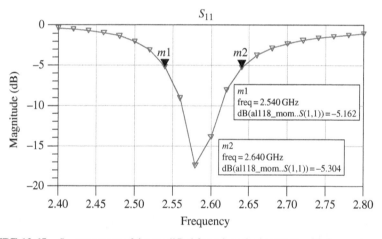

FIGURE 12.65 S_{11} parameter of the modified fractal stacked antenna with 8 mm air spacing.

The antenna bandwidth is around 10% for VSWR better than 3 : 1. The fractal stacked patch antenna radiation pattern is shown in Figure 12.66. The antenna beamwidth is around 76°, with 8dBi gain and 91.82% efficiency.

12.11.2   New Stacked Patch 2.5 GHz Fractal Printed Antennas

A new fractal microstrip antenna was designed as presented in Figure 12.67. The antenna was printed on duroid substrate 0.8 mm thick with 2.2 dielectric constant. The antenna dimensions are $45.8 \times 39.1 \times 0.08$ cm. The antenna was designed by using ADS software.

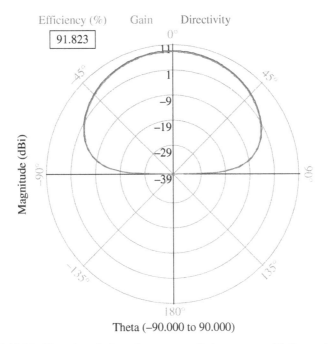

Theta (−90.000 to 90.000)

FIGURE 12.66 Fractal stacked patch antenna radiation pattern with 8 mm air spacing.

(a)   (b)

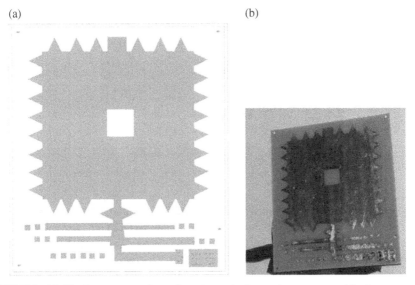

FIGURE 12.67 Resonator of a fractal stacked patch antenna. (a) Layout and (b) resonator photo.

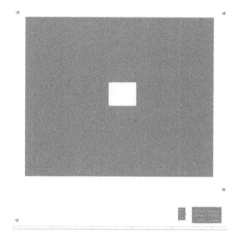

FIGURE 12.68   Radiator of a fractal stacked patch antenna.

FIGURE 12.69   Layout of the fractal stacked patch antenna.

The antenna resonator bandwidth is around 2% around 2.52 GHz for VSWR better than 3 : 1. The antenna bandwidth may be improved to 6%, for VSWR better than 3 : 1, by adding a second layer above the resonator. A patch radiator is printed on the second layer as presented in Figure 12.68. The radiator was printed on FR4 substrate 0.8 mm thick with 4.5 dielectric constant. The electromagnetic fields radiated by the resonator are electromagnetic coupled to the patch radiator. The patch radiator dimensions are $45.8 \times 39.1 \times 0.08$ cm. The stacked fractal antenna structure is shown in Figure 12.69. The spacing between the two layers may be

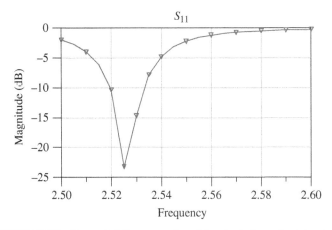

FIGURE 12.70   Computed S_{11} parameter of the single-layer fractal antenna.

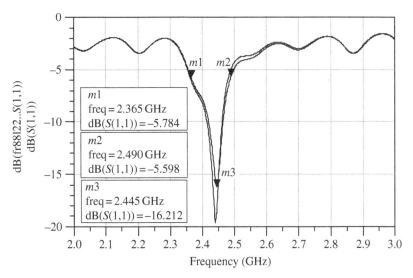

FIGURE 12.71   Measured and computed S_{11} parameter of the fractal stacked patch antenna with no air spacing between the layers.

varied to get wider bandwidth. The single-layer fractal antenna S_{11} parameter is presented in Figure 12.70.

A comparison of the computed and measured S_{11} parameter of the fractal stacked patch antenna with no air spacing is given in Figure 12.71. There is a good agreement between measured and computed results. The fractal stacked patch antenna radiation pattern is shown in Figure 12.72. The antenna beamwidth is around 82°, with 7.5dBi gain and 97.2% efficiency. A photo of the fractal stacked patch antenna is shown in Figure 12.73. The antenna resonator is shown in Figure 12.74a. The antenna radiator is shown in Figure 12.74b.

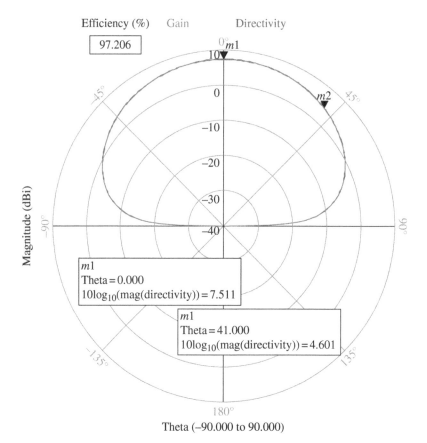

Efficiency (%)   Gain   Directivity

97.206

Magnitude (dBi)

$m1$
Theta = 0.000
$10\log_{10}(\mathrm{mag(directivity)}) = 7.511$

$m1$
Theta = 41.000
$10\log_{10}(\mathrm{mag(directivity)}) = 4.601$

Theta (−90.000 to 90.000)

FIGURE 12.72 Fractal stacked patch antenna radiation pattern with 8 mm air spacing.

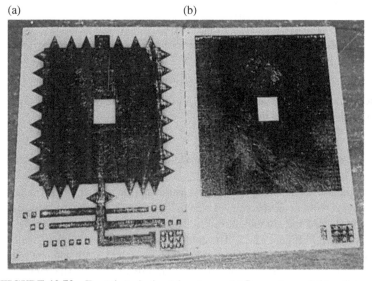

(a)   (b)

FIGURE 12.73 Fractal stacked patch antenna. (a) Resonator and (b) radiator.

(a)   (b)

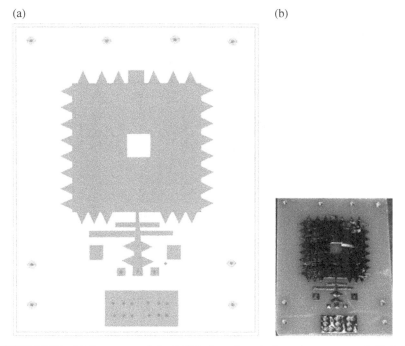

FIGURE 12.74 (a) Resonator of the 8 GHz fractal stacked patch antenna and (b) 8 GHz fractal resonator photo.

12.11.3 New 8 GHz Fractal Printed Antennas with Space-Filling Perimeter of the Conducting Sheet

A new fractal microstrip antenna was designed as presented in Figure 12.74. The antenna was printed on duroid substrate 0.8 mm thick with 2.2 dielectric constant. The antenna dimensions are $17.2 \times 21.8 \times 0.08$ cm. The antenna was designed by using ADS software.

The antenna resonator bandwidth is around 3% around 7. 2 GHz for VSWR better than 2 : 1. The antenna bandwidth is improved to 22%, for VSWR better than 3 : 1, by adding a second layer above the resonator. A patch radiator is printed on the second layer as presented in Figure 12.75. The radiator was printed on FR4 substrate 0.8 mm thick with 4.5 dielectric constant. The electromagnetic fields radiated by the resonator are electromagnetic coupled to the patch radiator. The patch radiator dimensions are $17.2 \times 21.8 \times 0.08$ cm. The stacked fractal antenna structure is shown in Figure 12.76. The spacing between the two layers may be varied to get wider bandwidth. The single-layer fractal antenna S_{11} parameter is presented in Figure 12.77. The computed S_{11} parameter of the fractal stacked patch antenna with 2 mm air spacing is given in Figure 12.78. The fractal stacked patch antenna radiation pattern at 7.5 GHz is shown

(a)   (b)

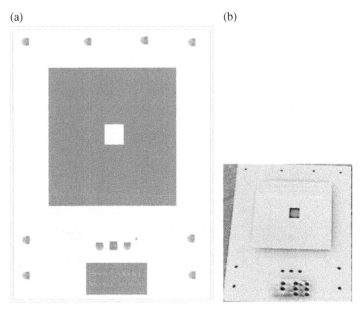

FIGURE 12.75 (a) Radiator of the 8 GHz fractal stacked patch antenna and (b) radiator photo.

FIGURE 12.76 Layout of the 8 GHz fractal stacked patch antenna.

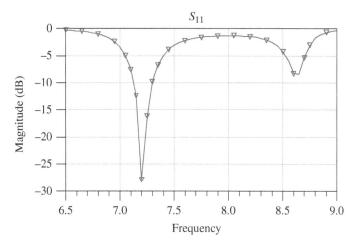

FIGURE 12.77 Computed S_{11} of the 8 GHz fractal stacked patch antenna.

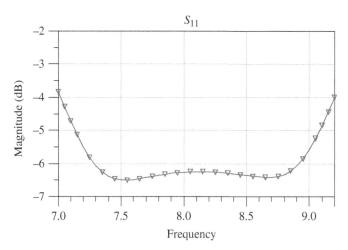

FIGURE 12.78 Computed S_{11} of the 8 GHz fractal stacked patch antenna with 2 mm air spacing.

in Figure 12.79. The antenna beamwidth is around 82°, with 7.8dBi gain and 82.2% efficiency. The fractal stacked patch antenna radiation pattern at 8 GHz is shown in Figure 12.80. The antenna beamwidth is around 82°, with 7.5dBi gain and 95.3% efficiency. A photo of the fractal stacked patch antenna is shown in Figure 12.81. The antenna resonator is shown in Figure 12.81a. The antenna radiator is shown in Figure 12.81b.

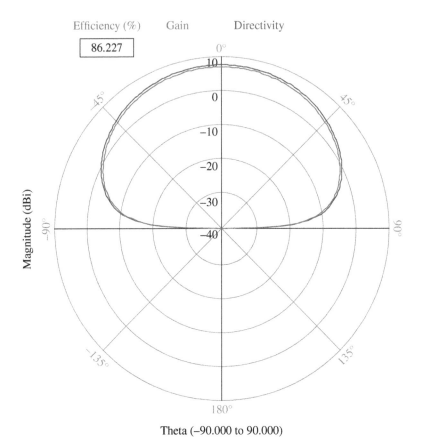

Theta (−90.000 to 90.000)

FIGURE 12.79   Fractal stacked patch antenna radiation pattern with 2 mm air spacing at 7.5 GHz.

12.11.4   New Stacked Patch 7.4 GHz Fractal Printed Antennas

A new fractal microstrip antenna was designed as presented in Figure 12.82. The antenna was printed on duroid substrate 0.8 mm thick with 2.2 dielectric constant. The antenna dimensions are $18 \times 12 \times 0.08$ cm. The antenna was designed by using ADS software.

The antenna resonator bandwidth is around 2% around 7. 4 GHz for VSWR better than 2 : 1. The antenna bandwidth is improved to 10%, for VSWR better than 3 : 1, by adding a second layer above the resonator. A patch radiator is printed on the second layer as presented in Figure 12.83. The radiator was printed on FR4 substrate 0.8 mm thick with 4.5 dielectric constant. The electromagnetic fields radiated by the resonator are electromagnetic coupled to the patch radiator. The fractal stacked patch dimensions are $18 \times 12 \times 0.08$ cm. The stacked fractal antenna structure is shown in

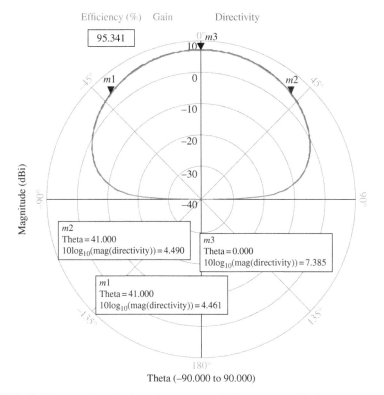

FIGURE 12.80 Fractal stacked patch antenna radiation pattern with 2 mm air spacing at 8 GHz.

FIGURE 12.81 A photo of the fractal stacked patch antenna. (a) Resonator and (b) radiator.

FIGURE 12.82   Layout of the 7.4 GHz fractal resonator.

FIGURE 12.83   Radiator of a 7.4 GHz fractal stacked patch antenna.

FIGURE 12.84   Layout of the 8 GHz fractal stacked patch antenna.

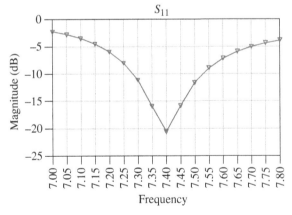

FIGURE 12.85   Computed S_{11} of the 7.4 GHz modified fractal antenna with 3 mm air spacing.

Figure 12.84. The spacing between the two layers may be varied to get wider bandwidth. The computed S_{11} parameter of the fractal stacked patch antenna with 3 mm air spacing is given in Figure 12.85. The fractal stacked patch antenna radiation pattern at 7.5 GHz is shown in Figure 12.86. The antenna beamwidth is around 86°, with 7.9 dBi gain and 89.7% efficiency.

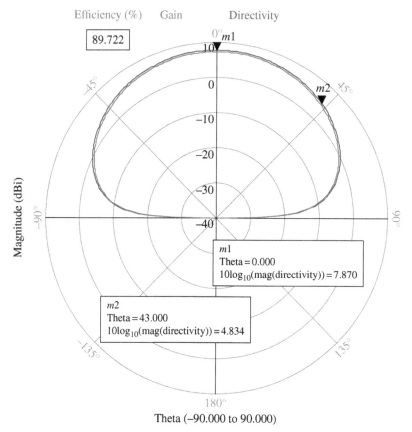

FIGURE 12.86 Fractal stacked patch antenna radiation pattern with 3 mm air spacing at 7.5 GHz.

A modified antenna structure is presented in Figure 12.87. The antenna matching network has been modified and S_{11} at 7.45 GHz is −23.5dB as shown in Figure 12.88. The computed S_{11} parameter of the fractal stacked patch antenna with 3 mm air spacing is given in Figure 12.88. The fractal stacked patch antenna bandwidth is around 9% for VSWR better than 3 : 1. The fractal stacked patch antenna radiation pattern at 7.5 GHz is shown in Figure 12.89. The antenna beamwidth is around 85°, with 7.8dBi gain and 86.2% efficiency.

12.12 CONCLUSIONS

Metamaterial technology is used to develop small antennas with high efficiency. A new class of printed metamaterial antennas with high efficiency is presented in this chapter. The bandwidth of the antenna with SRR and metallic strips is around 50% for

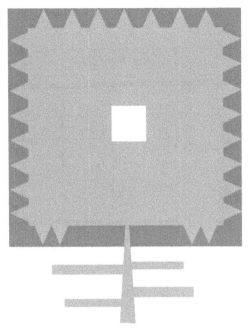

FIGURE 12.87 Layout of the modified 7.4 GHz fractal stacked patch antenna.

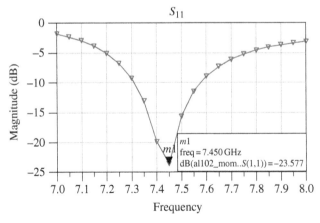

FIGURE 12.88 Computed S_{11} of the 7.4 GHz modified fractal antenna with 3 mm air spacing.

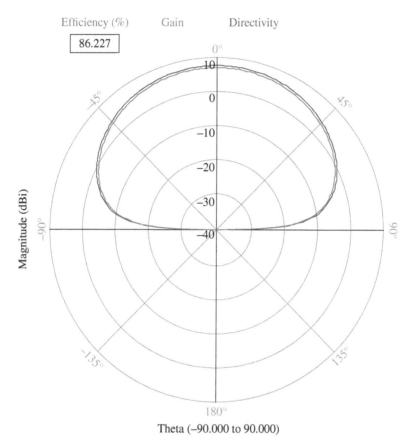

FIGURE 12.89   Modified fractal patch antenna radiation pattern with 3 mm air spacing at 7.5 GHz.

VSWR better than 2.3 : 1. Optimization of the number of the coupling stubs and the distance between the coupling stubs may be used to tune the antenna resonant frequency and number of resonant frequencies. The length of the antennas with SRR is smaller by 5% than the antennas without SRR. Moreover, the resonant frequency of the antennas with SRR is lower by 5–10% than the antennas without SRR. The gain and directivity of the patch antenna with SRR are higher by 2–3dB than the patch antenna without SRR. The resonant frequency of the antenna with SRR on human body is shifted by 3%.

Several designs of fractal printed antennas has been presented for communication applications. These fractal antennas are compact and efficient. Space-filling technique and Hilbert curves were employed to design the fractal antennas. The antenna bandwidth is around 10% with VSWR better than 3 : 1. The antenna gain is around 8dBi with 90% efficiency.

REFERENCES

[1] James JR, Hall PS, Wood C. *Microstrip Antenna Theory and Design*. New York: Peregrinus on behalf of the Institution of Electrical Engineers; 1981.

[2] Sabban A, Gupta KC. Characterization of radiation loss from microstrip discontinuities using a multiport network modeling approach. IEEE Trans Microwave Theory Tech 1991;39 (4):705–712.

[3] Sabban A. A new wideband stacked microstrip antenna. I.E.E.E Antenna and Propagation Symposium; June 1983; Houston, TX.

[4] Sabban A. Microstrip antenna arrays. In: Nasimuddin N, editor. *Microstrip Antennas*. Rijeka: InTech; 2011. p 361–384.

[5] Sabban A. New wideband printed antennas for medical applications. IEEE Trans Antennas Propag 2013;61 (1):84–91.

[6] Pendry JB, Holden AJ, Stewart WJ, Youngs I. Extremely low frequency plasmons in metallic mesostructures. Phys Rev Lett 1996;76:4773–4776.

[7] Pendry JB, Holden AJ, Robbins DJ, Stewart WJ. Magnetism from conductors and enhanced nonlinear phenomena. IEEE Trans Microwave Theory Tech 1999;47:2075–2084.

[8] Marque's R, Mesa F, Martel J, Medina F. Comparative analysis of edge and broadside coupled split ring resonators for metamaterial design. Theory and experiment. IEEE Trans Antennas Propag 2003;51:2572–2581.

[9] Marque's R, Baena JD, Martel J, Medina F, Falcone F, Sorolla M, and Martin F. Novel small resonant electromagnetic particles for metamaterial and filter design. Proceedings of ICEAA'03; Torino, Italy; 2003. p 439–442.

[10] Marque's R, Martel J, Mesa F, Medina F. Left-handed-media simulation and transmission of EM waves in resonator- subwavelength split-ring-loaded metallic waveguides. Phys Rev Lett 2002;89:183901.

[11] Baena JD, Marque's R, Martel J, Medina F. Experimental results on metamaterial simulation using SRR-loaded Waveguides. Proceedings of IEEE- AP/S International Symposium on Antennas and Propagation; June 2003; Ohio. p 106–109.

[12] Marque's R, Martel J, Mesa F, Medina F. A new 2-D isotropic left- handed metamaterial design: theory and experiment. Microwave Opt Technol Lett 2002;35:405–408.

[13] Shelby RA, Smith DR, Nemat-Nasser SC, Schultz S. Microwave transmission through a two-dimensional, isotropic, left-handed metamaterial. Appl Phys Lett 2001;78:489–491.

[14] Zhu J, Eleftheriades GV. A compact transmission-line metamaterial antenna with extended bandwidth. IEEE Antennas Wirel Propag Lett 2009;8:295–298.

[15] Chirwa LC, Hammond PA, Roy S, Cumming DRS. Electromagnetic radiation from ingested sources in the human intestine between 150 MHz and 1.2 GHz. IEEE Trans Biomed Eng 2003;50 (4):484–492.

[16] Werber D, Schwentner A, Biebl EM. Investigation of RF transmission properties of human tissues. Adv Radio Sci 2006;4:357–360.

[17] Gupta B, Sankaralingam S, Dhar S. Development of wearable and implantable antennas in the last decade. Microwave Symposium (MMS), 2010 Mediterranean; August 2010; Cyprus. p 251–267.

[18] Thalmann T, Popovic Z, Notaros BM, Mosig JR. Investigation and design of a multi-band wearable antenna. 3rd European Conference on Antennas and Propagation, EuCAP 2009; March 2009; Berlin, Germany. p 462–465.

[19] Salonen P, Rahmat-Samii Y, Kivikoski M. Wearable antennas in the vicinity of human body. IEEE Antennas Propag Soc Int Symp 2004;1:467–470.

[20] Kellomaki T, Heikkinen J, Kivikoski M. Wearable antennas for FM reception. First European Conference on Antennas and Propagation, EuCAP 2006; November 2006; Nice, France. p 1–6.

[21] Sabban A. Wideband printed antennas for medical applications. APMC 2009 Conference; December 2009; Singapore.

[22] Alomainy A, Sani A, et al. Transient characteristics of wearable antennas and radio propagation channels for ultrawideband Body-centric wireless communication. IEEE Trans Antennas Propag 2009;57 (4):875–884.

[23] Klemm M, Troester G. Textile UWB antenna for wireless body area networks. IEEE Trans Antennas Propag 2006;54 (11):3192–3197.

[24] Izdebski PM, Rajagoplan H, Rahmat-Sami Y. Conformal ingestible capsule antenna: a novel chandelier meandered design. IEEE Trans Antennas Propag 2009;57 (4):900–909.

[25] Sabban A. Wideband tunable printed antennas for medical applications. IEEE Antenna and Propagation Symposium; July 2012; Chicago, IL.

[26] Mandelbrot BB. *The Fractal Geometry of Nature*. San Francisco: W.H. Freeman and Company; 1983.

[27] Mandelbrot BB. How long is the coast of Britain? Statistical self-similarity and fractional dimension. Science 1967;156:636–638.

[28] Falkoner FJ. *The Geometry of Fractal Sets*. Cambridge: Cambridge University Press; 1990.

[29] Balanis CA. *Antenna Theory Analysis and Design*. Second ed. Hoboken: John Wiley & Sons, Inc.; 1997.

[30] Virga K, Rahmat-Samii Y. Low-profile enhanced-bandwidth PIFA antennas for wireless communications packaging. IEEE Trans Microwave Theory Tech 1997;45 (10):1879–1888.

[31] Skolnik MI. *Introduction to Radar Systems*. London: McGraw Hill; 1981.

[32] Minin I. Microwave and millimeter wave technologies from photonic bandgap devices to antenna and applications. In: Rusu MV, Baican R, editors. *Fractal Antenna Applications*. Rijeka: InTech; 2010. p 351–382.

[33] Baliarda CP. Anti-radar space-filling and/or multilevel chaff dispersers. European patent EP1317018 A2. November 27, 2002.

[34] Chiou T, Wong K. Design of compact microstrip antennas with a slotted ground plane. IEEE-APS Symposium; July 8–12, 2001; Boston, MA.

[35] Hansen RC. Fundamental limitations on antennas. Proc IEEE 1981;69 (2):170–182.

[36] Pozar D. *The Analysis and Design of Microstrip Antennas and Arrays*. Piscataway: IEEE Press; 1994.

[37] Zurcher JF, Gardiol FE. *Broadband Patch Antennas*. Boston: Artech House; 1995.

[38] Rusu MV, Hirvonen M, Rahimi H, Enoksson P, Rusu C, Pesonen N, Vermesan O, Rustad H. Minkowski fractal microstrip antenna for RFID tags. Proceedings of EuMW2008 Symposium; October 2008; Amsterdam.

[39] Rahimi H, Rusu M, Enoksson P, Sandström D, Rusu C. Small patch antenna based on fractal design for wireless sensors, MME07, 18th Workshop on Micromachining, Micromechanics, and Microsystems; September 16–18, 2007; Portugal.

13

WEARABLE COMMUNICATION AND MEDICAL SYSTEMS

Personal communication and biomedical industry is in continuous growth in the last few years. Due to the huge progress in development of communication systems in the last decade, the development of low-cost wearable communication systems is not risky. However, development of compact efficient wearable antennas is one of the major challenges in development of wearable communication and medical systems. Low-profile compact antennas and transceivers are crucial in the development of wearable human communication and biomedical systems. Development of wearable antennas and compact transceivers for communication and biomedical systems will be described in this chapter. Design considerations, computational results, and measured results of wearable antennas and compact transceivers will be presented in this chapter.

13.1 WEARABLE ANTENNAS FOR COMMUNICATION AND MEDICAL APPLICATIONS

Development of wearable printed antennas for communication and biomedical systems will be described in this section. Design considerations, computational results, and measured results on the human body of several compact wideband microstrip antennas with high efficiency at 434 MHz ± 5% are presented in this section. The proposed antenna may be used in communication and medicare RF systems. The antennas S_{11} results for different belt thickness, shirt thickness, and air spacing between the

Wideband RF Technologies and Antennas in Microwave Frequencies, First Edition. Dr. Albert Sabban.
© 2016 John Wiley & Sons, Inc. Published 2016 by John Wiley & Sons, Inc.

antennas and human body are presented in this section. If the air spacing between the new dually polarized antenna and the human body is increased from 0 to 5 mm, the antenna resonant frequency is shifted by about 5%.

13.1.1   Introduction

Microstrip antennas are widely used in wearable communication system. Microstrip antennas posse's attractive features that are crucial to wearable communication and medical systems. Microstrip antennas features are low profile, flexible, lightweight, and have low production cost. Microstrip antennas are widely presented in books and papers in the last decade [1–7]. However, the effect of human body on the electrical performance of wearable antennas at 434 MHz is not presented [8–13]. RF transmission properties of human tissues have been investigated in several articles [8, 9]. Several wearable antennas have been presented in the last decade [10–16]. A review of wearable and body-mounted antennas designed and developed for various applications at different frequency bands over the last decade can be found in Ref. [10]. In Ref. [11] meander wearable antennas in close proximity of a human body are presented in the frequency range between 800 and 2700 MHz. In Ref. [12] a textile antenna performance in the vicinity of the human body is presented at 2.4 GHz. In Ref. [13] the effect of human body on wearable 100 MHz portable radio antennas is studied. In Ref. [13] the authors concluded that wearable antennas need to be shorter by 15–25% from the antenna length in free space. Measurement of the antenna gain in Ref. [13] shows that a wide dipole (116 × 10 cm) has −13dBi gain. The antennas presented in Refs. [10–13] were developed mostly for cellular applications. Requirements and the frequency range for medical applications are different from those for cellular applications.

In this section, a new class of wideband compact wearable microstrip antennas for medical applications is presented. Numerical results with and without the presence of the human body are discussed. The antennas VSWR is better than $2:1$ at 434 MHz ± 5%. The antenna beamwidth is around 100°. The antennas gain is around 0–4dBi.

13.2   DUALLY POLARIZED WEARABLE 434 MHz PRINTED ANTENNA

Compact microstrip-loaded dipole antennas have been designed to provide horizontal polarization. The antenna dimensions have been optimized to operate on the human body by employing Agilent Advanced Design System (ADS) software [16]. The antenna consists of two layers. The first layer consists of RO3035 0.8 mm dielectric substrate. The second layer consists of RT/duroid 5880 0.8 mm dielectric substrate. The substrate thickness determines the antenna bandwidth. However, thinner antennas are flexible. Thicker antennas have been designed with wider bandwidth. The printed slot antenna provides a vertical polarization. In several medical systems, the required polarization may be vertical or horizontal. The proposed antenna is dually polarized. The printed dipole and the slot antenna provide dual orthogonal polarizations. The dimensions and current distribution of the dual-polarized wearable antenna are presented in Figure 13.1. The antenna dimensions are $26 \times 6 \times 0.16$ cm. The antenna may be used as a wearable antenna on a human body. The antenna may be attached to the patient shirt, patient stomach, or in the back zone. The antenna has been analyzed by using Agilent ADS software. There is a good agreement between measured and computed results. The antenna bandwidth is around 10% for VSWR better than 2 : 1. The antenna beamwidth is around 100°. The antenna gain is around 2dBi. The computed S_{11} and S_{22} parameters are presented in Figure 13.2. Figure 13.3 presents the antenna measured S_{11} parameters. The computed radiation

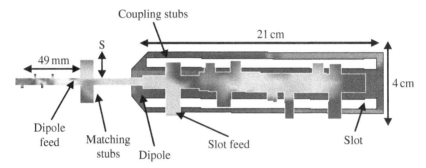

FIGURE 13.1　Current distribution of the wearable antenna.

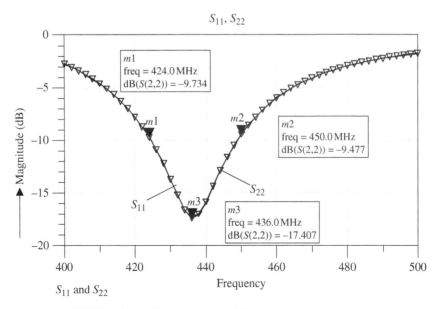

FIGURE 13.2　Computed S_{11} and S_{22} results on human body.

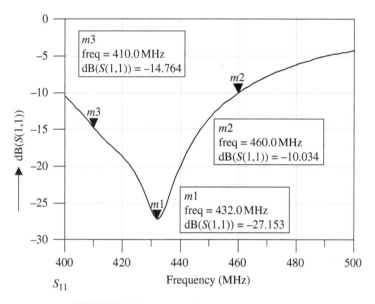

FIGURE 13.3   Measured S_{11} on human body.

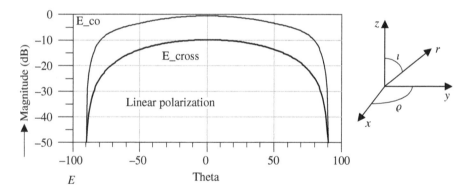

FIGURE 13.4   Antenna radiation patterns.

patterns are shown in Figure 13.4. The copolar radiation pattern belongs to the yz plane. The cross-polar radiation pattern belongs to the xz plane. The antenna cross-polarized field strength may be adjusted by varying the slot feed location. The dimensions and current distribution of the folded dually polarized antenna are presented in Figure 13.5. The antenna dimensions are $5.5 \times 4 \times 0.16$ cm. Figure 13.6 presents the antenna computed S_{11} and S_{22} parameters. The computed radiation patterns of the folded dipole are shown in Figure 13.7. The antenna's radiation characteristics on the human body have been measured by using a phantom. The phantom electrical characteristics represent the human body electrical characteristics.

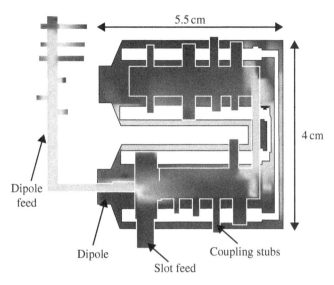

FIGURE 13.5   Current distribution of the folded dipole antenna, $7 \times 5 \times 0.16$ cm.

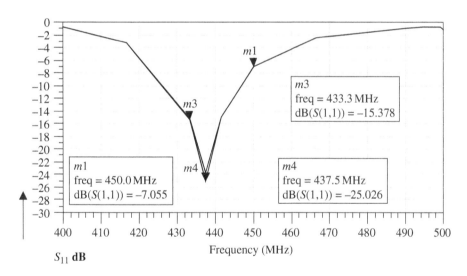

FIGURE 13.6   Folded antenna computed S_{11} and S_{22} results on human body.

The phantom has a cylindrical shape with a 40 cm diameter and a length of 1.5 m. The phantom contains a mix of 55% water, 44% sugar, and 1% salt. The antenna under test was placed on the phantom during the measurements of the antennas radiation characteristics. S_{11} and S_{12} parameters were measured directly on human body by using a network analyzer. The measured results were compared to a known reference antenna.

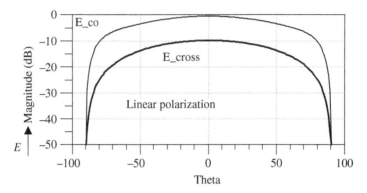

FIGURE 13.7   Folded antenna radiation patterns.

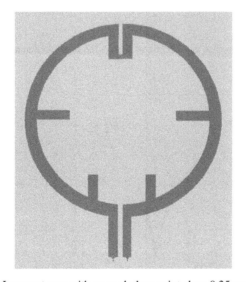

FIGURE 13.8   Loop antenna with ground plane printed on 0.25 mm thick substrate.

13.3   LOOP ANTENNA WITH GROUND PLANE

A new loop antenna with ground plane has been designed on Kapton substrates with thickness of 0.25 and 0.4 mm. The antenna without ground plane is shown in Figure 13.8. The loop antenna VSWR without the tuning capacitor was 4 : 1. This loop antenna may be tuned by adding a capacitor or varactor as shown in Figure 13.9. Tuning the antenna allow us to work in a wider bandwidth. Figure 13.10 presents the loop antenna computed S_{11} on human body. There is good agreement between measured and computed S_{11}. The computed 3D radiation pattern is shown in Figure 13.11.

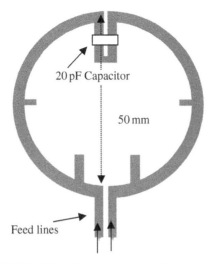

FIGURE 13.9 Tunable loop antenna without ground plane.

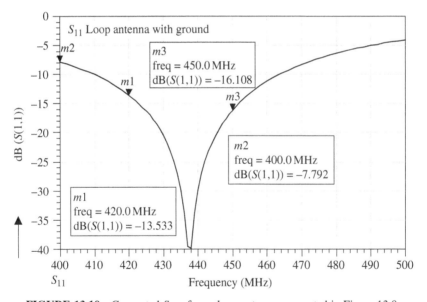

FIGURE 13.10 Computed S_{11} of new loop antenna, presented in Figure 13.8.

Table 13.1 compares the electrical performance of a loop antenna with ground plane with a loop antenna without ground plane. There is a good agreement between measured and computed results for several loop antennas electrical parameters on human body. The results presented in Table 13.1 indicate that the loop antenna with ground plane is matched to the human body environment, without the tuning capacitor, better than the loop antenna without ground plane.

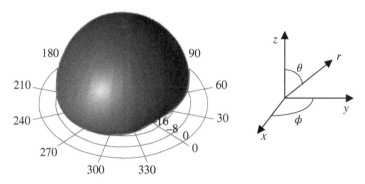

FIGURE 13.11   Loop antenna with ground, 3D radiation pattern.

TABLE 13.1   Electrical Performance of Several Loop Antenna Configurations

Antenna with No Tuning Capacitor	Beamwidth 3dB	Gain (dBi)	VSWR
Loop with no GND	100°	0	4 : 1
Loop with tuning capacitor (no GND)	100°	0	2 : 1
Wearable Loop with GND	100°	0–2	2 : 1
Loop with GND in free space	100°–110°	−3	5 : 1

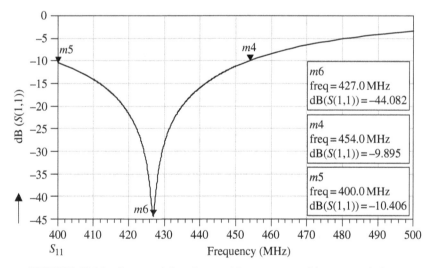

FIGURE 13.12   Computed S_{11} of a tuned loop antenna without ground plane.

The computed 3D radiation pattern and the coordinate used in this chapter are shown in Figure 13.11. Computed S_{11} of the loop antenna with a tuning capacitor is given in Figure 13.12. Figure 13.13 presents the radiation pattern of loop antenna without ground on human body. Figure 13.14 presents a loop antenna with ground

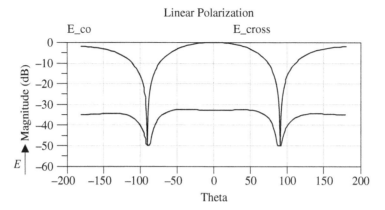

FIGURE 13.13   Radiation pattern of loop antenna without ground on human body.

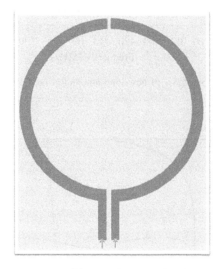

FIGURE 13.14   Loop antenna with ground plane printed on 0.4 mm thick substrate.

plane printed on 0.4 mm thick substrate. Figure 13.15 presents the loop antenna computed S_{11} on human body. Loop antennas printed on thicker substrate have wider bandwidth as presented in Figure 13.15. Figure 13.16 presents the loop antenna, printed on 0.4 mm thick substrate, radiation pattern.

13.4   ANTENNA S_{11} VARIATION AS FUNCTION OF DISTANCE FROM BODY

The antennas input impedance variation as function of distance from the body had been computed by employing ADS software. The analyzed structure is presented in Figure 13.17a. The antenna was placed inside a belt with thickness between 1

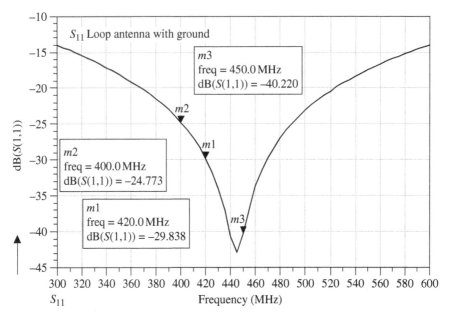

FIGURE 13.15   Computed S_{11} of new loop antenna printed on 0.4 mm thick substrate.

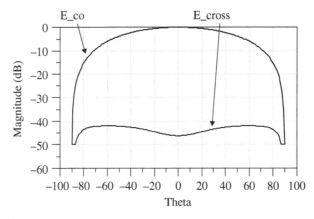

FIGURE 13.16   New loop antenna, printed on 0.4 mm thick substrate, radiation pattern.

and 4 mm as shown in Figure 13.17b. The patient body thickness was varied from 15 to 300 mm. The dielectric constant of the body was varied from 40 to 50. The antenna was placed inside a belt with thickness between 2 and 4 mm with dielectric constant from 2 to 4. The air layer between the belt and the patient shirt may vary from 0 to 8 mm. The shirt thickness was varied from 0.5 to 1 mm. The dielectric constant of the shirt was varied from 2 to 4. Properties of human body tissues are listed in Table 13.2 scc Ref. [8]. These properties were employed in the antenna design.

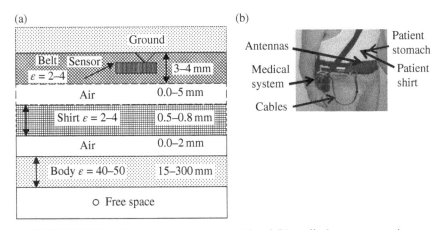

FIGURE 13.17   (a) Analyzed structure model and (b) medical system on patient.

TABLE 13.2   Properties of Human Body Tissues

Tissue	Property	434 MHz	800 MHz	1000 MHz
Prostate	σ	0.75	0.90	1.02
	ε	50.53	47.4	46.65
Stomach	σ	0.67	0.79	0.97
	ε	42.9	40.40	39.06
Colon, Heart	σ	0.98	1.15	1.28
	ε	63.6	60.74	59.96
Kidney	σ	0.88	0.88	0.88
	ε	117.43	117.43	117.43
Nerve	σ	0.49	0.58	0.63
	ε	35.71	33.68	33.15
Fat	σ	0.045	0.056	0.06
	ε	5.02	4.58	4.52
Lung	σ	0.27	0.27	0.27
	ε	38.4	38.4	38.4

Figure 13.18 presents S_{11} results (of the antenna shown in Fig. 13.1) for different belt thickness, shirt thickness, and air spacing between the antennas and human body. One may conclude from results shown in Figure 13.18 that the antenna has VSWR better than 2.5 : 1 for air spacing up to 8 mm between the antennas and patient body. For frequencies ranging from 415 to 445 MHz, the antenna has VSWR better than 2 : 1 when there is no air spacing between the antenna and the patient body. Results shown in Figure 13.19 indicates that the folded antenna (the antenna shown in Fig. 13.5) has VSWR better than 2.0 : 1 for air spacing up to 5 mm between the antennas and patient body. Figure 13.19 presents S_{11} results of the folded antenna results for different positions relative to the human body. Explanation of Figure 13.19 is given in Table 13.3. If the air spacing between the sensors and the human body is increased from 0 to 5 mm,

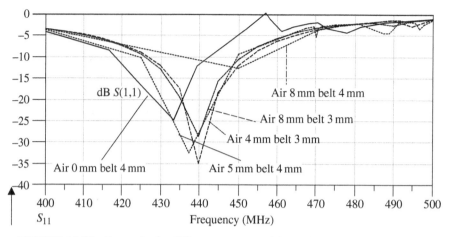

FIGURE 13.18 S_{11} results for different antenna positions relative to the human body.

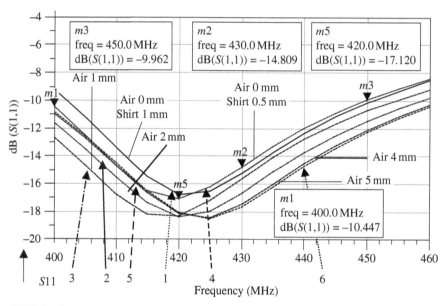

FIGURE 13.19 Folded antenna S_{11} results for different antenna positions relative to the human body.

TABLE 13.3 Explanation of Figure 13.19

Picture	Line Type	Sensor Position
1	Dot·············	Shirt thickness 0.5 mm
2	Line———	Shirt thickness 1 mm
3	Dash dot—·—·—	Air spacing 2 mm
4	Dash-----·	Air spacing 4 mm
5	Long dash— — —	Air spacing 1 mm
6	Big dots···········	Air spacing 5 mm

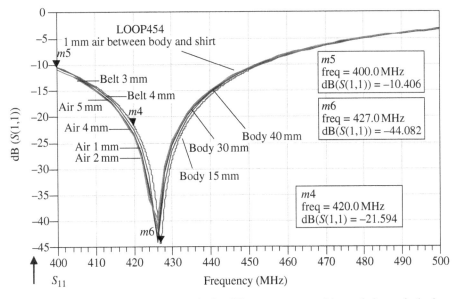

FIGURE 13.20 Loop antenna S_{11} results for different antenna positions relative to the body.

TABLE 13.4   Explanation of Figure 13.20

Plot Colure	Sensor Position
Red	Body 15 mm air spacing 0 mm
Blue	Air spacing 5 mm Body 15 mm
Pink	Body 40 mm air spacing 0 mm
Green	Body 30 mm air spacing 0 mm
Sky	Body 15 mm air spacing 2 mm
Purple	Body 15 mm air spacing 4 mm

the antenna resonant frequency is shifted by 5%. The loop antenna with ground plane has VSWR better than 2.0 : 1 for air spacing up to 5 mm between the antennas and patient body.

If the air spacing between the sensors and the human body is increased from 0 to 5 mm, the computed antenna resonant frequency is shifted by 2%. However, if the air spacing between the sensors and the human body is increased up to 5 mm, the measured loop antenna resonant frequency is shifted by 5%. Explanation of Figure 13.20 is given in Table 13.4.

13.5   WEARABLE ANTENNAS

An application of the proposed antenna is shown in Figure 13.21. Three to four folded dipole or loop antennas may be assembled in a belt and attached to the patient stomach. The cable from each antenna is connected to a recorder. The received signal is

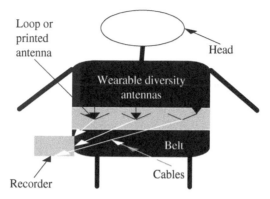

FIGURE 13.21 Printed wearable antenna.

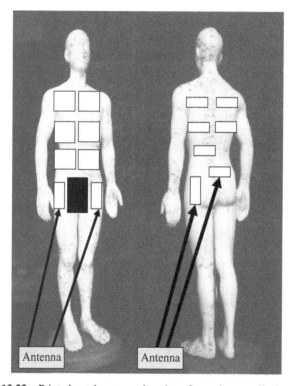

FIGURE 13.22 Printed patch antenna locations for various medical applications.

routed to a switching matrix. The signal with the highest level is selected during the medical test. The antennas receive a signal that is transmitted from various positions in the human body. Folded antennas may be also attached on the patient back in order to improve the level of the received signal from different locations in the human body. Figures 13.22 and 13.23 show various antenna locations on the back and front of the human body for different medical applications.

(a)

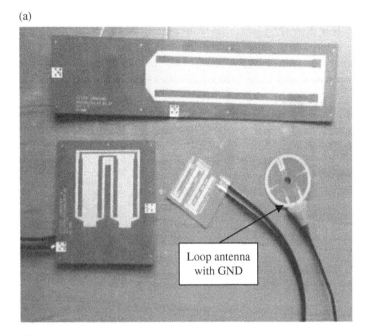

(b)

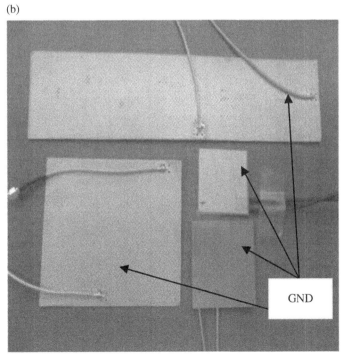

FIGURE 13.23   (a) Microstrip antennas for medical applications and (b) backside of the antennas.

In several applications the distance separating the transmitting and receiving antennas is less than $2D^2/\lambda$. D is the largest dimension of the radiator. In these applications the amplitude of the electromagnetic field close to the antenna may be quite powerful, but because of rapid falloff with distance, the antenna do not radiate

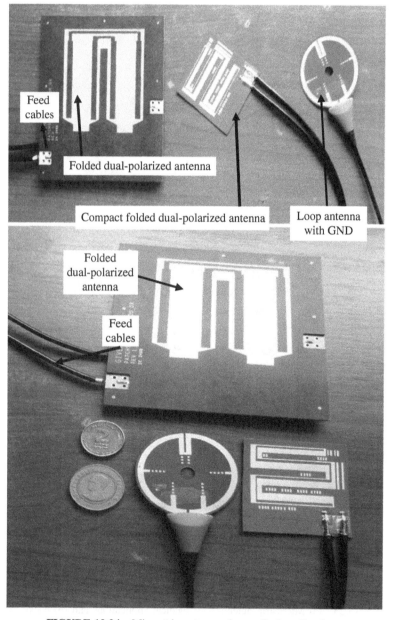

FIGURE 13.24 Microstrip antennas for medical applications.

COMPACT WEARABLE RFID ANTENNAS

energy to infinite distances but instead the radiated power remain trapped in the region near to the antenna. Thus, the near fields only transfer energy to close distances from the receivers. The receiving and transmitting antennas are magnetically coupled. Change in current flow through one wire induces a voltage across the ends of the other wire through electromagnetic induction. The amount of inductive coupling between two conductors is measured by their mutual inductance. In these applications we have to refer to the near field and not to the far-field radiation.

In Figures 13.23 and 13.24, several microstrip antennas for medical applications at 434 MHz are shown. The backside of the antennas is presented in Figure 13.23b. The diameter of the loop antenna presented in Figure 13.24 is 50 mm. The dimensions of the folded dipole antenna are $7 \times 6 \times 0.16$ cm. The dimensions of the compact folded dipole presented in Figure 13.24 are $5 \times 5 \times 0.5$ cm.

13.6   COMPACT DUAL-POLARIZED PRINTED ANTENNA

New compact microstrip-loaded dipole antennas have been designed. The antenna consists of two layers. The first layer consists of FR4 0.25 mm dielectric substrate. The second layer consists of Kapton 0.25 mm dielectric substrate. The substrate thickness determines the antenna bandwidth. However, with thinner substrate we may achieve better flexibility. The proposed antenna is dual polarized. The printed dipole and the slot antenna provide dual orthogonal polarizations. The dual-polarized antenna is shown in Figure 13.25. The antenna dimensions are $5 \times 5 \times 0.05$ cm.

The antenna may be attached to the patient shirt in the patient stomach or back zone. The antenna has been analyzed by using Agilent ADS software. There is a good agreement between measured and computed results. The antenna bandwidth is around 10% for VSWR better than 2 : 1. The antenna beamwidth is around 100°. The antenna gain is around 0 dBi. The computed S_{11} parameters are presented in Figure 13.26. Figure 13.27 presents the antenna measured S_{11} parameters. The antenna cross-polarized field strength may be adjusted by varying the slot feed location. The computed 3D radiation pattern of the antenna is shown in Figure 13.28. The computed radiation pattern is shown in Figure 13.29.

13.7   COMPACT WEARABLE RFID ANTENNAS

*Radio Frequency Id*entification (RFID) is an electronic method of exchanging data over radio frequency waves. There are three major components in RFID system: transponder (tag), antenna, and a controller. The RFID tag, antenna, and controller may be assembled on the same board. Microstrip antennas are widely presented in books and papers in the last decade [1–7]. However, compact wearable printed antennas are not widely used at 13.5 MHz RIFD systems. HF tags work best at close range but are more effective at penetrating nonmetal objects especially objects with high water content.

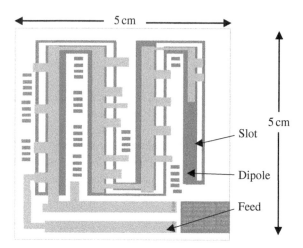

FIGURE 13.25   Printed compact dual-polarized antenna.

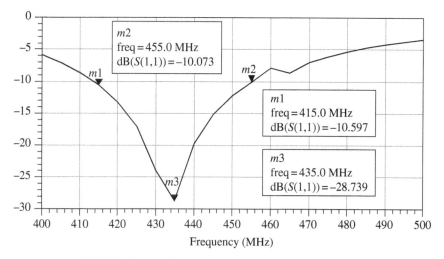

FIGURE 13.26   Computed S_{11} results of compact antenna.

A new class of wideband compact printed and microstrip antennas for RFID applications is presented in this chapter. RF transmission properties of human tissues have been investigated in several papers [8, 9]. The effect of human body on the antenna performance is investigated in this chapter. The proposed antennas may be used as wearable antennas on persons or animals. The proposed antennas may be attached to cars, trucks, containers, and other various objects.

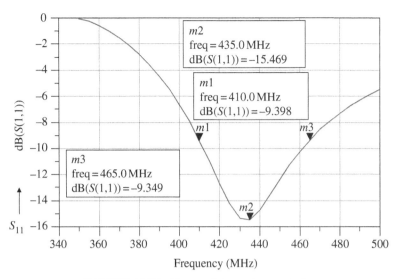

FIGURE 13.27 Measured S_{11} on human body.

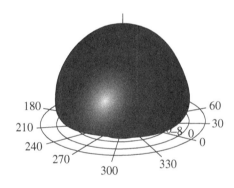

FIGURE 13.28 Compact antenna 3D radiation pattern.

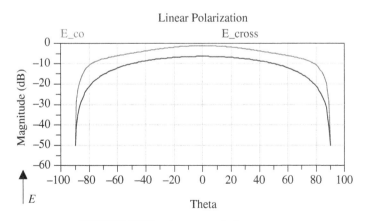

FIGURE 13.29 Antenna radiation pattern.

13.7.1   Dual-Polarized 13.5 MHz Compact Printed Antenna

One of the most critical elements of any RFID system is the electrical performance of its antenna. The antenna is the main component for transferring energy from the transmitter to the passive RFID tags, receiving the transponder's replying signal and avoiding in-band interference from electrical noise and other nearby RFID components. Low-profile compact printed antennas are crucial in the development of RIFD systems.

New compact microstrip-loaded dipole antennas have been designed at 13.5 MHz to provide horizontal polarization. The antenna consists of two layers. The first layer consists of FR4 0.8 mm dielectric substrate. The second layer consists of Kapton 0.8 mm dielectric substrate. The substrate thickness determines the antenna bandwidth. A printed slot antenna provides a vertical polarization. The proposed antenna is dual polarized. The printed dipole and the slot antenna provide dual orthogonal polarizations. The dual-polarized RFID antenna is shown in Figure 13.30. The antenna dimensions are $6.4 \times 6.4 \times 0.16$ cm. The antenna may be attached to the customer shirt in the customer stomach or back zone. The antenna has been analyzed by using Agilent ADS software.

The antenna S_{11} parameter is better than -21dB at 13.5 MHz. The antenna gain is around -10dBi. The antenna beamwidth is around $160°$. The computed S_{11} parameters are presented in Figure 13.31. There is a good agreement between measured and computed results. The antenna cross- polarized field strength may be adjusted by varying the slot feed location. The computed radiation pattern is shown in Figure 13.32.

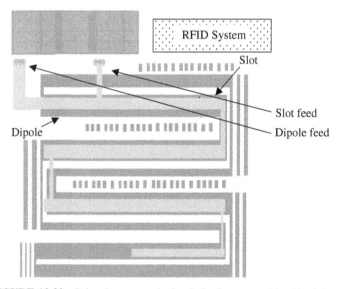

FIGURE 13.30   Printed compact dual-polarized antenna, $64 \times 64 \times 1.6$ mm.

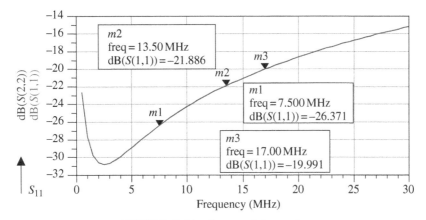

FIGURE 13.31 Computed S_{11} results.

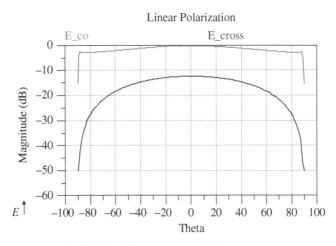

FIGURE 13.32 Antenna radiation pattern.

13.7.2 Varying the Antenna Feed Network

Several designs with different feed networks have been developed. A compact antenna with different feed networks is shown in Figure 13.33. The antenna dimensions are $8.4 \times 6.4 \times 0.16$ cm. Figure 13.34 presents the antenna computed S_{11} on human body. There is a good agreement between measured and computed results. The computed radiation pattern is shown in Figure 13.35. Table 13.5 compares the electrical performance of a loop antenna with the compact dual-polarized antenna.

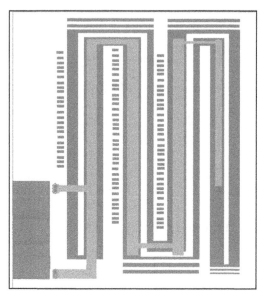

FIGURE 13.33   RFID printed antenna, $8.4 \times 6.4 \times 0.16$ cm.

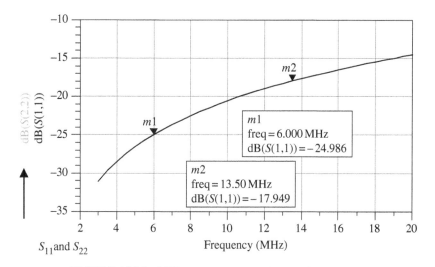

FIGURE 13.34   RFID antenna computed S_{11} and S_{22} results.

13.7.3 RFID Wearable Loop Antennas

Several RFID loop antennas are presented in Ref. [15]. The disadvantages of loop antennas with number of turns are low efficiency and narrow bandwidth. The real part of the loop antenna impedance approaches 0.5 Ω. The image part of the loop antenna impedance may be represented as high inductance. A matching network may be used

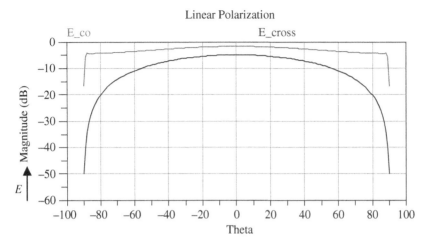

FIGURE 13.35   Compact antenna, 8.4 × 6.4 × 0.16 cm, radiation pattern.

TABLE 13.5   Comparison of Loop Antenna and Microstrip Antenna Parameters

Antenna	Beamwidth 3dB	Gain (dBi)	VSWR
Loop antenna	140°	−25	2 : 1
Microstrip antenna	160°	−10	1.2 : 1

to match the antenna to 50 Ω. The matching network consists of an RLC matching network. This matching network has narrow bandwidth. The loop antenna efficiency is lower than 1%.

A square four-turn loop antenna has been designed at 13.5 MHz by using Agilent ADS software. The antenna is printed on a FR4 substrate. The antenna dimensions are 32 × 52.4 × 0.25 mm. The antenna layout is shown in Figure 13.36a. The antenna photo is shown in Figure 13.36b. S_{11} results of the printed loop antenna are shown in Figure 13.37. The antenna S_{11} parameter is better than −9.5dB without an external matching network. The computed radiation pattern is shown in Figure 13.38. The computed radiation pattern takes into account an infinite ground plane.

The microstrip antenna input impedance variation as function of distance from the body has been computed by employing ADS software. The analyzed structure is presented in Figure 13.17. Properties of human body tissues are listed in Table 13.2 (see Ref. [8]). These properties were used in the antenna design. S_{11} parameters for different human body thicknesses have been computed. We may note that the differences in the results for body thickness of 15–100 mm are negligible. S_{11} parameters for different positions relative to the human body have been computed. If the air spacing between the antenna and the human body is increased from 0 to 10 mm, the antenna S_{11} parameters may change by less than 1%. The VSWR is better than 1.5 : 1.

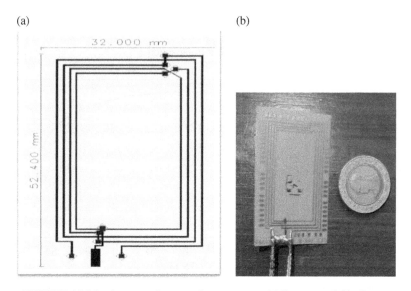

FIGURE 13.36 A square four-turn loop antenna. (a) Layout and (b) photo.

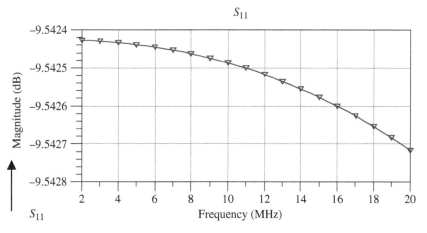

FIGURE 13.37 Loop antenna computed S_{11} results.

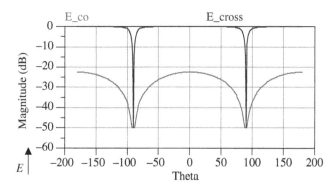

FIGURE 13.38 Loop antenna radiation patterns for an infinite ground plane.

13.7.4 Proposed Antenna Applications

An application of the proposed antenna is shown in Figure 13.39. The RFID antennas may be assembled in a belt and attached to the customer stomach. The antennas may be employed as transmitting or as receiving antennas. The antennas may receive or transmit information to medical systems.

In RFID systems the distance between the transmitting and receiving antennas is less than $2D^2/\lambda$, where D is the largest dimension of the antenna. The receiving and transmitting antennas are magnetically coupled. In these applications we refer to the near field and not to the far-field radiation pattern.

Figures 13.40 and 13.41 present compact printed antenna for RFID applications. The presented antennas may be assembled in a belt and attached to the patient stomach or back.

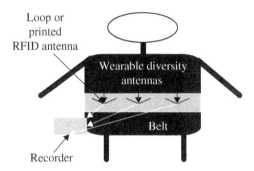

FIGURE 13.39 Wearable RFID antenna.

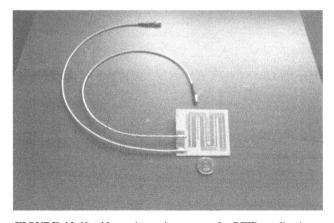

FIGURE 13.40 New microstrip antenna for RFID applications.

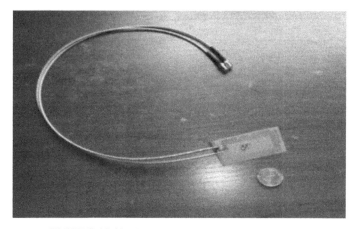

FIGURE 13.41   Loop antenna for RFID applications.

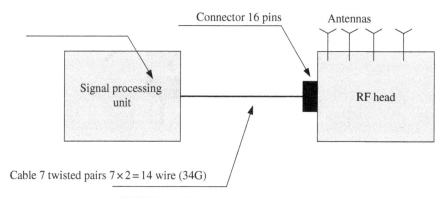

FIGURE 13.42   Recorder block diagram.

13.8   434 MHz RECEIVING CHANNEL FOR COMMUNICATION AND MEDICAL SYSTEMS

A medical system may be implanted or inserted to the human body as swallowed capsule. The medical device will transmit medical data to a recorder. The medical data may be analyzed by the medical stuff online or stored as medical data about the patient. The receiving channel is part of the recorder and consists of receiving wearable antennas, RF head, and a signal processing unit. A block diagram of the receiver is shown in Figure 13.42.

Receiving channel main specifications:

Requirement	Specification
Frequency range up link	430–440 MHz
Return loss (dB)	−9
Group delay	Max. 50 ns for 12 MHz BW
Input power	−30 to −60dBm
SNR	>20dB
Current consumption (mA)	50
Dimensions (cm)	$12 \times 12 \times 5$
Frequency range down link	13.56 MHz
Received power down link	−8dBm ± 1dB
Current consumption (mA)	100–110

A block diagram of the receiving channel is shown in Figure 13.43. The receiving channel consists of an uplink channel at 434 MHz and a downlink at 13.56 MHz.

The uplink channel consists of a switching matrix, low-noise amplifier (LNA), and filter. The switching matrix losses are around 2dB. The LNA noise figure is around 1dB with 21dB gain. The downlink channel consists of a transmitting antenna, antenna matching network, and differential amplifier. The downlink channel transmits commands to the medical system. The receiving channel gain and noise figure budget is shown in Figure 13.44. The receiving channel noise figure is around 3.5dB. A receiving channel with lower noise figure values is shown in Figure 13.45. The LNA is connected to the receiving antenna.

Receiving channel gain and noise figure budget is shown in Figure 13.46. The receiving channel noise figure is around 3.5dB.

Four folded dipole or loop antennas may be assembled in a belt and attached to the patient stomach as shown in Figure 13.47a and b. The cable from each antenna is connected to a recorder. The received signal is routed to a switching matrix. The signal with the highest level is selected during the medical test. The antennas receive a signal that is transmitted from various positions in the human body. Wearable antennas may be also attached on the patient back in order to improve the level of the received signal from different locations in the human body.

13.9 CONCLUSIONS

Development of wearable antennas and compact transceiver for communication and biomedical systems is presented in this chapter. This chapter presents wideband microstrip antennas with high efficiency for medical applications. The antenna dimensions may vary from $26 \times 6 \times 0.16$ cm to $5 \times 5 \times 0.05$ cm according to the medical system specification. The antennas bandwidth is around 10% for VSWR better than 2 : 1.

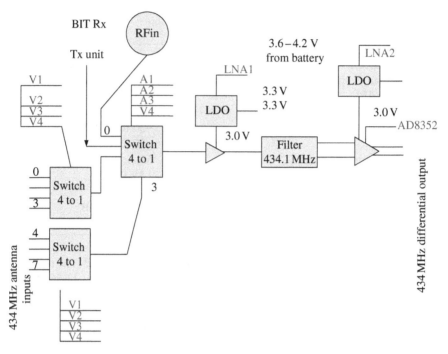

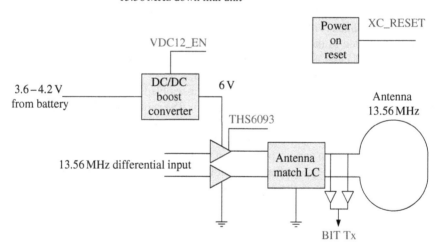

FIGURE 13.43   Receiving channel block diagram.

The antenna beamwidth is around 100°. The antennas gain varies from 0 to 4dBi. The antenna S_{11} results for different belt thickness, shirt thickness, and air spacing between the antennas and human body are presented in this chapter. If the air spacing between the new dual-polarized antenna and the human body is increased from 0 to 5 mm, the

System1

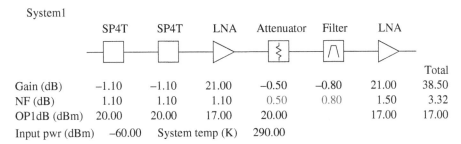

	SP4T	SP4T	LNA	Attenuator	Filter	LNA	Total
Gain (dB)	−1.10	−1.10	21.00	−0.50	−0.80	21.00	38.50
NF (dB)	1.10	1.10	1.10	0.50	0.80	1.50	3.32
OP1dB (dBm)	20.00	20.00	17.00	20.00		17.00	17.00
Input pwr (dBm)	−60.00	System temp (K)	290.00				

FIGURE 13.44 Receiving channel gain and noise figure budget.

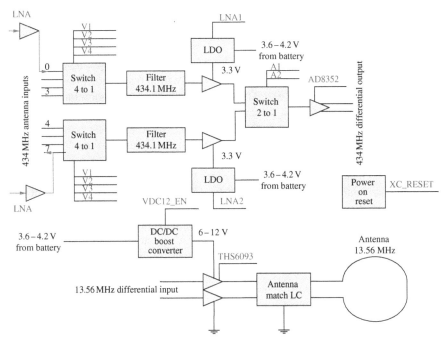

FIGURE 13.45 Receiving channel block diagram with LNA connected to the antennas.

System1

	LNA	SP4T	Filter	SP2T	LNA	Attenuator	LNA	Total
Gain (dB)	21.00	−1.20	-0.80	−1.00	17.00	−1.00	15.50	49.50
NF (dB)	1.10	1.20	0.80	2.00	1.10	1.00	2.00	1.16
OP1dB (dBm)	17.00	20.00		20.00	17.00	20.00	17.00	17.00
Input pwr (dBm)	−60.00	System temp (K)	290.00					

FIGURE 13.46 Receiving channel with LNA connected to the antennas, gain and noise figure budget.

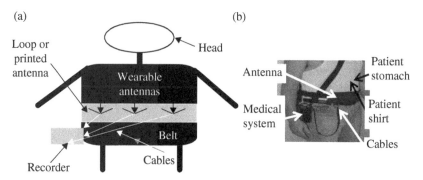

FIGURE 13.47   (a) Wearable medical system and (b) medical system on patient.

antenna resonant frequency is shifted by 5%. However, if the air spacing between the helix antenna and the human body is increased only from 0 to 2 mm, the antenna resonant frequency is shifted by 5%. The effect of the antenna location on the human body should be considered in the antenna design process. The proposed antenna may be used in Medicare RF systems.

A wideband tunable microstrip antennas with high efficiency for medical applications has been presented in this chapter. The antenna dimensions may vary from $26 \times 6 \times 0.16$ cm to $5 \times 5 \times 0.05$ cm according to the medical system specification. The antennas bandwidth is around 10% for VSWR better than 2 : 1. The antenna beamwidth is around 100°. The antennas gain varies from 0 to 2dBi. If the air spacing between the dual-polarized antenna and the human body is increased from 0 to 5 mm, the antenna resonant frequency is shifted by 5%. A varactor is employed to compensate variations in the antenna resonant frequency at different locations on the human body.

This chapter presents also wideband compact printed antennas, microstrip, and loop antennas for RFID applications. The antenna beamwidth is around 160°. The antenna gain is around −10dBi. The proposed antennas may be used as wearable antennas on persons or animals. The proposed antennas may be attached to cars, trucks, and other various objects. If the air spacing between the antenna and the human body is increased from 0 to 10 mm, the antenna S_{11} parameters may change by less than 1%. The antenna VSWR is better than 1.5 : 1 for all tested environments.

REFERENCES

[1] James JR, Hall PS, Wood C. *Microstrip Antenna Theory and Design*. New York: Peregrinus on behalf of the Institution of Electrical Engineers; 1981.

[2] Sabban A, Gupta KC. Characterization of radiation loss from microstrip discontinuities using a multiport network modeling approach. IEEE Trans Microwave Theory Tech 1991;39 (4):705 712.

[3] Sabban A. A new wideband stacked microstrip antenna. IEEE Antenna and Propagation Symposium; June 1983; Houston, TX.

[4] Sabban A, Navon E. A mm-waves microstrip antenna array. IEEE Symposium; March 1983; Tel-Aviv.

[5] Kastner R, Heyman E, Sabban A. Spectral domain iterative analysis of single and double-layered microstrip antennas using the conjugate gradient algorithm. IEEE Trans Antennas Propag 1988;36 (9):1204–1212.

[6] Sabban A. Wideband microstrip antenna arrays. IEEE Antenna and Propagation Symposium MELCOM; Tel-Aviv; 1981.

[7] Sabban A. Microstrip antenna arrays. In: Nasimuddin N, editor. *Microstrip Antennas.* Croatia: InTech; 2011. p 361–384.

[8] Chirwa LC, Hammond PA, Roy S, Cumming DRS. Electromagnetic radiation from ingested sources in the human intestine between 150 MHz and 1.2 GHz. IEEE Trans Biomed Eng 2003;50 (4):484–492.

[9] Werber D, Schwentner A, Biebl EM. Investigation of RF transmission properties of human tissues. Adv Radio Sci 2006;4:357–360.

[10] Gupta B, Sankaralingam S, Dhar S. Development of wearable and implantable antennas in the last decade. Microwave Symposium (MMS), 2010 Mediterranean; August 2010; Cyprus. p 251–267.

[11] Thalmann T, Popovic Z, Notaros BM, Mosig JR. Investigation and design of a multi-band wearable antenna. 3rd European Conference on Antennas and Propagation, EuCAP 2009; March 2009; Berlin, Germany. p 462–465.

[12] Salonen P, Rahmat-Samii Y, Kivikoski M. Wearable antennas in the vicinity of human body. IEEE Antennas Propag Soc Int Symp 2004;1:467–470.

[13] Kellomaki T, Heikkinen J, Kivikoski M. Wearable antennas for FM reception. First European Conference on Antennas and Propagation, EuCAP 2006; November 2006; Nice, France. p 1–6.

[14] Sabban A. Wideband printed antennas for medical applications. APMC 2009 Conference; December 2009; Singapore.

[15] Lee Y. Antenna circuit design for RFID applications. Microchip Technology Inc., Microchip AN 710c.ADS software, Agilent. Available at http://www.home.agilent.com/agilent/product.jspx?cc=IL&lc=eng&ckey=1297113&nid=-34346.0.00&id=1297113. Accessed February 9, 2016.

[16] Sabban A. *Low Visibility Antennas for Communication Systems.* Boca Raton: Taylor & Francis Group; 2015.

14

RF MEASUREMENTS

14.1 INTRODUCTION

This chapter describes electromagnetics, microwave engineering and antenna measurements. Basic RF measurement theory is presented in Sections 14.2 and 14.3.

S-parameters measurements are the first stage in electromagnetics, microwave engineering, and antenna measurements. Setups for microwave engineering measurements will be presented in this chapter. Maximum input and output measurements of communication system are also presented in this chapter. Intermodulation measurements, IP_2 and IP_3, are discussed in this chapter.

Antenna measurements will discussed in this chapter. It is more convenient to measure antennas in the receiving mode. If the measured antenna is reciprocal, the antenna radiation characteristics are identical for the receiving and transmitting modes. Active antennas are not reciprocal. Radiation characteristics of antennas are usually measured in the far field. Far-field antenna measurements suffer from some disadvantages. A long free-space area is needed. Reflection from the ground from walls affect measured results and add errors to measured results. It is difficult and almost impossible to measure the antenna on the antenna operating environment, such as airplane or satellite. Antenna measurement facilities are expensive. Some of these drawbacks may be solved by near-field and indoor measurements. Near-field measurements are presented in Ref. [1]. Small communication companies do not own antenna measurement facilities. However, there are several companies around the world that provides antenna measurement services, near-field and far-field measurements. One day near-field measurements may cost around US$5000 and far-field measurements may cost around US$2000.

Wideband RF Technologies and Antennas in Microwave Frequencies, First Edition. Dr. Albert Sabban.
© 2016 John Wiley & Sons, Inc. Published 2016 by John Wiley & Sons, Inc.

14.2   MULTIPORT NETWORKS WITH *N*-PORTS

Antenna systems and communication systems may be represented as multiport networks with *N*-ports as shown in Figure 14.1. We may assume that only one mode propagate in each port. The electromagnetic fields in each port represents incident and reflected waves. The electromagnetic fields may be represented by equivalent voltages and currents as given in Equations 14.1 and 14.2 (see Refs. [1–5]):

$$V_n^- = Z_n I_n^-$$
$$V_n^+ = Z_n I_n^+$$

(14.1)

$$I_n^- = Y_n V_n^-$$
$$I_n^+ = Y_n V_n^+$$

(14.2)

The voltages and currents in each port are given in Equation 14.3:

$$I_n = I_n^+ - I_n^-$$
$$V_n = V_n^+ + V_n^-$$

(14.3)

The relations between the voltages and currents may be represented by the *Z* matrix as given in Equation 14.4. The relations between the currents and voltages may be represented by the *Y* matrix as given in Equation 14.5. The *Y* matrix is the inverse of the *Z* matrix:

$$[V] = \begin{bmatrix} V_1 \\ V_2 \\ \vdots \\ V_N \end{bmatrix} = \begin{bmatrix} Z_{11} & Z_{12} & \cdots & Z_{1N} \\ Z_{21} & Z_{22} & \cdots & Z_{2N} \\ \vdots & \vdots & \ddots & \vdots \\ Z_{N1} & Z_{N2} & \cdots & Z_{NN} \end{bmatrix} \begin{bmatrix} I_1 \\ I_2 \\ \vdots \\ I_N \end{bmatrix} = [Z][I]$$

(14.4)

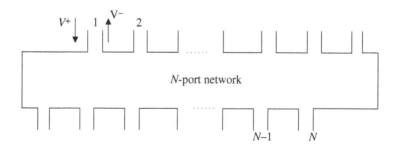

FIGURE 14.1   Multiport networks with *N*-ports.

$$
[I] = \begin{bmatrix} I_1 \\ I_2 \\ \vdots \\ I_N \end{bmatrix} = \begin{bmatrix} Y_{11} & Y_{12} & \cdots & Y_{1N} \\ Y_{21} & Y_{22} & \cdots & Y_{2N} \\ \vdots & \vdots & \ddots & \vdots \\ Y_{N1} & Y_{N2} & \cdots & Y_{NN} \end{bmatrix} \begin{bmatrix} V_1 \\ V_2 \\ \vdots \\ V_N \end{bmatrix} = [Y][V] \tag{14.5}
$$

14.3   SCATTERING MATRIX

We cannot measure voltages and currents in microwave networks. However, we can measure power, VSWR, and the location of the minimum field strength. We can calculate the reflection coefficient from these data. The scattering matrix is a mathematical presentation that describes how electromagnetic energy propagates through a multiport network. The S-matrix allows us to accurately describe the properties of complicated networks. S-parameters are defined for a given frequency and system impedance and vary as a function of frequency for any nonideal network. The scattering S-matrix describes the relation between the forward and reflected waves as written in Equation 14.6. S-parameters describe the response of an N-port network to voltage signals at each port. The first number in the subscript refers to the responding port, while the second number refers to the incident port. Thus S_{21} means the response at port 2 due to a signal at port 1:

$$
[V^-] = \begin{bmatrix} V_1^- \\ V_2^- \\ \vdots \\ V_2^+ \end{bmatrix} = \begin{bmatrix} S_{11} & S_{12} & \cdots & S_{1N} \\ S_{21} & S_{22} & \cdots & S_{2N} \\ \vdots & \vdots & \ddots & \vdots \\ S_{N1} & S_{N2} & \cdots & S_{NN} \end{bmatrix} \begin{bmatrix} V_1^+ \\ V_2^+ \\ \vdots \\ V_2^+ \end{bmatrix} = [S][V^+] \tag{14.6}
$$

The S_{nn} elements represent reflection coefficients. The S_{nm} elements represent transmission coefficients as written in Equation 14.7, where a_i represents the forward voltage in the i port:

$$
S_{nn} = \frac{V_n^-}{V_n^+} \Big|_{a_i = 0} \quad i \neq n
$$

$$
S_{nm} = \frac{V_n^-}{V_m^+} \Big|_{a_i = 0} \quad i \neq m
$$

$$\tag{14.7}$$

By normalizing the S-matrix, we can represent the forward and reflected voltages as written in Equation 14.8. S-parameters depend on the frequency and are given as function of frequency. In a reciprocal microwave network, $S_{nm} = S_{mn}$ and $[S]^t = [S]$:

$$I = I^+ - I^- = V^+ - V^-$$

$$V = V^+ + V^-$$

$$V^+ = \frac{1}{2}(V + I)$$

$$V^- = \frac{1}{2}(V - I)$$

(14.8)

The relation between Z- and S-matrix is derived by using Equations 14.8 and 14.9 and is given in Equation 14.10 and 14.11:

$$I_n = I_n^+ - I_n^- = V_n^+ - V_n^-$$
$$V_n = V_n^+ + V_n^-$$

(14.9)

$$[V] = [V^+] + [V^-] = [Z][I] = [Z][V^+] - [Z][V^-]$$
$$([Z] + [U])[V^-] = ([Z] - [U])[V^+]$$
$$[V^-] = ([Z] + [U])^{-1}([Z] - [U])[V^+]$$
$$[V^-] = [S][V^+]$$
$$[S] = ([Z] + [U])^{-1}([Z] - [U])$$

(14.10)

$$[V^+] = \frac{1}{2}([V] + [I]) = \frac{1}{2}([Z] + [U])[I]$$
$$[V^-] = \frac{1}{2}([V] - [I]) = \frac{1}{2}([Z] - [U])[I]$$
$$\frac{1}{2}[I] = ([Z] + [U])^{-1}[V^+]$$
$$[V^-] = ([Z] - [U])([Z] + [U])^{-1}[V^+]$$
$$[S] = ([Z] - [U])([Z] + [U])^{-1}$$

(14.11)

A network analyzer is employed to measure S-parameters, as shown in Figure 14.2(a). A network analyzer may have 2–16 ports.

14.4 S-PARAMETERS MEASUREMENTS

Antenna S-parameters measurement is usually a one-port measurement. First we calibrate the network analyzer to the desired frequency range. One port, S_{1P}, calibration process consists of three steps:

(a)   (b)

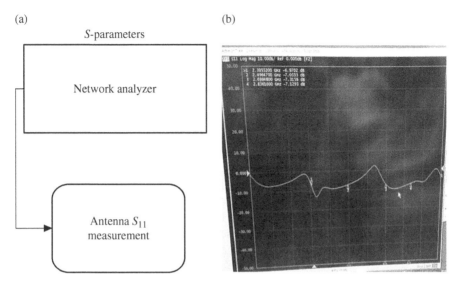

FIGURE 14.2   (a) Antenna *S*-parameter measurements and (b) measured antenna S_{11}.

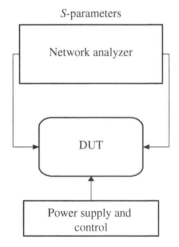

FIGURE 14.3   Two-port *S*-parameter measurements.

1. Short calibration
2. Open calibration
3. Load calibration

Connect the antenna to the network analyzer and measure S_{11} parameter. Save and plot S_{11} results. Antenna *S*-parameters measured result is shown in Figure 14.2. A setup for *S*-parameters measurement is shown in Figure 14.3. Two-port *S*-parameter measurement setup is shown in Figure 14.3. Two-port, S_{2P}, calibration process consists of four steps:

1. Short calibration
2. Open calibration
3. Load calibration
4. Through calibration

Measure S-parameters S_{11}, S_{22}, S_{12}, and S_{21} for N channels. RF head gain is given by S_{21} parameter. Gain flatness and phase balance between channels can be measured by comparing S_{21} magnitude and phase measured values. RF head gain and flatness measurement setup is presented in Figure 14.4(a). A two-port network analyzer is shown in Figure 14.4(b). Table 14.1 presents a typical table of measured S-parameter results. S-parameters in decibel may be calculated by using Equation 14.12:

$$S_{ij}(\text{dB}) = 20 \ \log\left[S_{ij}(\text{magnitude})\right] \tag{14.12}$$

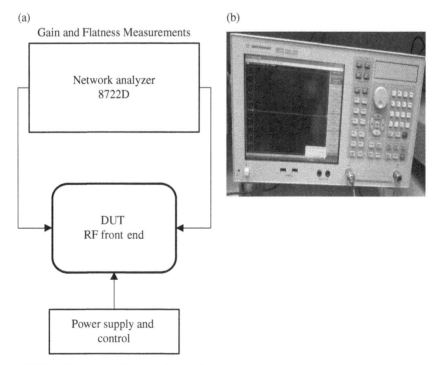

FIGURE 14.4   (a) RF head gain and flatness measurements and (b) network analyzer.

TABLE 14.1 S-Parameter Results

Channel	$S_{11}(E/T)$ dB	S_{22} (E/T) dB	S_{12} (E/T) dB	S_{21} (E/T) dB
1	−10	−10	−20	30
2	−10.5	−11	−21	29
3	−11	−10	−20	29
$n-1$	−10	−9	−20	29
n	−9	−10	−19	30

14.4.1 Types of S-parameters Measurements

Small-signal S-parameters measurements—In small-signal S-parameter measurements, the signals have only linear effects on the network so that gain compression does not take place. Passive networks are linear at any power level.

Large-signal S-parameters measurements—S-parameters are measured for different power levels. The S-matrix will vary with input signal strength.

14.5 TRANSMISSION MEASUREMENTS

A block diagram of transmission measurement setup is shown in Figure 14.5(a). The transmission measurement setup consists of a sweep generator, device under test (DUT), transmitting and receiving antennas, and spectrum analyzer. Measured transmission results by using a spectrum analyzer are shown in Figure 14.5(b).

The received power may be calculated by using Friis equation as given in Equations 14.13 and 14.14. The receiving antenna may be a standard gain antenna with a known gain. Where r represents the distance between the antennas,

$$P_R = P_T G_T G_R \left(\frac{\lambda}{4\pi r}\right)^2$$

$$\text{For}: \ G_T = G_R = G \tag{14.13}$$

$$G = \sqrt{\frac{P_R}{P_T}}\left(\frac{4\pi r}{\lambda}\right)$$

$$P_R = P_T G_T G_R \left(\frac{\lambda}{4\pi r}\right)^2$$

$$\text{For}: \ G_T \neq G_R \tag{14.14}$$

$$G_T = \frac{1}{G_R}\frac{P_R}{P_T}\left(\frac{4\pi r}{\lambda}\right)^2$$

Transmission measurements results may be summarized as in Table 14.2.

(a)

Transmitting channel measurements

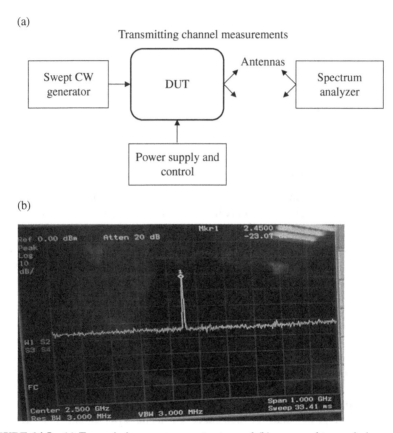

FIGURE 14.5 (a) Transmission measurement setup and (b) measured transmission results.

TABLE 14.2 Transmission Measurements Results

Transmission Results for Antennas Under Test (AUT) dBm

Antenna	F_1 (MHz)	F_1 (MHz)	F_1 (MHz)	Remarks
1	10	9	8	
2	9	8	7	
3	9.5	8.5	7.5	
4	10	9	8	
5	9	8	7	
6	10.5	9.5	8.5	
7	9	8	7	
8	11	10	9	

14.6 OUTPUT POWER AND LINEARITY MEASUREMENTS

A block diagram of output power and linearity measurement setup is shown in Figure 14.6. The output power and linearity measurement setup consists of a sweep generator, DUT, and a spectrum analyzer or power meter. In output power and linearity measurements, we increase the synthesizer power in 1dB steps and measure the output power level and linearity.

14.7 POWER INPUT PROTECTION MEASUREMENT

A block diagram of power input protection measurement setup is shown in Figure 14.7. The output power and linearity measurement setup consists of a sweep generator, power amplifier, DUT, attenuator, and a spectrum analyzer or power meter. In power input protection measurements, we increase the synthesizer power to 1dB steps from 0dBm and measure the output power level and observe that the DUT functions with no damage.

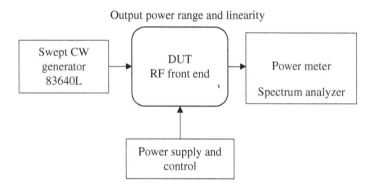

FIGURE 14.6 Output power and linearity measurement setup.

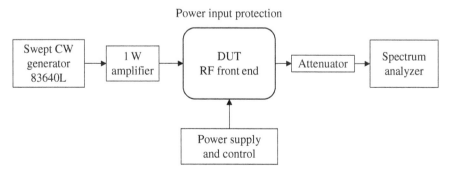

FIGURE 14.7 Power input protection measurement.

(a)

Non harmonic spurious measurements

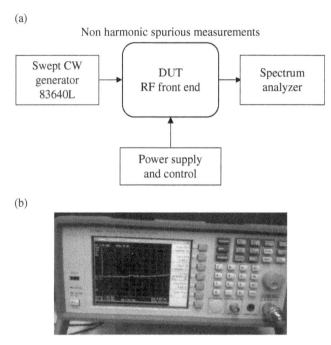

FIGURE 14.8   (a) Nonharmonic spurious measurements and (b) spectrum analyzer.

14.8   NONHARMONIC SPURIOUS MEASUREMENTS

A block diagram of nonharmonic spurious measurement setup is shown in Figure (a). A spectrum analyzer is shown in Figure 14.8(b).

The nonharmonic spurious measurement setup consists of a sweep generator, DUT, and a spectrum analyzer. In nonharmonic spurious measurements, we increase the synthesizer power in 1dB steps, up to 1dBc point, and measure the spurious level.

14.9   SWITCHING TIME MEASUREMENTS

A block diagram of switching time measurement setup is shown in Figure 14.9. The switching time measurement setup consists of a sweep generator, DUT, detector, pulse generator, and oscilloscope. In switching time measurements, we transmit an RF signal through the DUT. We inject a pulse via the switch control port. The pulse envelope may be observed on the oscilloscope. Switching time may be measured by using the oscilloscope.

14.10   IP$_2$ MEASUREMENTS

The setup for IP$_2$ and IP$_3$ measurements is shown in Figure 14.10. Second-order inter-modulation results are shown in Figure 14.11. IP$_2$ may be computed by Equation 14.15:

$$IP_{2[dBm]} = P_{out\ [dBm]} + \Delta_{[dB]} \qquad (14.15)$$

Second-order intermodulation results are shown in Figure 14.11.

Third-order intermodulation results are shown in Figure 14.12. IP$_3$ may be computed by Equation 14.16:

$$IP_{3[dBm]} = P_{out\ [dBm]} + \frac{\Delta_{[dB]}}{2} \qquad (14.16)$$

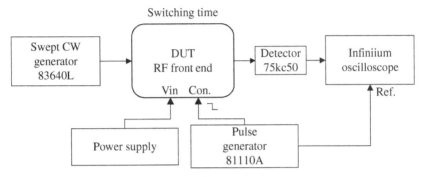

FIGURE 14.9   Switching time measurement setup.

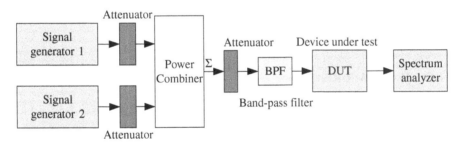

FIGURE 14.10   Setup for IP$_2$ and IP$_3$ measurements.

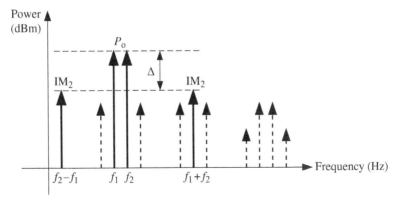

FIGURE 14.11   Second-order intermodulation results.

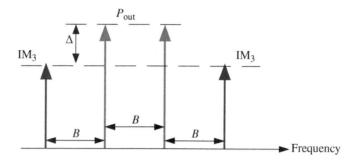

FIGURE 14.12   Third-order intermodulation results.

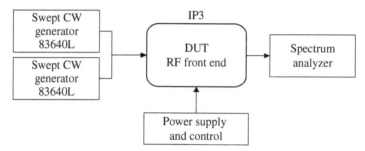

FIGURE 14.13   Two-tone measurements.

14.11   IP₃ MEASUREMENTS

A block diagram of output IP$_3$ measurement setup is shown in Figure 14.13. IP$_3$ setup consists of two sweep generators, DUT, and a spectrum analyzer. In IP$_3$ test we inject two signals to the DUT and measure the intermodulation signals. Test results listed in Table 14.3 present IP$_3$ measurements of a receiving channel. Input- and output-measured IP$_3$ may be calculated by using Equations 14.16–14.18:

$$IP_{3_{out}} = P_{F1} + \frac{((P_{F1} + P_{F2})/2) - P_{IM_1}}{2} \tag{14.17}$$

$$IP_{3_{in}} = \frac{(P_{F1} + P_{F2})}{2} + \frac{(((P_{F1} + P_{F2})/2) - P_{IM_2})}{2} - \left(\frac{(P_{F1} + P_{F2})}{2} - P_{in}\right) \tag{14.18}$$

For example, the first signal is at 20 GHz with a power level of 10dBm. The second signal is at 20.001 GHz with a power level of 10dBm. The first intermodulation signal is at 19.999 GHz with a power level of −10.8dBm. The second intermodulation signal is at 20.001 GHz with a power level of −13.8dBm. The power level of IP$_3$ output is 20.4dBm. The power level of IP$_3$ at the input is −5.1dBm.

TABLE 14.3 IP$_3$ Measurements

Pin (dBm) B	19.999 (GHz) IM1	F$_1$ = 20 (GHz)	F$_2$ = 20.001 (GHz)	20.002 (GHz) IM2	IP$_3$ OUT	IP$_3$ INPUT
	C (dBm)	D (dBm)	E (dBm)	F (dBm)	(dBm)	(dBm)
-17	-10.8	10	10	-13.8	20.4	-5.1
Pin (dBm)	29.999 (GHz) IM1	F$_1$ = 30 (GHz)	F$_2$ = 30.001 (GHz)	30.002 (GHz) IM2	IP$_3$ OUT	IP$_3$ INPUT
-14.50	-17	10	10	-18.8	23.5	-0.1
Pin (dBm)	39.998 (GHz) IM1	F$_1$ = 39.999 (GHz)	F$_2$ = 40 (GHz)	40.001 (GHz) IM2	IP$_3$ OUT	IP$_3$ INPUT
-14.5	-12	10	10	-11.2	21	-3.9

14.12   NOISE FIGURE MEASUREMENTS

A block diagram of noise figure measurement setup is shown in Figure 14.14. The noise figure measurement setup consists of a noise source, DUT, amplifier, and a spectrum analyzer. The noise level is measured without the DUT as a calibration level. We measure the difference, delta (Δ) value, in the noise figure when the noise source is on to the measured noise figure when the noise source is off:

$$NF = \frac{10\log\left(10^{0.1*ENR}\right)}{\left(10^{0.1*\Delta}\right)-1}\qquad(14.19)$$

where ENR is listed on the noise source for a given frequency. Delta, Δ, is the difference in the noise figure measurement when the noise source is on to the noise figure measurement when the noise source is off. Measured NF is calculated by using Equation 14.19. Noise Figure measurements are listed in Table 14.4.

14.13   ANTENNA MEASUREMENTS

Typical parameters of antennas are radiation pattern, gain, directivity, beamwidth, polarization, and impedance. During antenna measurements we ensure that the antenna meets the required specifications, and we can characterize the antenna parameters.

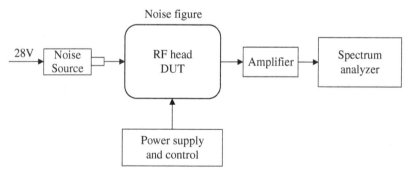

FIGURE 14.14   Noise figure measurement setup.

TABLE 14.4   Noise Figure Measurements

Parameter	Measurement 1	Measurement 2	Measurement 3
ENR	23.62	23.62	24
Delta (dB)	14	15	14
NF (dB)	9.777	8.740	10.159

14.13.1   Radiation Pattern Measurements

A radiation pattern is the antenna radiated field as function of the direction in space. The radiated field is measured at various angles at a constant distance from the antenna. The radiation pattern of an antenna can be defined as the locus of all points where the emitted power per unit surface is the same. The radiated power per unit surface is proportional to the square of the electric field of the electromagnetic wave. The radiation pattern is the locus of points with the same electrical field strength. Usually the antenna radiation pattern is measured in a far-field antenna range. The antenna under test is placed in the far-field distance from the transmitting antenna. Due to the size required to create a far-field range for large antennas, near-field techniques are employed. Near-field techniques allow to measure the fields on a surface close to the antenna (usually 3–10 wavelength). Near field is transferred to far field by using Fourier transform.

The far-field distance or Fraunhofer distance, R, is given in Equation 14.20:

$$R = \frac{2D^2}{\lambda} \tag{14.20}$$

where D is the maximum antenna dimension and λ is the antenna wavelength.

The radiation pattern graphs can be drawn using Cartesian (rectangular) coordinates as shown in Figure 14.15, see Refs. [1–5]. A polar plot is shown in Figure 14.16. The polar plot is useful to measure the beamwidth, which is the angle at the −3dB points around the maximum gain. A 3D radiation pattern is shown in Figure 14.17.

Main beam—Main beam is the region around the direction of maximum radiation, usually the region that is within 3dB of the peak of the main lobe.

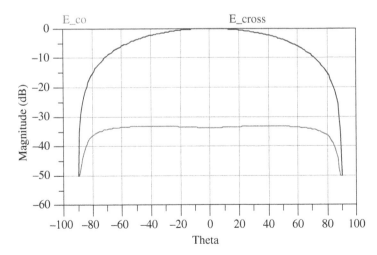

FIGURE 14.15   Radiation pattern of loop antenna with ground plane rectangular plot.

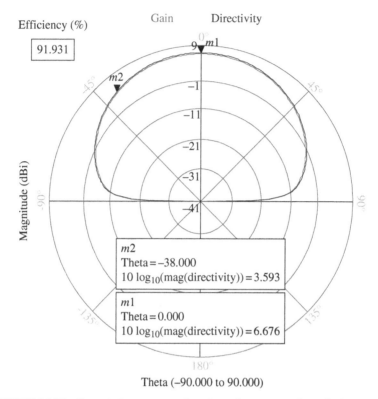

FIGURE 14.16   Grounded quarter wavelength patch antenna polar radiation pattern.

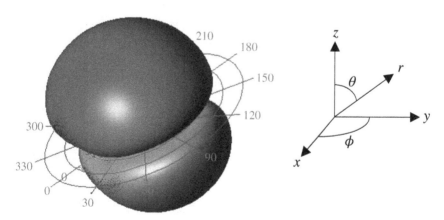

FIGURE 14.17   Loop antenna 3D radiation pattern.

Beamwidth—Beamwidth is the angular range of the antenna pattern in which at least half of the maximum power is emitted. This angular range, of the major lobe, is defined as the points at which the field strength falls around 3dB regarding to the maximum field strength.

Side lobes level—Side lobes are smaller beams that are away from the main beam. Side lobes present radiation in undesired directions. The side lobe level is a parameter used to characterize the antenna radiation pattern. It is the maximum value of the side lobes away from the main beam and is expressed usually in decibels.

Radiated power—Total radiated power when the antenna is excited by a current or voltage of known intensity.

14.13.2 Directivity and Antenna Effective Area

Antenna directivity is the ratio between the amounts of energy propagating in a certain direction compared to the average energy radiated to all directions over a sphere as given in Equation 14.21, see Refs. [1–4]:

$$D = \frac{P(\theta,\phi)\text{maximal}}{P(\theta,\phi)\text{average}} = 4\pi\frac{P(\theta,\phi)\text{maximal}}{P\,\text{rad}} \tag{14.21}$$

where $P(\theta,\phi)\text{average} = \frac{1}{4}P(\theta,\phi)\sin\theta\,d\theta\,d\phi = \frac{P\,\text{rad}}{4\pi}$

An approximation used to calculate antenna directivity is given in Equation 14.22:

$$D \sim \frac{4\pi}{\theta E \times \theta H} \tag{14.22}$$

θE – Measured beamwidth in radian in EL plane

θH – Measured beamwidth in AZ plane

Measured beamwidth in *radian in AZ plane and in EL plane allows us to calculate antenna directivity*.

Antenna effective area (A_{eff})—The antenna area which contributes to the antenna directivity is given in Equation 14.23:

$$A_{\text{eff}} = \frac{D\lambda^2}{4\pi} \sim \frac{\lambda^2}{\theta E \times \theta H} \tag{14.23}$$

14.13.3 Radiation Efficiency (α)

Radiation efficiency is the ratio of power radiated to the total input power, $\alpha = G/D$. The efficiency of an antenna takes into account losses and is equal to the total radiated power divided by the radiated power of an ideal lossless antenna. Efficiency is equal to the radiation resistance divided by total resistance (real part) of the feed-point impedance. Efficiency is defined as the ratio of the power that is radiated to the total power used by the antenna as given in Equation 14.24. Total power equals to power radiated plus power loss:

$$\alpha = \frac{P_r}{P_r + P_l} \qquad (14.24)$$

E and *H* plane 3D radiation pattern of a wire loop antenna in free space is shown in Figure 14.18.

14.13.4   Typical Antenna Radiation Pattern

A typical antenna radiation pattern is shown in Figure 14.19. The antenna main beam is measured between the points that the maximum relative field intensity *E* decays to 0.707E. Half of the radiated power is concentrated in the antenna main beam. The antenna main beam is called 3dB beamwidth. Radiation to undesired direction is concentrated in the antenna side lobes.

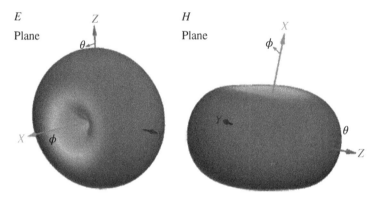

FIGURE 14.18   *E* and *H* plane radiation pattern of loop antenna in free space.

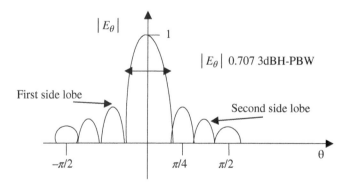

FIGURE 14.19   Antenna typical radiation pattern.

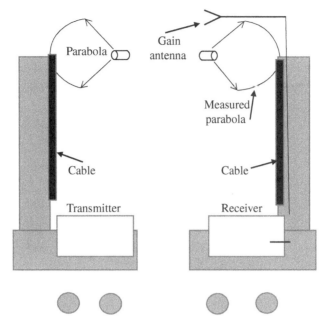

FIGURE 14.20   Antenna range for radiation pattern measurements.

Antenna radiation pattern is usually measured in free-space ranges. An elevated free-space range is shown in Figure 14.20. An anechoic chamber is shown in Figure 14.21.

14.13.5   Gain Measurements

Antenna Gain (G)—The ratio between the amounts of energy propagating in a certain direction compared to the energy that would be propagating in the same direction if the antenna were not directional. Isotropic radiator is known as its gain.

Figure 14.20 presents antenna far-field range for radiation pattern measurements. Antenna gain is measured by comparing the field strength measured by the antenna under test to the field strength measured by a standard gain horn as shown in Figure 14.20. The gain as function of frequency of the standard gain horn is supplied by the standard gain horn manufacturer. Figure 14.21 presents an anechoic chamber used to indoor antenna measurements. The chamber metallic walls are covered with absorbing materials.

14.14   ANTENNA RANGE SETUP

Antenna rage setup is shown in Figure 14.20. Antenna rage setup consists of the following instruments:

- Transmitting system that consists of a wideband signal generator and transmitting antenna

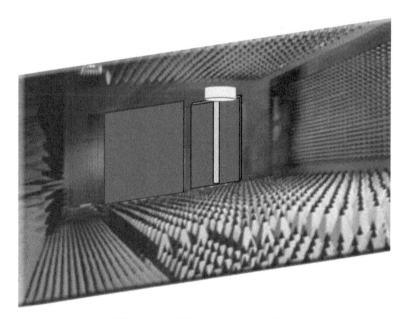

FIGURE 14.21 Anechoic chamber.

- Measured receiving antenna
- Receiver
- Positioning system
- Recorder and plotter
- Computer and data processing system

The signal generator should be stable with controlled frequency value, good spectral purity, and controlled power level. Low-cost receiving system consists of a detector and amplifiers. Several companies sale antenna measurement setups such as Agilent, Tektronix, and Anritsu.

REFERENCES

[1] Balanis CA. *Antenna Theory: Analysis and Design*. 2nd ed. Chichester: John Wiley & Sons, Ltd; 1996.

[2] Godara LC, editor. *Handbook of Antennas in Wireless Communications*. Boca Raton: CRC Press LLC; 2002.

[3] Kraus JD, Marhefka RJ. *Antennas for All Applications*. third ed. New York: McGraw Hill; 2002.

[4] Sabban A. *Low visibility Antennas for Communication Systems*. Boca Raton: Taylor & Francis Group; 2015.

[5] Sabban A. *RF Engineering, Microwave and Antennas*. Tel Aviv, Israel: Saar Publication; 2014.

INDEX

Wideband RF Technologies and Antennas in Microwave Frequencies, First Edition. Dr. Albert Sabban.
© 2016 John Wiley & Sons, Inc. Published 2016 by John Wiley & Sons, Inc.